HYDRAULIC AND ENVIRONMENTAL MODELLING:

ESTUARINE AND RIVER WATERS

Hydraulic and Environmental Modelling: Estuarine and River Waters

Proceedings of the Second International Conference on Hydraulic and Environmental Modelling of Coastal, Estuarine and River Waters Volume 2

Edited by

R. A. Falconer
K. Shiono
R. G. S. Matthew

Routledge
Taylor & Francis Group

LONDON AND NEW YORK

First published 1992 by Ashgate Publishing

Reissued 2018 by Routledge
2 Park Square, Milton Park, Abingdon, Oxon, OX14 4RN
52 Vanderbilt Avenue, New York, NY 10017

Routledge is an imprint of the Taylor & Francis Group, an informa business

A Library of Congress record exists under LC control number:

ISBN 13: 978-1-138-38613-6 (hbk)
ISBN 13: 978-1-138-38614-3 (pbk)
ISBN 13: 978-0-429-42693-3 (ebk)

Contents

v

vii

Editors' foreword

In recent years there has continued to be an increasing interest in the development and application of numerical hydraulic models as design and management tools for flow and pollutant and sediment transport simulation studies in hydraulic and environmental engineering. Such models also depend heavily on laboratory and/or field measuring programmes for calibration, verification and interpretation of flow phenomena and water quality evaluation. This increasing use of deterministic models by engineers and scientists, in both private and public organisations, can be partly attributed to:- (i) the growing public perception and concern over the pollution of the coastal, estuarine and river environment, (ii) the adoption of a more rational basis for coastal defence works due to an improved understanding of surf zone hydrodynamics, (iii) the more widespread use of sophisticated field and laboratory data analysis techniques and recording equipment, and (iv) the availability of cheaper and more powerful personal computers and work stations and the increasing availability of computer software relating to coastal, estuarine and river flow and solute transport processes.

In continuing to recognise this increasing interest in the use of hydraulic, water quality and sediment transport models, and following the success of the first international conference on Hydraulic and Environmental Modelling of Coastal, Estuarine and River Waters, held in September 1989 and opened by HRH The Princess Royal, the Computational Hydraulics and Environmental Modelling Research Group at the University of Bradford have organised a similar conference to discuss recent advances in the field. The main aim of the conference was to provide a forum whereby engineers, scientists and planners involved in the application, research and development of multi-disciplinery models could share experiences.

The excellent response to the call for papers resulted in a large number of contributions, with the conference proceedings of the edited papers being included in two volumes namely:- Hydraulic and Environmental Modelling : Coastal Waters and Hydraulic and Environmental Modelling : Estuarine and River Waters. This volume contains the papers relating to estuarine and river waters and has been subdivided into the following parts:- Flow Processes, Flow Modelling, Salinity Intrusion Modelling, Water Quality Modelling, Sediment Transport Modelling and Expert Systems.

In organising the conference and in the preparation of this volume the editors would like to thank the following:-

.   The keynote speakers, including Dr Stuart Reed, Director of Environmental Protection, Hong Kong Government Environmental Protection Department and Prof Donald Harleman, Ford Professor of Engineering (Emeritus), Massachusetts Institute of Technology, USA

.   The authors for their time and effort in preparing the papers for this volume and the delegates and all those who participated in the organisation of the conference.

.   The members of the UK and Overseas Technical Advisory committees, the paper referees and Mrs Christine Dove (Conference Co-ordinator)

.   The Co-sponsors including: Institution of Civil Engineers, Institution of Water and Environmental Management and International Association for Hydraulic Research

.   The staff of the publishers - particularly Mr John Hindley - for their ready help and encouragement

.   Staff of the University of Bradford who assisted with the organisation of the conference.

RAF, KS, RGSM
Bradford 1992

Acknowledgments

The editors' wish to acknowledge the valuable assistance of the other members of the Organising Committees, the Conference Co-ordinator and the Panel of Referees.

U K ORGANISING COMMITTEE

Prof R A Falconer (Chairman)	University of Bradford
Mr P Ackers	Private Consultant
Dr S N Chandler-Wilde	University of Bradford
Mr Y Chen	University of Bradford
Mr N E Denman	ABP Research & Consultancy Ltd
Mr C Evans	Wallace Evans Ltd
Dr G P Evans	WRc
Dr C A Fleming	Sir William Halcrow & Partners
Mr A Hooper	W S Atkins Engineering Sciences
Dr M W Horner	Bullen & Partners
Dr D W Knight	University of Birmingham
Prof D M McDowell	Private Consultant
Mr A McLean	Yorkshire Water plc
Dr R G S Matthew	University of Bradford
Dr K Shiono	University of Bradford
Mr D V Smith	Wessex Water
Mr G Thompson	Binnie & Partners
Mr M F C Thorn	HR Wallingford

OVERSEAS COMMITTEE

Prof M B Abbott (Chairman)	International Institute for Hydraulic and Environmental Engineering, The Netherlands
Dr J A Cunge	Laboratoire d'Hydraulique de France, France
Dr P Goodwin	Philip Williams & Associates, USA
Mr M Hartnett	MCS International, Eire
Prof Dr Ing P K Holz	University of Hannover, Germany
Dr C W Li	Hong Kong Polytechnic, Hong Kong
Prof S Q Liu	Tongji University, Peoples Republic of China
Prof K Nakatsuji	Osaka University, Japan
Prof R E Nece	University of Washington, USA
Mrs C H Dove	Conference Co-ordinator, University of Bradford

List of contributors

Aburatani, S	The Ministry of Transport, Japan
Alam, M K	Bangladesh University of Engineering and Technology, Bangladesh
Alcrudo, F	University of Zaragoza, Spain
Al-Hamid, A A I	King Saud University, Saudi Arabia
Almassy, A	W S Atkins Consultants Ltd, UK
Aristodemou, E	Wessex Water Technologies, UK
Arnold, U	IKW Consultants for Public and Environmental Affairs, Germany
Awaya, Y	Kyushu Kyuritsu University, Japan
Ban, M	Kochi University, Japan
Bennett, J M	National Rivers Authority Thames Region, UK
Bijker, E W	Technical University Delft, The Netherlands
Biro, I	Water Resources Research Centre (VITUKI), Hungary
Brockhaus, T	Aachen University of Technology, Germany
Chawdhary, K S	University of Reading, UK
Clark, K J	WRc, UK
Clifford, N J	University of Hull, UK
Costanza, R	University of Maryland, USA
Crowther, J M	University of Strathclyde, UK
Curran, J C	Clyde River Purification Board, UK
Darby, S E	University of Nottingham, UK
Day Jr, J W	Louisiana State University, USA

Deursen van, W	University of Utrecht, The Netherlands
Dittrich, A	University of Karlsruhe, Germany
Ervine, D A	University of Glasgow, UK
Feldhaus, R	Aachen University of Technology, Germany
Findlay, J S	William Grant Ltd, UK
French, J R	University College London, UK
Fujisaki, K	Kyushu Institute of Technology, Japan
Garcia-Navarro, P	University of Zaragoza, Spain
Gasser, M M	Hydraulics and Sediment Research Institute, Egypt
Gaweesh, M T K	Hydraulics and Sediment Research Institute, Egypt
Ghosh, L K	Central Water and Power Research Station, India
Goodwin, P	Philip Williams & Associates Ltd, USA
Han, Z	Zhejiang Provincial Institute of Estuarine and Coastal Engineering Research, China
Harbott, B	National Rivers Authority Anglian Region, UK
Hassan, R M	Faculty of Engineering Cairo University, Egypt
Helsloot, I C M	Ministry of Transport, The Netherlands
Hiramatsu, K	Kyoto University, Japan
Hottges, J	Aachen University of Technology, Germany
Janssen, J P F M	Ministry of Transport, The Netherlands
Jensen, O K	Danish Hydraulic Institute, Denmark
Johnsen, J	Danish Hydraulic Institute, Denmark
Johnson, H K	Danish Hydraulic Institute, Denmark
Jolankai, G	Water Reseources Research Centre (VITUKI), Hungary
Kawachi, T	Kyoto University, Japan
Khan, L A	Bangladesh University of Engineering and Technology, Bangladesh
Kleinschmidt, D G	University of Maine, USA
Knight, D W	University of Birmingham, UK
Komatsu, T	Kyushu University of Japan
Kwadijk, J	University of Utrecht, The Netherlands
Larsen, H G H	Danish Hydraulic Institute, Denmark
Lee, M O	National Fisheries University of Yosu, Korea
Lewandowski, J	Philip Williams & Associates Ltd, USA
Lewis, R E	ICI Group Environmental Laboratory, UK
Li, B	Wuhan Institute of Urban Construction, China
Lin, B	University of Bradford, UK
Liu, S Q	Tongji University, China
Liu, Y	Shanghai University of Technology, China
Lorena, M L	University of Glasgow, UK
Lu, X	Zhejiang Provincial Institute of Estuarine and Coastal Engineering Research, China

Matthew, R G S	University of Bradford, UK
Matsunaga, N	Kyushu University, Japan
Maxwell, T	University of Maryland, USA
McKemey, M D	Environmental Assessment Services Ltd, UK
McIlvaine, H	Atlantic Electric Company, USA
Mironovsky, A L	Technical University of St Petersburg, Russia
Mollowney, B M	WRc, UK
Muraoka, K	Osaka University, Japan
Murota, A	Osaka University, Japan
Nakatsuji, K	Osaka University, Japan
Ohki, H	Kyushu Institute of Technology, Japan
Olesen, K W	Danish Hydraulic Institute, Denmark
Parfitt, A J	Rendel Palmer and Tritton, UK
Parker, W R	Blackdown Consultants Ltd, UK
Pearce, B R	University of Maine, USA
Porter, J H	Mersey Barrage Company, UK
Prinos, P	University of Thessaloniki, Greece
Rijn van, L C	Delft Hydraulics, The Netherlands
Rouve, G	Aachen University of Technology, Germany
Ruland, P	Aachen University of Technology, Germany
Saad, M B A	Hydraulics and Sediment Research Institute, Egypt
Sadek, E El	Suez Canal Authority, Egypt
Samuels, P G	H R Wallingford, UK
Shao, Y	Zhejiang Provincial Institute of Estuarine and Coastal Engineering Research, China
Shiono, K	University of Bradford, UK
Simek, E	Environmental Resources Management, USA
Singh, C B	Central Water and Power Research Station, India
Sklar, F	University of South Carolina, USA
Slow, T M	University of Bradford, UK
Smith, D V	Wessex Water plc, UK
Sobey, R J	University of California, Berkeley, USA
Sucsy, P	University of Maine, USA
Sugihara, Y	Kyushu University, Japan
Suszka, L	Institute of Hydroengineering Polish Academy of Sciences, Poland
Thorne, C R	University of Nottingham, UK
Vivek, V	University of Maine, USA
Vodslon, J	Technical University of Prague, Czechoslovakia
Wallis, S G	Heriot-Watt University, UK
Wang, Z	Institute of Water Conservancy and Hydroelectric Power Research, China

Warren, R   Danish Hydraulic Institute, Denmark

White, M L   Louisiana State University, USA

Whitlow, C D   Sir William Halcrow & Partners Ltd, UK

Wilson, E A   Mersey Barrage Company, UK

Wood, B J B   University of Strathclyde, UK

Yuen, K W H   Consultant, Canada

Part 1
FLOW PROCESSES

1 Boundary shear in differentially roughened trapezoidal channels

D. W. Knight, A. A. I. Al-Hamid and K. W. H. Yuen

ABSTRACT

Experimental data on flow in homogeneously and heterogeneously roughened trapezoidal channels are presented. The walls were differentially roughened with respect to the bed and two series of experiments undertaken (k_{sw}/k_{sb} = 679 & 419, 0.85 < B/H < 10.0). Equations are presented for the percentage of the total shear force which acts on the walls and the associated mean and maximum boundary shear stresses on both bed and walls. These equations include the effects of aspect ratio, wall side slope angle and relative roughness.

1 Introduction

There are many instances of open channel flow in which the walls of the channel are rougher than the bed. However the majority of research has often been directed at channels in which the bed is differentially rougher than the walls, since this case commonly occurs in sediment flume experiments for which sidewall correction procedures are needed. In order to redress this balance, and in order to study lateral eddy diffusivity, a series of experiments has recently been completed in which large lateral variations of velocity and boundary shear stress were induced in various trapezoidal channels by means of adding roughness to the channel walls.

2 Experimental apparatus, procedure and results

2.1 Experimental apparatus and procedure

The experiments were conducted at the University of Birmingham in a 22m long rectangular tilting flume, 0.615m wide and 0.365m deep [1]. Three trapezoidal channels with side slopes s = 1 (1:s,vertical:horizontal) and nominal base widths of 150, 300 and 450mm were constructed inside the flume and roughened with gravel. Two gravel roughness sizes, denoted R1 and R2, with d_{84} values of 18.0 and 9.3mm respectively were used to roughen the channels. The walls of the channels were differentially roughened with respect to the smooth bed and a series of 38 experiments conducted in all three channels with both roughnesses between aspect ratios $0.85 < B/H < 10.0$, where B = base width of channel and H = depth of flow. The corresponding wetted perimeter ratio range was $0.3 < P_b/P_w < 3.53$, where P=wetted perimeter of bed (= B) and P_w=wetted perimeter of walls (= $2H\sqrt{(1+s^2)}$). All flows were subcritical with the Froude number, Fr, varying between $0.39 < Fr < 0.89$ and the Reynolds number, Re, varying between $3.4 \times 10^4 < Re < 1.6 \times 10^5$. Two series of experiments were conducted with uniformly roughened channels, with the bed roughened in a similar manner to the walls, in order to obtain the Nikuradse equivalent roughness size for each gravel.

The procedure for setting normal depth flow was that originally proposed by the senior author [5]. This procedure ensures that sufficiently accurate stage discharge and tailgate height discharge curves are obtained. These curves were later used to interpolate discharge and tailgate settings corresponding to the desired normal depth at preselected B/H or P_b/P_w values. Velocities were measured locally using a 4.0mm diameter Pitot-static tube, mounted on a precision rotating pointer gauge so that velocity profiles could be obtained accurately normal to any surface. Readings were generally taken at 5mm intervals normal to the surface and at 10 to 20mm intervals laterally. For each experiment the overall cross section discharge was evaluated by integration of the local velocities and compared with the discharge measured independently by calibrated orifice plates in the suppy line to the flume. Boundary shear stresses were measured with a 4.7mm Preston tube on smooth surfaces and by logarithmic plotting of the velocity data for rough surfaces. For each experiment the mean overall boundary shear stress was computed by integration of the local boundary shear stresses and compared with the mean overall value, τ_0, computed from the energy gradient(= $\rho g R S_f$).

2.2 Location of datum planes

For flow over rough boundaries a datum plane is required to which all measurements should be related. This plane is known to depend upon the shape, size, distribution and concentration of the roughness elements, but no generally agreed definition or method of calculation is available [1,8]. In this study an attempt was made to determine the appropriate datum planes for both velocity and boundary shear stress. Since the gravel roughness had been glued to marine plywood panels prior to insertion on the walls or bed of the trapezoidal channels, the geometric datum plane, n_0, measured relative to the backing panel, could be readily determined by detailed measurement. The values of n_0 were found to be 11.7mm (0.65d_{84}) for R1 and 7.03mm (0.76d_{84}) for R2. Appropriate values for datum planes, n_1 for velocity and n_2 for boundary shear, were computed using a minimisation technique on the errors between the integrated point velocities or local boundary shear stresses and their section mean values.

The discharge error between the integrated point velocities and the orifice plate were evaluated for each experiment and n_1 varied until the error was a minimum. For differentially roughened channels with rough walls and smooth beds, the average discharge error using n_0 was +1.43%, with a standard deviation of 3.09%. The %Q_{err} values were generally within the ±5.0% band for all P_b/P_w values tested. Because the mean error of +1.43% was so small, and in the absence of any significant trend, the value of n_1 was taken to be the same as n_0.

The boundary shear stress data was treated in a similar manner, minimising errors between the integrated mean and the energy slope mean values until the appropriate value of n_2 was found. During the minimisation process due regard was also taken of the influence of n_2 on the basic geometrical parameters of the channel. The results of this analysis are shown in Fig.1 where a surprisingly consistent pattern emerges. It is apparent that the boundary shear displacement length n_2 differs somewhat from n_0, being located at 9.8mm ($0.54d_{84}$) for R1 and 5.84mm ($0.63d_{84}$) for R2. The standard deviations in both cases were extremely small, being 0.52mm and 0.37mm respectively. The discharge datum plane therefore appears to be about 17% above the shear stress datum plane, based on these two series of experiments.

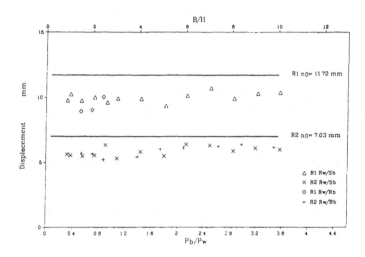

Figure 1 *Boundary shear datum plane n_2*

2.3 Nikuradse equivalent roughness sizes

In differentially roughened channels in which the Nikuradse equivalent sand roughness size, k_s, is used to distinguish between the roughness on the walls and the bed, it is important that the absolute values, k_{sw} and k_{sb} respectively, are prescribed accurately rather than just the ratio, k_{sw}/k_{sb}. This is not always as straightforward as it appears since it is known that the shape of the channel cross section, as well as the roughness height, affects the flow resistance [1,3,6,10,12]. Therefore in determining equivalent k_s values from measurements of Darcy-Weisbach f and hydraulic radius R, some account should strictly be taken of the channel shape. However in this particular series of experiments since all the channels were of a simple trapezoidal cross section and such large roughnesses were employed, the standard Colebrook-White equation was used without correction to the hydraulic radius.

Values of overall k_s were evaluated from the experimental data by altering the value of $u*k_s/v$ for the uniformly roughened channels and $k_s/4R$ for the differentially roughened channels until

5

a minimum error was achieved. Lines of constant $u*k_s/v$ were ensured in the experimental programme by keeping the product of the shear velocity and k_s nearly constant for each sequential depth change. For roughnesses R1 and R2 this analysis gave $u*k_s/v$ values of 1378.2 and 778.8 respectively. Individual values of k_s were then back calculated for each experiment. The homogeneously roughened channel data are shown in **Fig.2**. Since this figure shows that there was only a slight reduction in k_s with Reynolds number, mean values of 32.98, 20.37, and 0.0486 mm were taken for channels with R1, R2 and smooth surfaces respectively. For differentially roughened channels, $k_s/4R$ values were obtained for each channel and best fit lines drawn through the data. **Fig.3** shows some data for roughness R1. Using the mean values of $k_s/4R$, individual k_s values could be computed.

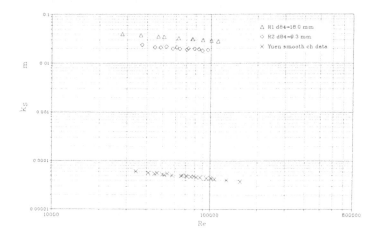

Figure 2 *Nikuradse equivalent roughness sizes*

Because the absolute values of k_s for a given gravel size would be used later as a key parameter in the analysis of both composite resistance and boundary shear stress data, the homogeneously roughened channel data were examined further using velocity profiles normal to each surface roughness. Using the standard two dimensional logarithmic law for a rough wall, individual k_s values were determined using the geometric datum plane n_0. The average k_s value for roughness R2, based on 9 experiments with a total of 234 individual velocity profiles with 5 data points nearest to the boundary in each profile, was 19.88mm, only 2.4% less than the value obtained by using the Colebrook White equation. For roughness R1, using 36 individual profiles, the average k_s value was 30.22mm, 8.8% less than the Colebrook-White value. For smooth channels, using $k_s/4R = 0.00026$, the average k_s was found to be 0.0516mm, some 6.2% higher than the value based on a mean $u*k_s/v$ value of 0.89.

6

If the boundary shear datum plane, n_2, had been used instead of the geometric datum plane, n_0, then the aspect ratio, Reynolds number, Nikuradse size and shear force values all change. As a check the equivalent roughness sizes were computed using n_2 and found to be 39.8mm and 23.84mm for roughnesses R1 and R2 respectively. The consequent changes in wall shear forces for uniformly roughened channels were within ±1.0% and for differentially roughened channels were within 0.5% to 4.0% for both R1 and R2. In order to standardise the shear force results which follow, the finally adopted Nikuradse equivalent roughness sizes were based on the geometric datum plane, n_0, and the overall channel resistance data since this averages the net effect of the roughness and also introduces a smoothing procedure implied in constant $u*k_s/v$ values. The consequent k_{sw}/k_{sb} values required for shear force analysis in differentially roughened channels were therefore taken as 679 and 419 for R1 and R2 on the channel walls respectively. See **Table 1**.

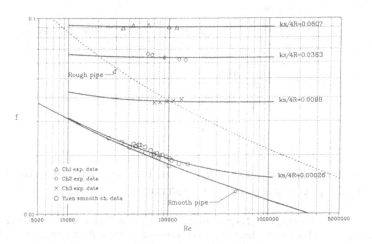

Figure 3 *Darcy-Weisbach friction factors*

TABLE 1 *Gravel size, roughness and displacement values*

Roughness	d_{84} (mm)	d_{50} (mm)	n_0 (mm)	n_2 (mm)	k_s (mm)
R1	18.0	14.2	11.7	9.8	32.98
R2	9.3	7.5	7.03	5.84	20.37
Smooth	-	-	-	-	0.0486

3 Boundary shear in homogeneously roughened channels

The total shear force per unit streamwise length, SF_T, acting on the wetted perimeter, P, of a prismatic channel is given by

$$SF_T = \tau_o P = (\rho g R S_f) P = \rho g A S_f \qquad (1)$$

where τ_o is the mean boundary shear stress, R is the hydraulic radius (=A/P), A is the cross section area of the channel and S_f is the energy slope. For a trapezoidal channel the overall wetted perimeter, P, may be decomposed into two elements, P_b and P_w, where P_b(=B) relates to the bed and P_w (=$2H\sqrt{(1+s^2)}$) relates to the walls. The breadth/depth ratio, B/H, or the wetted perimeter ratio, P_b/P_w, is frequently used to characterise the shape of a trapezoidal channel with side slopes 1:s (vertical:horizontal). The percentage of the total shear force which acts on the walls of such a channel may be related to the mean boundary shear stresses $\bar{\tau}_w$ and $\bar{\tau}_b$, acting on walls and bed by the equation

$$\%SF_w = 100 SF_w/SF_T = 100(P_w\bar{\tau}_w)/(P_w\bar{\tau}_w + P_b\bar{\tau}_b) \qquad (2)$$

$$\therefore \ \%SF_w = \frac{100}{\left[1 + \dfrac{P_b}{P_w}\dfrac{\bar{\tau}_b}{\bar{\tau}_w}\right]} \qquad (3)$$

Equation (3) thus shows that $\%SF_w$ will only vary in the same way as the geometry of the channel, i.e. in proportion to the perimeter length ratio P_b/P_w, if $\bar{\tau}_b = \bar{\tau}_w$.

There is much evidence [2,4,5,6,10,12] to suggest that even in homogeneously roughened channels the mean shear stresses differ markedly due to three dimensional turbulence effects and streamwise vorticity [7,9,11]. Recent work by the authors [6] suggests that $\%SF_w$ varies exponentially with P_b/P_w according to the equations

$$\%SF_w = C_{sf}\, e^\alpha \qquad (4)$$

$$\alpha = -3.23 \log_{10}[(P_b/(P_w C_2)) + 1.0] + 4.6052 \qquad (5)$$

where $\quad C_{sf} = $ 1.0 for $P_b/P_w < 6.546$

else $\quad C_{sf} = 0.5857(P_b/P_w)^{0.28471}$

and $\quad C_2 = $ 1.50 for subcritical flow

Fig.4 illustrates this exponential variation together with a wide range of data, drawn from various sources. Equations (4) and (5) appear to represent the experimental data reasonably well provided that the channel is homogeneously roughened. Fig.5 illustrates, to an enlarged

8

scale, how the data from two differently roughened channels, but each with the same roughness on the walls and the bed (symbols ◊ and + for roughnesses R1 and R2 respectively), merge together to form a single curve. This curve, as **Fig.4** has already shown, indicates that homogeneously smooth and rough trapezoidal channels of any side slope may be defined by equations (4) and (5).

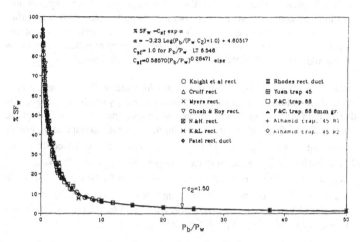

Figure 4 *Percentage shear force on channel walls for smooth and homogeneously roughened channels together with equations (4) & (5)*

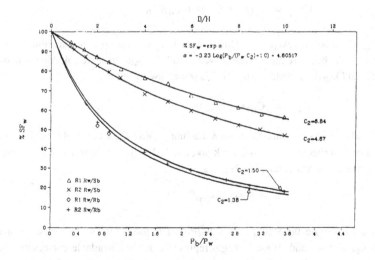

Figure 5 *Percentage shear force values for homogeneously and differentially roughened channels*

9

4   Boundary shear in heterogeneously roughened channels

For trapezoidal channels in which the bed or the walls are differentially roughened with respect to each other, it is obvious from equation (3) that as $\bar\tau_w$ and $\bar\tau_b$ depart further from each other, then the %SF_w versus P_b/P_w relationship will depart further from a simple geometrical proportioning law. Although it has already been shown that such a law is invalid for homogeneously roughened channels, and should be replaced by an exponential law, it nevertheless follows by analogy that equations (4) and (5) will require further modification as the roughness becomes more heterogeneous. The precise way in which these equations should be modified is open to debate. Given the paucity of experimental data with which to test hypotheses, the authors put forward the following ideas somewhat tentatively.

Fig.5 shows data from two series of experiments in which the walls of a trapezoidal channel were roughened differentially with respect to a smooth bed (symbols Δ and X). The rough wall/smooth bed data, annotated Rw/Sb, clearly lie above the rough wall/rough bed data, annotated Rw/Rb. The %SF_w values are clearly higher than the corresponding homogeneous values due to the higher values of $\bar\tau_w$. Also shown in **Fig.5** are lines based on equations (4) and (5) with the coefficient C_2 varied from its homogeneous value of 1.50 ($k_{sw}/k_{sb} = 1.0$) to 4.87 for $k_{sw}/k_{sb} = 419$ and 6.84 for $k_{sw}/k_{sb} = 679$ in order to fit the data. Various attempts at fitting the data either by changing the form of equations (4) and (**5**) or by adjusting the one remaining constant 3.23 in equation (5) were also tried [1,6] but abandoned in favour of the alternative strategy of simply changing the constant C_2.

The relationship between C_2 and $k_{sw}/k_{sb} = 1$ clearly cannot be accurately determined on the basis of these three series of experiments giving just three k_{sw}/k_{sb} values, but if a linear relationship was assumed then this relationship might be expressed as

$$C_2 \quad = \quad 1.492 + 0.008 \ (k_{sw}/k_{sb}) \tag{6}$$

giving $C_2 = 1.50$ for $k_{sw}/k_{sb} = 1$. However such a relationship is not sutiable for those cases in which the bed is rougher than the walls and $k_{sb} > k_{sw}$. A more suitable relationship between C_2 and k_{sw}/k_{sb} would obviously be a power law equation

$$C_2 \quad = \quad C_o(k_{sw}/k_{sb})^n \tag{7}$$

where C_o and n are coefficients. Flintham & Carling [2] give $C_o = 1.50$ and n = 0.203, based on 4 series of experiments in which $2.1 < k_{sb}/k_{sw} < 91.1$. Using all 6 series of data, together with homogeneous series of data [1,12] gives

$$C_2 \quad = \quad 1.50(k_{sw}/k_{sb})^{0.2115} \tag{8}$$

Figs. 6 and 7 illustrate the %SF_w versus P_b/P_w relationship for the 7 series of available data with $k_{sw}/k_{sb} > 1$ and also < 1. **Fig.6** represents the best available correlations using equation (6) for the data with $k_{sw}/k_{sb} \geq 1$ and equation (7) with n = 0.203 for the data with $k_{sb}/k_{sw} > 1$. **Fig.7** shows the same data sets, but with a unified equation for C_2 in the form

10

of equation (8). It is evident from a comparison between Figs. 7 and 8 that the power law relationship does not fit the more recent data as well as the linear relationship. However until more data become available it is not sensible to speculate further or refine the two alternative strategies represented by equations (6) and (8).

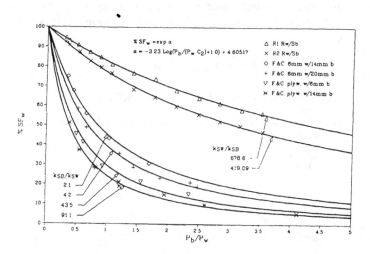

Figure 6 *Percentage shear force values for differentially roughened channels together with equations (4),(5),(6) & (8)*

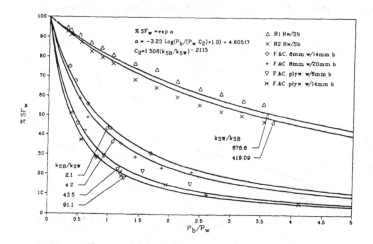

Figure 7 *Percentage shear force values for differentially roughened channels together with equations (4),(5) & (8)*

5 Mean and maximum boundary shear stresses

The mean wall and bed shear stresses $\bar{\tau}_w$ and $\bar{\tau}_b$, were determined from the experimental data for the two differentially roughened and the two homogeneously roughened channels. It may be shown [1,4] that the mean wall and bed shear stresses may be expressed by equations (9) & (10), with the %SF$_w$ versus P$_b$/P$_w$ relationship supplied via equations (4) and (5). These equations, with C$_2$ given by equation (6), together with the experimental data, are shown in **Figs. 8** and **9**.

$$\bar{\tau}_w/(\rho gRS_f) \quad = \quad 0.01\%SF_w(1.0 + P_b/P_w) \tag{9}$$

$$\bar{\tau}_b/(\rho gRS_f) \quad = \quad (1.0 - 0.01\%SF_w)(1 + 1/(P_b/P_w)) \tag{10}$$

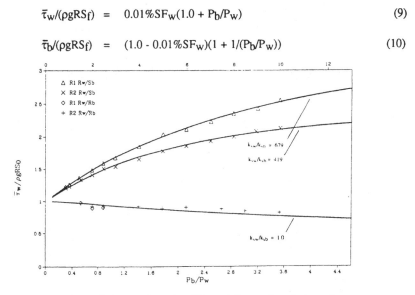

Figure 8 *Mean wall shear stresses for differentially roughened channels*

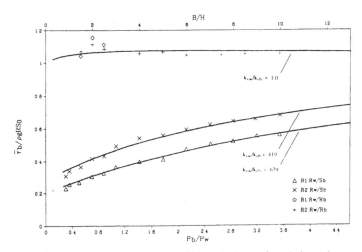

Figure 9 *Mean bed shear stresses for differentially roughened channels*

12

The maximum wall, τ_{wm}, and bed, τ_{bm}, shear stresses were also determined from the experimental data and plotted in a similar format to **Figs. 8** and **9**. These are shown in **Figs. 10** and **11** together with some best fit equations [1]. The maximum boundary shear stress data is less consistent than the mean boundary shear stress data due to perturbations caused by secondary flow cells. These distort the lateral variation of boundary shear stress in a complex way and will inevitably cause the maxima to increase in a non uniform manner [1,4,6,7,12]. The best fit equations therefore only represent the general trend of the data in these figures.

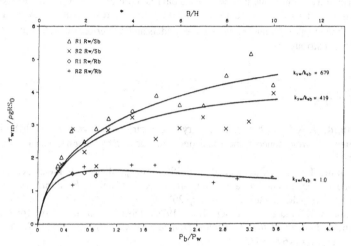

Figure 10 *Maximum wall shear stresses for differentially roughened channels*

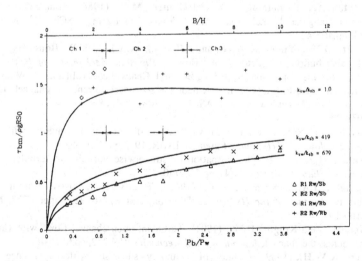

Figure 11 *Maximum bed shear stresses for differentially roughened channels*

13

6 Concluding remarks

The boundary shear force results presented in **Figs. 4-7** indicate the general reduction in $\%SF_w$ that occurs with increasing P_b/P_w and k_{sb}/k_{sw} values. Equations (4) and (5) represent` these distributions quite well but are dependent upon an ancillary equation linking C_2 with k_{sb}/k_{sw}. Equation (6) is shown to fit the data best for $k_{sw}/k_{sb} > 1$. The mean bed and wall boundary shear stresses may be readily derived from these basic equations as **Figs. 8** and **9** indicate. The maximum boundary shear stress data is more difficult to fit due to secondary flow structures. The $\%SF_w$ and mean boundary shear stress results provide a useful benchmark for the calibration of mathematical models and numerical modellers are invited to simulate these distributions.

REFERENCES

1   **Alhamid, A.A.I.**, (1991), Boundary shear stress and velocity distributions in differentially roughened trapezoidal open channels, *PhD. Thesis*, The University of Birmingham, UK.

2   **Flintham, T.P.** and **Carling, P.A.** (1988), The prediction of mean bed and wall boundary shear in uniform and compositely rough channels, *Proc.Int.Conf. on River Regime*, (Ed.W.R. White), J. Wiley, pp.267-287.

3   **Kartha, V.C.**, and **Leutheusser, H.J.** (1970), Distribution of tractive force in open channels, *Journal of the Hydraulics Division*, ASCE, Vol.96, No.HY7, Proc.Paper 7415, July, pp.1469-1483.

4   **Knight, D.W., Patel, H.S., Demetriou, J.D.** and **Hamed, M.E.**, (1982), Boundary shear stress distributions in open channel and closed conduit flows, *Proceedings of Euromech 156-Mechanics of sediment transport*, (Ed.B.Mutlu Sumer and A.Muller), Istanbul, July, A.A. Balkema, Rotterdam, Netherlands, pp.33-40.

5   **Knight, D.W., Demetriou, J.D.** and **Hamed, M.E.**, (1984), Boundary shear in smooth rectangular channels, *Journal of Hydraulic Engineering*, ASCE, Vol.110, No.4, April, pp.405-422.

6   **Knight, D.W., Yuen, K.W.H.**, and **Alhamid, A.A.I**, (1992), Boundary shear stress distributions in open channel flow, in *Physical Mechanisms of Mixing and Transport in the Environment*, (Ed. K. Beven, P.Chatwin & J.Millbank), J. Wiley.

7   **Naot, N.** and **Rodi, W.**, (1982), Calculation of secondary currents in channel flow, *Journal of the Hydraulics Division*, ASCE, Vol.108, No.HY8, Proc. Paper 17269, Aug., pp.938-968.

8   **Pyle, R.**, and **Novak, P.**, (1981), Coefficient of friction in conduits with large roughness, *Journal of Hydraulic Research*, IAHR, 19, 2, pp.119-140.

9   **Perkins, H.J.**, (1970), The formation of streamwise vorticity in turbulent flow, *Journal of Fluid Mechanics*, 44(4), pp 721-740.

10   **Replogle, J.A.**, and **Chow, V.T.**, (1966), Tractive-force distribution in open channels, *Journal of the Hydraulics Division*, ASCE, Vol.92, Paper 4727, Mar., pp.169-191.

11   **Shiono, K.** and **Knight, D.W.**, (1991), Turbulent open channel flows with variable depth across the channel, *Journal of Fluid Mechanics*, Vol.222, pp.617-646.

12   **Yuen, K.W.H.**, (1989), A study of boundary shear stress, flow resistance and momentum transfer in open channels with simple and compound trapezoidal cross sections, PhD thesis, The University of Birmingham, UK.

2 An experimental study for estimating hydraulic roughness of alluvial beds

M. B. A. Saad, M. T. K. Gaweesh and M. M. Gasser

ABSTRACT

Intensive measurements have been conducted experimently to get an accurate assessment for hydraulic roughness of the alluvial beds which is considered to be an essential parameter in determining of many hydraulic relations such as stage discharge, water level, design of water ways, degradation and aggradation. The experimental tests include six different types of alluvial materials, each type has a grain size distribution which can be characterized by its median diameter and geometric standard deviation. Those two parameters, for the involved experimental materials, are as follows: 0.089 mm, 1.42 & 0.210 mm, 1.3 & 0.235 mm, 1.85 & 0.390 mm, 1.65 & 1.070 mm, 3.31 & 1.960 mm, 2.95 & 4.0mm, 2.0.

In all tests the main hydraulic parameters such as: flow discharge, depth, slope and bed configuration were measured. The hydraulic roughness is expressed in this research by Chezy coefficient. It has been found that the Chezy coefficient is highly affected by the following three parameters: relative roughness, bed material characteristics and the geometry of the bed forms. Also, the Chezy coefficient was found to be proportional to the relative roughness in the case of the planed beds, while for the formed bed the adverse result was found.

1   Introduction

The resistance to flow over alluvial beds have received considerable attention nowadays. The prediction of an accurate assessment of the hydraulic roughness helps in solving the problems associated with alluvial channels such as degradation, aggradation and channel

15

design. The resistance of a channel with rigid bed has been accurately defined by relationships given by Keulegan and Manning. However extending the data collected on alluvial channels, in recent years, have indicated considerable variations in the hydraulic roughness.This factor is expressed in this paper by Chezy coefficient.

In alluvial channels the movement of bed material is accompanied by an undulating forms of the bed. The nature and size of these bed formations have been found to change appreciably with changes in flow conditions. Naturally, the resistance coefficient would vary with the change in the bed conditions. The main objective of this paper is to develop empirical equations to predict the Chezy coefficient for alluvial channels based on experimental flume data.

2   Data

The data used in the present analysis were quoted from different sources for different bed configuration as follows:

2.1   *Planed beds*

Saad [4] in 1986 conducted a series of armouring tests using different size bed materials and flow conditions. These tests resulted in a planed beds. Although small bed forms were initially developed, they were eroded at the end of the test.

2.2   *Rippled and dunned beds*

Simons and Richardson [5] in 1962 conducted a series of laboratory experiments and field measurments to study the formes of alluvial beds with respect to the flow parameters. Gaweesh [1],[2] and [3] in 1988, 1989 and 1991 conducted a series of experiments to study the sediment movement, as well as sediment flow discharge measurements. The bed deformation resulted in rippled and dunned beds.

The data of the mentioned references used in this paper are: the geometric mean diameter of the grain size (D_{50}), the geometric standard deviation (σ_g) and the flow parameters of the different tests such as flow depth (d) , discharge (Q) and slope (S). All the employed data are listed in Tables 1, 2 and 3 for planed beds, rippled beds and dunned beds, respectively.

3   Results and analysis

The hydraulic roughness, expressed in the present analysis as the Chezy coefficient (C) can be obtained from the Chezy equation,

$$V = C \sqrt{RS} \tag{1}$$

in which (V) is the average flow velocity obtained from dividing the flow discharge (Q) by the flow cross section area (A) ; (R) is the hydraulic radius obtained from dividing the flow cross section area (A) by the wetted perimeter (P) and (S) is the flow surface slope.The obtained results of the Chezy coefficient (C) is shown in Tables 1, 2 and 3.

16

TABLE 1 *Experimental data and results of planed beds*

No	Aut	D_{50}	σ_g	d	Q	S	A	R	V	C_m	C_p
1	s	1.07	3.31	0.04	0.005	0.00257	0.015	0.036	0.34	35.69	36.08
2	s	1.07	3.31	0.05	0.007	0.00227	0.018	0.042	0.40	40.84	39.45
3	s	1.07	3.31	0.06	0.010	0.00298	0.021	0.049	0.47	39.01	38.97
4	s	1.07	3.31	0.06	0.012	0.00311	0.024	0.054	0.52	40.04	39.91
5	s	1.07	3.31	0.07	0.015	0.00321	0.026	0.059	0.55	40.24	40.40
6	s	1.07	3.31	0.07	0.015	0.00322	0.027	0.061	0.56	40.10	40.51
7	s	1.07	3.31	0.07	0.017	0.00323	0.029	0.064	0.59	40.74	41.04
8	s	1.07	3.31	0.08	0.020	0.00354	0.031	0.068	0.64	41.38	41.50
9	s	1.96	2.95	0.03	0.005	0.00498	0.012	0.030	0.41	33.92	33.82
10	s	1.96	2.95	0.04	0.006	0.00500	0.014	0.034	0.46	35.28	34.99
11	s	1.96	2.95	0.04	0.008	0.00504	0.015	0.037	0.49	35.53	35.54
12	s	1.96	2.95	0.05	0.009	0.00486	0.017	0.041	0.52	36.78	36.71
13	s	1.96	2.95	0.05	0.010	0.00518	0.018	0.042	0.54	36.57	36.56
14	s	1.96	2.95	0.05	0.011	0.00509	0.020	0.046	0.58	37.49	37.52
15	s	1.96	2.95	0.06	0.011	0.00379	0.021	0.050	0.52	37.62	38.45
16	s	4.00	2.00	0.06	0.010	0.00364	0.022	0.052	0.45	33.50	33.70
17	s	4.00	2.00	0.07	0.013	0.00368	0.025	0.059	0.51	34.51	34.64
18	s	4.00	2.00	0.08	0.017	0.00381	0.030	0.068	0.57	35.27	35.64
19	s	4.00	2.00	0.08	0.021	0.00451	0.032	0.073	0.66	36.38	36.16
20	s	4.00	2.00	0.08	0.017	0.00388	0.029	0.068	0.58	35.64	35.72
21	s	4.00	2.00	0.08	0.021	0.00462	0.032	0.073	0.66	36.13	36.01
22	s	4.00	2.00	0.09	0.023	0.00501	0.033	0.075	0.71	36.40	36.17
23	s	4.00	2.00	0.09	0.027	0.00548	0.035	0.080	0.78	37.00	36.58
24	s	4.00	2.00	0.10	0.031	0.00502	0.039	0.088	0.80	37.86	37.61

17

TABLE 2 *Experimental data and results of rippled beds*

No	Aut	D_{50}	σ_g	d	Q	S	Δ	λ	A	R	V	C_m	C_p
1	G	0.24	1.85	.12	0.091	0.0009	0.030	0.120	0.320	0.12	0.28	26.94	28.28
2	G	0.24	1.85	.10	0.091	0.0010	0.040	0.150	0.270	0.10	0.34	34.00	30.81
3	G	0.09	1.45	0.27	0.026	0.00064	0.020	0.146	0.081	0.19	0.32	29.22	27.34
4	G	0.09	1.45	0.27	0.031	0.00082	0.021	0.153	0.081	0.19	0.38	30.32	27.79
5	G	0.09	1.45	0.27	0.020	0.00046	0.013	0.114	0.081	0.20	0.24	25.26	25.68
6	G	0.21	1.30	0.27	0.026	0.00082	0.019	0.145	0.081	0.21	0.32	24.81	26.81
7	G	0.21	1.30	0.27	0.031	0.00091	0.020	0.145	0.081	0.20	0.38	28.17	28.31
8	G	0.21	1.30	0.21	0.031	0.00190	0.021	0.170	0.063	0.17	0.50	27.90	27.47
9	G	0.39	1.65	0.49	0.210	0.00035	0.015	0.177	0.490	0.39	0.43	36.81	40.70
10	G	0.39	1.65	0.49	0.210	0.00042	0.021	0.198	0.485	0.40	0.43	33.41	38.89
11	S&R	0.45	1.60	0.30	0.176	0.00016	0.012	0.235	0.721	0.24	0.24	39.38	39.45
12	S&R	0.45	1.60	0.25	0.145	0.00017	0.018	0.210	0.602	0.21	0.24	40.82	39.02
13	S&R	0.45	1.60	0.18	0.103	0.00031	0.012	0.216	0.433	0.16	0.24	34.33	34.37
14	S&R	0.45	1.60	0.25	0.244	0.00036	0.021	0.369	0.612	0.21	0.37	42.40	39.89
15	S&R	0.45	1.60	0.26	0.224	0.00039	0.027	0.415	0.633	0.21	0.35	38.75	38.31
16	S&R	0.45	1.60	0.17	0.109	0.00040	0.018	0.223	0.407	0.15	0.27	34.83	34.25
17	S&R	0.45	1.60	0.24	0.222	0.00042	0.030	0.610	0.592	0.20	0.38	40.61	38.97
18	S&R	0.45	1.60	0.23	0.225	0.00047	0.021	0.305	0.600	0.19	0.40	42.21	39.29
19	S&R	0.45	1.60	0.11	0.055	0.00049	0.018	0.244	0.258	0.10	0.21	30.74	30.20
20	S&R	0.45	1.60	0.16	0.108	0.00060	0.018	0.223	0.382	0.14	0.28	31.10	32.01
21	S&R	0.45	1.60	0.10	0.055	0.00088	0.024	0.253	0.240	0.09	0.23	25.31	27.13
22	S&R	0.45	1.60	0.14	0.110	0.00088	0.030	0.475	0.337	0.13	0.33	30.96	31.52
23	S&R	0.45	1.60	0.09	0.055	0.00106	0.018	0.271	0.212	0.08	0.26	27.78	27.92

18

TABLE 3 *Experimental data and results of dunned beds*

No.	Aut	D50	σ_g	d	Q	S	Δ	λ	A	R	V	C_m	C_p
1	G	0.39	1.65	0.445	0.21	0.0052	0.054	1.34	0.445	0.369	0.49	33.93	33.93
2	S&R	0.45	1.6	0.21	0.225	0.00057	0.046	1.268	0.513	0.174	0.439	43.46	43.45
3	S&R	0.45	1.6	0.213	0.226	0.00078	0.061	1.344	0.518	0.197	0.436	35.17	35.16
4	S&R	0.45	1.6	0.125	0.120	0.00112	0.079	1.957	0.303	0.113	0.396	35.20	35.19
5	S&R	0.45	1.6	0.293	0.343	0.00114	0.094	1.469	0.712	0.236	0.482	29.39	29.38
6	S&R	0.45	1.6	0.305	0.383	0.00124	0.158	2.286	0.739	0.420	0.518	29.78	29.77
7	S&R	0.45	1.6	0.128	0.139	0.00189	0.079	1.917	0.31	0.116	0.448	30.26	30.25
8	S&R	0.45	1.6	0.186	0.23	0.00193	0.11	1.637	0.449	0.161	0.512	29.05	29.03
9	S&R	0.45	1.6	0.198	0.378	0.00247	0.125	2.012	0.483	0.170	0.783	38.21	38.20
10	S&R	0.45	1.6	0.189	0.247	0.00289	0.094	1.679	0.461	0.164	0.536	24.62	24.61
11	S&R	0.45	1.6	0.247	0.606	0.00301	0.094	2.24	0.599	0.205	1.012	40.74	40.73

3.1　Planed beds

Figure 1 indicates the variation of the Chezy coefficient (C) with the parameter (R/D_{50}) for different size distributions of bed material used in case of planed beds. The figure shows the effect of the geometric standard deviation on the Chezy coefficient. The Chezy coefficient (C) increases as the geometric standard deviation (σ_g) decreases. This means that the roughness of the planed beds increases with increasing the (σ_g). In mean time, the Chezy coefficient (C) increases as the relative roughness of the planed beds (R/D_{50}) increases for a given geometric standard deviation (σ_g). The variation of Chezy coefficient with the average flow velocity of different grain sizes is shown in Figure 2. For a given average velocity the Chezy coefficient increases as the geometric standard deviation of the bed material distribution increases.

3.2　Rippled beds

Figure 3 and Figure 4 show the variation of the Chezy coefficient (C) with the parameter (R/D_{50}) and the flow velocity V, respectively. These Figures indicate the scattering of the observations of each grain size distribution. Accordingly, σ_g is not an effective parameter on the Chezy coefficient.

3.3　Dunned beds

Figures 5 and 6 show the variation of Chezy coefficient (C) with (R/D_{50}) and the average flow velocity, respectively. Since only two types of bed material were used in the investigation of dunned bed, with almost the same value of the geometric standard deviation (σ_g); therefore the effect of (σ_g) on the Chezy (C) can not be distinguished.

A multi-regression program was employed to find out a correlation between the Chezy coefficient (C) and other parameter such as: relative roughness (R/D_{50}); geometric standard deviation (σ_g); flow surface slope (S); average flow velocity (V); and the ratio (λ/Δ) of bed form length (λ) to the bed form height (Δ). Many trials were made, consequently, to test the influence of each of the mentioned parameters as well as to develop the best fitting equation for predicting the Chezy coefficient. Table 4 shows the results obtained from these trials. Table 4 indicates that the best estimated Chezy coefficient (C) can obtained when all parameters are considered which corresponds to the smallest value of the standard deviation. Thus, the following equations were concluded to predict (C) according to the form of the alluvial beds:

a) for planed beds:

$$C = 9.73 \; \sigma_g^{0.411} \; S^{-0.28} \; V^{0.466} \; (R\backslash D_{50})^{-0.1} \tag{2}$$

with standard deviation = 1.03 %

b) for Rippled beds:

$$C = 13.75 \; \sigma_g^{0.26} \; S^{-.249} \; V^{0.494} \; (R\backslash D_{50})^{-0.091} \; (\lambda\backslash\Delta)^{0.008} \tag{3}$$

with standard deviation = 6.27 %

20

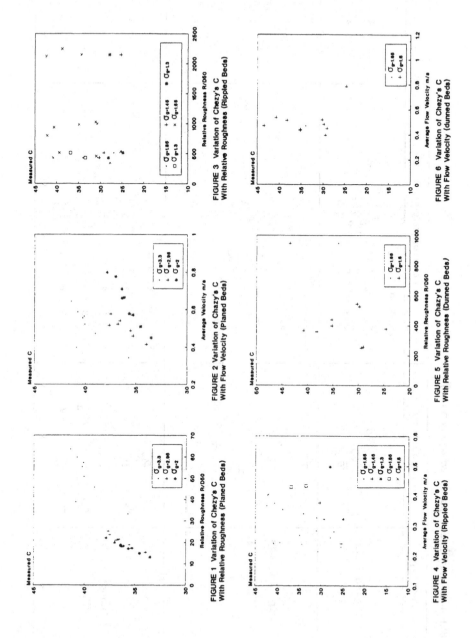

FIGURE 1 Variation of Chezy's C
With Relative Roughness (Planed Beds)

FIGURE 2 Variation of Chezy's C
With Flow Velocity (Planed Beds)

FIGURE 3 Variation of Chezy's C
With Relative Roughness (Rippled Beds)

FIGURE 4 Variation of Chezy's C
With Flow Velocity (Rippled Beds)

FIGURE 5 Variation of Chezy's C
With Relative Roughness (Dunned Beds)

FIGURE 6 Variation of Chezy's C
With Flow Velocity (dunned Beds)

21

TABLE 4 *Effect of different parameters on Chezy coefficient.*

Bed type	Parameters	Developed equation	Stand. dev.
Planed beds	$\sigma_g, \dfrac{R}{D_{50}}$	$C=25.73\ \sigma_g^{.053}\left(\dfrac{R}{D_{50}}\right)^{.129}$	2.24 %
	$\sigma_g, \dfrac{R}{D_{50}}, S$	$C=31.16\ \sigma_g^{-.054}\ S^{0.042}\left(\dfrac{R}{D_{50}}\right)^{.142}$	2.05 %
	$\sigma_g, \dfrac{R}{D_{50}}, S, V$	$C=9.73\ \sigma_g^{.411}\ S^{-.28}\ V^{0.466}\left(\dfrac{R}{D_{50}}\right)^{-.100}$	1.03 %
Rippled beds	$\sigma_g, \dfrac{R}{D_{50}}$	$C=23.43\ \sigma_g^{.747}\left(\dfrac{R}{D_{50}}\right)^{-.001}$	16.16 %
	$\sigma_g, \dfrac{R}{D_{50}}, V$	$C=39.5\ \sigma_g^{0.785}\left(\dfrac{R}{D_{50}}\right)^{.274}\ V^{.036}$	14.76 %
	$\sigma_g, \dfrac{R}{D_{50}}, S$	$C=7.87\ \sigma_g^{.317}\ S^{-.189}\left(\dfrac{R}{D_{50}}\right)^{.023}$	11.85 %
	$\sigma_g, \dfrac{R}{D_{50}}, \dfrac{\lambda}{\Delta}$	$C=6.18\ \sigma_g^{1.070}\left(\dfrac{R}{D_{50}}\right)^{0.089}\left(\dfrac{\lambda}{\Delta}\right)^{.262}$	11.95 %
	$\sigma_g, \dfrac{R}{D_{50}}, \dfrac{\lambda}{\Delta}, S$	$C=5.79\ \sigma_g^{0.680}\ S^{-.11}\left(\dfrac{R}{D_{50}}\right)^{-.037}\left(\dfrac{\lambda}{\Delta}\right)^{0.150}$	11.17 %
	$\sigma_g, \dfrac{R}{D_{50}}, \dfrac{\lambda}{\Delta}, S, V$	$C=13.75\ \sigma_g^{0.260}\ S^{-.249}\ V^{.494}\left(\dfrac{R}{D_{50}}\right)^{.091}\left(\dfrac{\lambda}{\Delta}\right)^{0.008}$	6.27 %
Dunned beds	$\sigma_g, \dfrac{R}{D_{50}}$	$C=26.42\ \sigma_g^{.243}\left(\dfrac{R}{D_{50}}\right)^{.019}$	17.05 %
	$\sigma_g, \dfrac{R}{D_{50}}, S$	$C=57.48\ \sigma_g^{-2.485}\ S^{-.106}\left(\dfrac{R}{D_{50}}\right)^{.013}$	15.48 %
	$\sigma_g, \dfrac{R}{D_{50}}, S, V$	$C=15.77\ \sigma_g^{2.330}\ S^{-0.5}\ V^{-0.500}\left(\dfrac{R}{D_{50}}\right)$	0.0 %
	$\sigma_g, \dfrac{R}{D_{50}}, S, V, \dfrac{\lambda}{\Delta}$	$C=15.8\ \sigma_g^{2.300}\ S^{-0.5}\ V^{-0.500}\left(\dfrac{R}{D_{50}}\right)$	0.0 %
	$\sigma_g, \dfrac{R}{D_{50}}, V$	$C=11.595\ \sigma_g^{3.188}\ V^{-0.197}\left(\dfrac{R}{D_{50}}\right)^{-.055}$	16.22 %

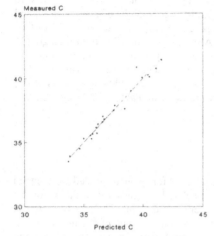

FIGURE 7 Predicted and Measured Chezy's C (Planed Beds)

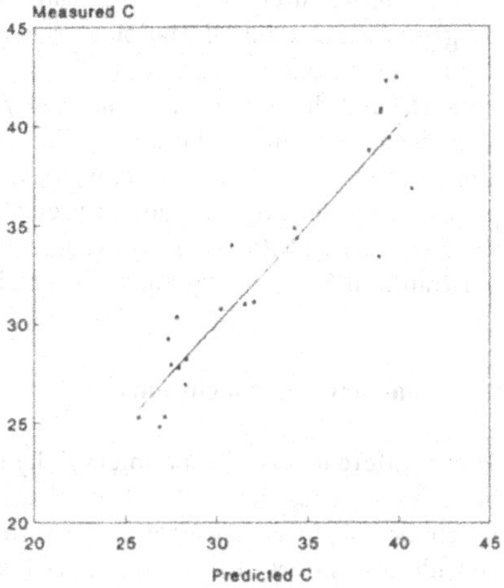

FIGURE 8 Predicted and Measured Chezy's C (Rippled Beds)

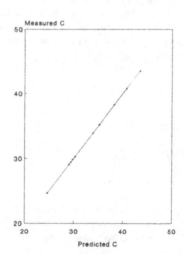

FIGURE 9 Predicted and Measyred Chezy'C (Dunned Beds)

c) for dunned beds:

$$C = 15.8 \ \sigma_g^{2.325} \ S^{-0.5} \ V \ (R \backslash D_{50})^{-0.50} \tag{4}$$

with standard deviation = 0.0 %

In the above three developed equations, the Chezy coefficient (C) is in proportion to the geometric standard deviation of the grain size distribution of the bed material(σ_g) and the average velocity of the flow (V). While it is adversely proportional to the flow surface slope (S) and the relative roughness (R/D_{50}). Also, the ratio of the bed form length (λ) to its height (Δ) has a little effect in case of rippled beds and has no effect in case of dunned beds. The three developed equations were applied to the experimental data presented in the current study to predict the value of (C). The obtained results of (C) were compared with the measured values as indicated in Tables 1, 2 and 3 and are illustrated in Figures 7, 8, and 9, respectively.

4 Summary and conclusions

Three different sets of experimental data were adopted for planed; rippled and dunned beds in the present study. These data were analyzed to study the variation of the hydraulic roughness parameter (C) with the different bed forms. The study showed a considerable effect of the bed material characteristics on the value of (C) in the case of planed beds only.Three empirical equations 2, 3 and 4 were developed to predict the coefficient (C) for different forms of the alluvial beds. These three equations involved the bed and flow characteristics. They were applied to the experimental data and the obtained results of the coefficient (C) and were found to be in a good agreement with the measured values. The developed three equations indicated the proportionality of (C) values with the geometric standard deviation of the bed material distributions (σ_g), and with the average flow velocity (V), while it is in reverse proportion with the flow surface slope and with the relative roughness(R/D_{50}).

Symbol definitions

A   Flow cross sectional area (m^2);
b   Flume width (m);
C   Chezy coefficient ($m^{1/2}/s$);
C_m Measured Chezy coefficient ($m^{1/2}$)/s);
C_p Predicted Chezy Coefficient ($m^{1/2}/s$);
d   Flow depth (m);
D_{50} Geometric mean diameter of the grain size distribution of the bed material (mm);
Q   Flow discharge rate (m^3/s);
P   Wetted perimeter of the flume cross section (m).
R   Hydraulic radius of the cross section (m);
S   Flow surface slope;
V   Flow average velocity (m/s);
Δ Height of the bed form (m);
λ Length of the bed form (m); and
σ_g Geometric standard deviation of the grain size distribution of the bed material.

Aut   Author
No   Number of run

REFERENCES

[1] Gaweesh, M.T.K. (1988), "Investigation of Sediment Behavior in a Channel with Flood plains", Ph.D. Thesis, Civil Eng. Dept., University of Southampton, Southampton, England.

[2] Gaweesh, M.T.K. (1989), Hydraulic Studies on the Nile River and its Structures. Netherlands Technical Assistance Project, Report on Training in the Netherlands on Sediment Transport.

[3] Gaweesh, M.T.K. (1991), "Calibration of the Delft-Nile Sampler",Report, the Hydraulics and Sediment Research Institute, Delta Barrage, Egypt.

[4] Saad, M.B.E. (1986), "The Armouring of Alluvial Channel Beds and the Evaluation of the Hydraulic Characteristics of the Armour coat", Ph.D. Thesis, Civil Eng. Dept. University of Southampton, Southampton, England.

[5] Simons, D.B and Richardson, E.V. (1962), "Resistance to Flow in Alluvial Channels", Transactions, ASCE, paper No. 3360, Vol. 127, Part 1, pp. 927-1006.

3 Conveyance of meandering compound channels

M. L. Lorena and D. A. Ervine

ABSTRACT

In the period 1988 to 1991, a programme of research, entitled S.E.R.C. Flood Channels Series B, has been carried out at Hydraulics Research Ltd, Wallingford. The main thrust of this work has been the experimental study of meandering compound channels with overbank flow and meandering two-stage channels. The work at Wallingford used the large experimental flume (50*10m), investigated two sinuosities of meandering main channel, a range of flood plain roughness, a range of cross-sectional geometry, and for various depths of inbank and overbank flow. This study is a brief summary of some of the experimental work above, concentrating on flow resistance and flow conveyance. This study begins with an analysis of the effect of several parameters on the stage-discharge curve and on the flow of the meandering compound flow. These parameters, include the sinuosity(curved meander length/straight length)of the meandering main channel (low/high), the cross-section geometry(trapezoidal/natural), the boundary reference(smooth/rough), together with relative flow depth. This paper includes an analysis of how skin friction, together with other additional energy losses such as secondary currents expansion-contraction phenomena and lateral shear can affect the conveyance of meandering channels. The analysis is performed by a non-dimensional function F_5 (actual discharge/computed discharge), and includes data from S.E.R.C. Series A (the straight compound channel) and as well as data from model studies from Willetts and Hardwick [7].

1 Introduction

Many parameters affect the conveyance of meandering compound channels. Previous research has shown that:

- Increasing the main channel sinuosity diminishes conveyance significantly (U.S.A. Corps of Engineers [6]). Typically the discharge reduces 30%-40% for increasing sinuosity from 1.0 to 1.57.
- Floodplain roughness reduces conveyance substantially (Sellin and Giles [5]). Typically 30% reduction in the total conveyance comparing compound channels with cut and uncut plain vegetation.
- The geometry of the main channels cross-section seems to have a considerable

effect on the conveyance. This is true both for the shape of the main channel (Willetts and Hardwick [7]) but also for the aspect ratio of the main channel (Ervine and Jasem [3]).
- The relative depth of flow (Kiely [4]) is also a significant parameter. For instance the strength of the secondary currents that develop in the main channel are much greater for an overbank case than for an inbank case. These secondary currents as shown by Ervine and Ellis [2], introduce significant energy losses and reduce the conveyance of meandering compound channels.

The experimental programme using the S.E.R.C. flume Series B at Hydraulics Research Ltd, Wallingford, lasted for a two year period and included detailed measurements of stage(H), discharge(Q), velocity(U), stream angles(θ), boundary shear stress(τ_0), local water levels, dispersion tests, turbulent velocities and turbulent shear stress measurements. One of the aims of the experimental programme was to carry out an investigation of the parameters that affect the conveyance of meander compound channels. In this programme the following parameters were varied: stage(inbank and overbank in the range 50<H<300mm); sinuosity (1.37 and 2.04); geometry of the main channel cross-section (natural section for both sinuosities and trapezoidal for sinuosity 1.37); floodplain roughness (smooth and roughened). The bend apex cross-section of the natural meander was designed based on averaged values of 17 bend apices of natural rivers. In this cross-section as it happens in nature, the deepest part of the section is near the outer bank of the bend and the shallower part of the section is near the inner bank of the bend. It was assumed as well that the cross-section area of the natural main channel between bend apex and the trapezoidal cross-over was maintained constant. This avoids the distortion of the velocity field. Fig. 1 shows both sinuosities used in S.E.R.C. flume.

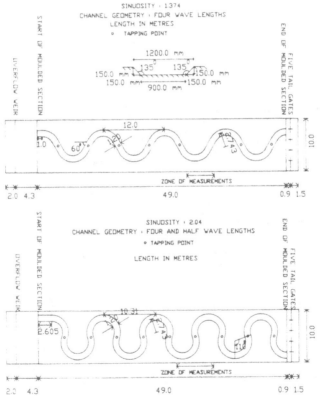

FIGURE 1   Plan view of S.E.R.C. flume Series B with two sinuosities used in the Flood Channel Programme.

The parametric analysis of the meander channels will be made in terms of the effect of

28

the above parameters on the stage-discharge curve and on the flow resistance. The parametric analysis is concluded through the application of a non-dimensional function F_S(actual discharge/computed discharge) which will allow to separate the skin friction losses from other additional losses that occur in meandering compound channels such as secondary currents, expansion-contraction phenomena and lateral shear (Ervine and Ellis [2]).

2   Stage-discharge Curve

The stage-discharge curve was determined for the steady state and quasi uniform flow conditions. In this case the uniform flow conditions were defined as having the same depth in each tapping point of the cross-over. The location of the tapping points in each cross-over is shown in Fig. 1. The channel slope was $0.996*10^{-3}$ for sinuosity 1.37 and $1.02*10^{-3}$ for sinuosity 2.04. For the fully roughened floodplain case, the roughness was produced through 25.0mm diameter rods, with a density of 12 rods per square meter, always piercing the water surface. The cross-section area of the trapezoidal section ($0.1575\ m^2$) was 40% greater than the natural cross-section for sinuosity 1.37. However, both cross-sections had the same top width (1.2 m). The area of the natural cross-section for sinuosity 1.37 ($0.0941\ m^2$) and 2.04 ($0.098\ m^2$) was approximately the same. The stage-discharge curve for the S.E.R.C. flume Series B are plotted in Fig. 2.

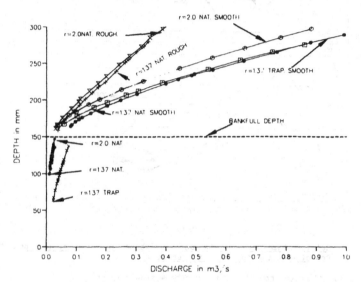

FIGURE 2   Stage-discharge curve of S.E.R.C. flume Series B

This figure shows that for the overbank stage 200.0 mm, for instance:
- As the sinuosity increases from 1.37 to 2.04, the discharge is reduced by 23% (for the case of smooth floodplain and natural main channel cross-section).
- As the floodplain roughness changes from the smooth to the fully roughened case, the discharges by 58% for the case of sinuosity 1.37 and natural cross-section.

Fig. 2 reveals also the effect of the cross-sectional geometry on the conveyance of the meander channel for the case of sinuosity 1.37 and smooth floodplains. For low stages, the meander channel with trapezoidal cross-section delivers more flow than the natural cross-section geometry. This is the result of the larger area of the trapezoidal cross-section (40% larger) compared with the natural cross-section. However, for higher stages the stage-discharge curves show a trend that the opposite will occur and the natural cross-sectional geometry conveys more flow.

These findings are evidence that not only the sinuosity and the floodplain roughness can affect the conveyance of meander compound channels, but also the cross-sectional geometry of the main channel seems to have an important role. A further parameter was

29

discovered by the authors to be important, namely the aspect ratio (top width/bankfull depth) of the main channel. This was discovered by comparing S.E.C.R. flume data with other smaller scale data (with smaller aspect ratios) and will be discussed in section 4.

3   Flow resistance

The flow resistance analysis was carried out by computing the Darcy-Weisbach friction factor(f) and plotting with Reynolds number as in Fig. 3. The calculation of the friction factor was based on the assumptions that: both main channel and floodplain were treated as a single channel; the friction factor was calculated for bend apex cross-section only; the bed slope for the inbank case was the floodplain slope divided by the sinuosity; and for overbank cases the bed slope was taken as the floodplain slope throughout. In essence this is a crude definition of the friction factor but at least gives a qualitative picture of the effect of each parameter.

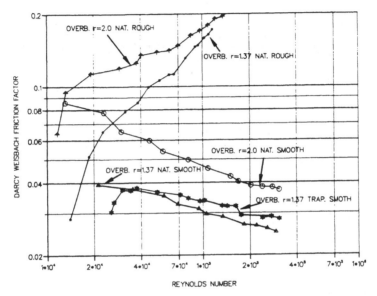

FIGURE 3   Relationship between the Darcy-Weisbach friction factor and the Reynolds Number for S.E.R.C. Series B.

The following can be noted from Fig. 3 for overbank flow data only:
    - The friction factor decreases with stage, for the smooth floodplain case and increases with stage for the rough floodplain cases.
    - Flow resistance increases with sinuosity, particularly with sinuosity for lower Reynolds Numbers or for lower values of stage.
    - Flow resistance varies with cross-sectional shape with the natural geometry, producing lower resistance than the trapezoidal geometry, at least at higher values of stage.

4   Parametric analysis of flow conveyance

Following on from the work of Ackers [1], an initial parametric analysis of flow conveyance in meandering compound channels was carried using a non-dimensional function F_5. This function is defined as the ratio of actual discharge (total) divided by the theoretical discharge which was calculated by the horizontal division wall at the bankfull level, a vertical division wall at the edge of the meander belt width, but the only energy loss taken was that produced by the skin friction. This method was developed by Ervine and Ellis [2]. Therefore for a straight open-channel the value of function F_5

30

will be near 1.0. In compound meandering channels because of range of other additional energy loss, the value of function F_5 is smaller than 1.0. The degree that F_5 is smaller than 1.0, depends on the additional energy loss other than skin friction. The skin friction term was computed used the smooth turbulent law for the S.E.R.C. flume recommended by Ackers [1]:

$$\frac{1}{\sqrt{f}} = 2.02 \log(Re \sqrt{f}) - 1.38 \qquad (1)$$

where f is the Darcy-Weisbach friction factor and Re is the Reynolds Number. Equation 1 was based in S.E.R.C. Series A data for inbank case. For the case of fully roughened floodplain the theory devised by Ackers [1], considering the drag coefficient of vertical was employed. The analysis carried out with function F_5 involved not only data from S.E.R.C. Series B but also S.E.R.C. Series A (the straight compound channel sinuosity 1.0) data, together with data from Willetts and Hardwick [7] (the meander compound channel with aspect ratio 3.48).

The variation of F_5 with sinuosity for the smooth floodplain case and for the fully roughened floodplain case are presented in Figs 4 and 5. This is shown for a range of

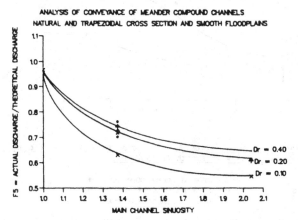

ANALYSIS OF CONVEYANCE OF MEANDER COMPOUND CHANNELS
NATURAL AND TRAPEZOIDAL CROSS SECTION AND SMOOTH FLOODPLAINS

FIGURE 4   Variation of function F_5 with main channel sinuosity for smooth floodplains.

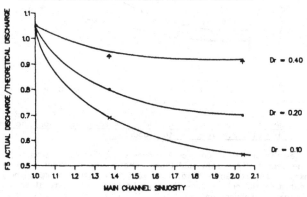

ANALYSIS OF CONVEYANCE OF MEANDER COMPOUND CHANNELS
NATURAL AND TRAPEZOIDAL CROSS SECTION WITH ROUGH. FLOODPLAINS

FIGURE 5   Variation of function F_5 with main channel sinuosity for fully roughened floodplains.

depth ratios ((H-h)/H) up to 0.4. It is evident the reduction of F_5 with increasing sinuosity. However, as depth ratio increase this effect is diminished. It can be seen in both figs 4 and 5 that for low depth ratios, the discharge can be reduced by almost 50% by increasing sinuosity from 1.0 to 2.0. This is an important finding. the effect of sinuosity is felt more in smooth floodplain cases compared with rough floodplains where considerable energy
loss was due to the vertical rods.
The effect of aspect ratio of the main channel (Bc/h) on the variation of function F_5 for sinuosities 1.37 and 2.04 is shown in figs. 6 and 7. As the aspect ratio of the main

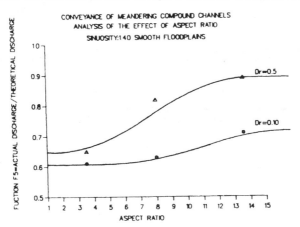

FIGURE 6   Variation of function F_5 with aspect ratio for the main channel sinuosity 1.4 with smooth floodplains.

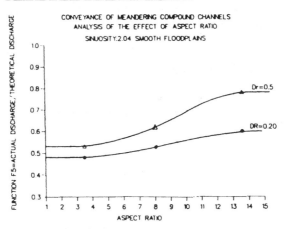

FIGURE 7   Variation of function F_5 with aspect ratio for the main channel sinuosity 2.0 with smooth floodplains.

channel increases the value of function F_5 moves closer to 1. This means, that with an increase of the aspect ratio, the additional energy losses other than skin friction diminish. Consequently, it can be said, that most model studies employing small aspect ratios, over·estimate the energy loss which would exist in a natural channel where the aspect ratio is often greater than 10.

32

5   Conclusions

In this parametric analysis of the conveyance of meandering compound channels the following conclusions can be drawn:

- The conveyance of meandering compound channels is not only affected by the skin friction but also by other losses produced by: floodplain roughness (for instance, as the floodplain roughness changes from the smooth to the fully roughened case, the discharge reduces by 58% for the case of sinuosity 1.37 and natural cross-section); sinuosity(for instance, as sinuosity increases from 1.37 to 2.04 the friction factor may raise by 50% and give discharge reductions of 20-25%); main channel cross-section geometry (for higher stages, the natural cross-section conveys more flow than the trapezoidal cross-section); aspect ratio (main channel cross-sections with larger aspect ratios produce less resistance to the flow).
- Model studies of meandering compound flows will give misleading results when the model is distorted and the aspect ratio is much lower than nature.
- In meandering channels, for low depth ratios on the floodplain, the additional loss of energy produced by secondary currents, expansion-contraction phenomena and lateral shear can reach approximately 50% of the total loss of energy. As the depth ratio increases the role of additional loss of energy diminish and most of energy loss is then produced by skin friction.

Acknowledgements

The research described in this paper was sponsored by the Scientific Research Council (S.E.R.C.). The authors are thankful to Prof. B. Willetts and Mr. R. Hardwick from Aberdeen University, Prof. R. Sellin and Ms. R. Greenhill from Bristol University, and Dr. D. Knight, Dr. Y. Fares from Birmingham University which were part of the same team of S.E.R.C. Series B Flood Channels project.

References

[1]   Ackers,P., (1991), "Hydraulic Design of Straight Compound Channels", H.R. Wallingford, Report SR 281, October.
[2]   Ervine, D. A. and Ellis, J.,(1987), "Experimental and Computational Aspects of Overbank Flood Plain Flow", Transactions of the Royal Society of Edinburgh, Earth Sciences, Vol78, pp.315-325.
[3]   Ervine, D. A. and Jasem, H. K., (1992), "Two-Stage Channels with Skew Compound Flow" (for publication in A.S.C.E.).
[4]   Kiely, G.K., (1989), "An Experimental Study of Overbank Flow in Straight and Meandering Compound Channels", Ph.D. Thesis. Department of Civil Engineering, U.C. Cork, Ireland, September.
[5]   Sellin R. and Giles A., (1988), "Two Stage Channel Flow", Publication of University of Bristol, department of Civil Engineering, July.
[6]   U.S.A. Corps of Engineers, (1956), "Hydraulic Capacity of Meandering Channels in Straight Floodways", Technical Memorandum, No2-249, March.
[7]   Willetts, B.B. and Hardwick R.I., (1990), "Model Studies of Overbank Flow from a Meandering Channel", International Conference on River Hydraulics edited by W.R. White Hydraulics Research Ltd, Published by John Wiley & Sons Ltd, pp 197-205, September.

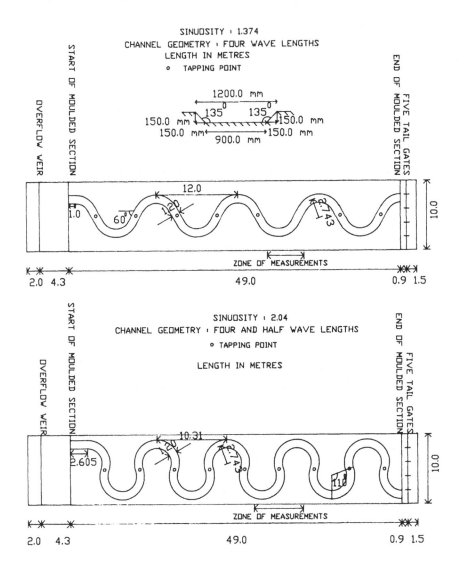

SINUOSITY : 1.374
CHANNEL GEOMETRY : FOUR WAVE LENGTHS
LENGTH IN METRES
o   TAPPING POINT

SINUOSITY : 2.04
CHANNEL GEOMETRY : FOUR AND HALF WAVE LENGTHS
o TAPPING POINT

LENGTH IN METRES

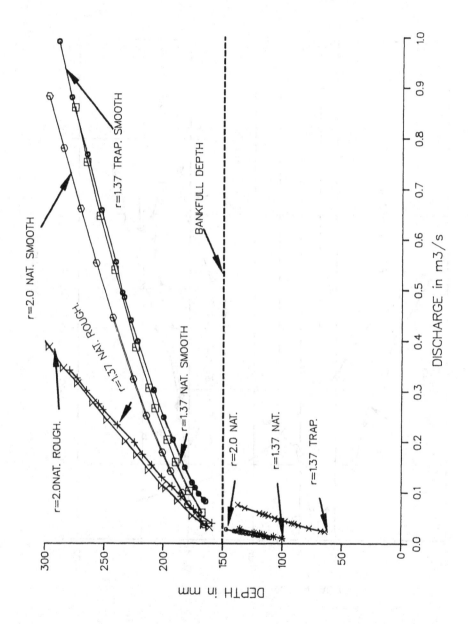

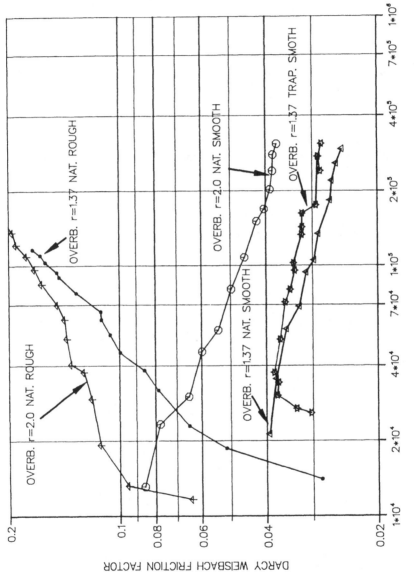

35

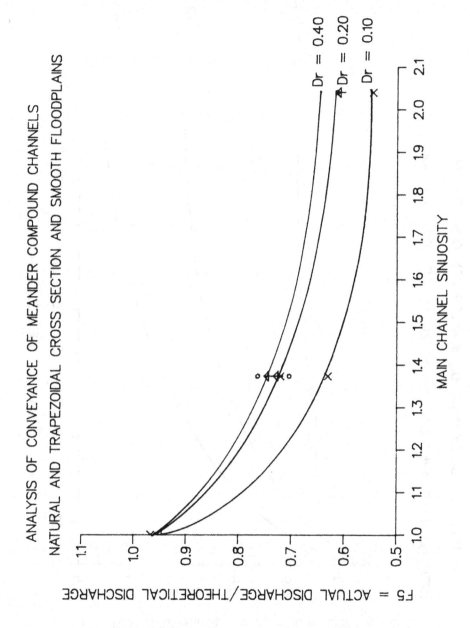

ANALYSIS OF CONVEYANCE OF MEANDER COMPOUND CHANNELS
NATURAL AND TRAPEZOIDAL CROSS SECTION AND SMOOTH FLOODPLAINS

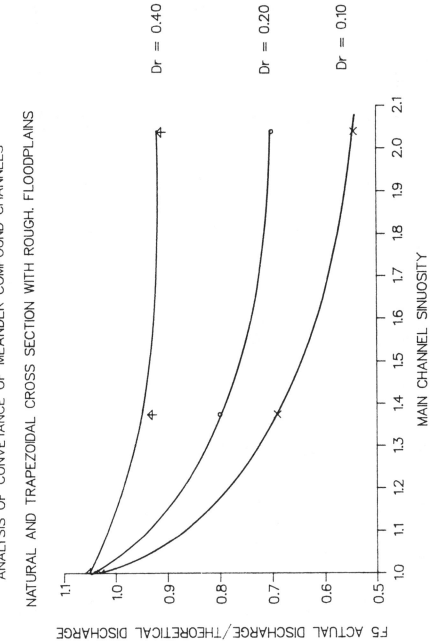

ANALYSIS OF CONVEYANCE OF MEANDER COMPOUND CHANNELS

NATURAL AND TRAPEZOIDAL CROSS SECTION WITH ROUGH. FLOODPLAINS

Dr = 0.40

Dr = 0.20

Dr = 0.10

F5 ACTUAL DISCHARGE/THEORETICAL DISCHARGE

MAIN CHANNEL SINUOSITY

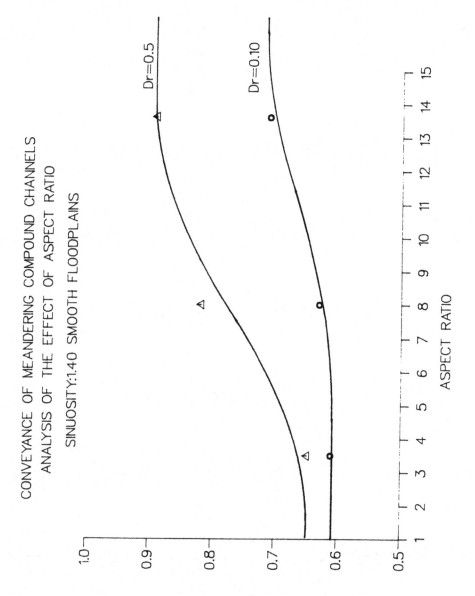

CONVEYANCE OF MEANDERING COMPOUND CHANNELS
ANALYSIS OF THE EFFECT OF ASPECT RATIO

SINUOSITY:1:40 SMOOTH FLOODPLAINS

Dr=0.5

Dr=0.10

ASPECT RATIO

FUCTION F5=ACTUAL DISCHARGE/THEORETICAL DISCHARGE

39

4 Estimation of turbulence parameters within intertidal saltmarsh channels

J. R. French and N. J. Clifford

ABSTRACT

Interest in sedimentary processes within coastal marshes has been stimulated by an increasing awareness of their delicate adjustment to tidal elevations, and of their obvious engineering significance in the context of coastal protection schemes. In tide-dominated marshes, typical of the UK, a large proportion of the tidal exchange of water and materials occurs via well-defined channel systems which frequently drain completely at low water. Flows within these are correspondingly complex. In particular, estimation of turbulence parameters known to be important in the entrainment and suspension of fine sediment (channel sand and externally derived silt) is rendered problematic by virtue of the need to eliminate severe nonstationarity due to flow unsteadiness over a range of non-turbulent scales. In addition to tidal variation, large velocity transients occur on both flood and ebb phases as the marsh surface undergoes inundation and drainage. After careful attention to series de-trending, however, a large number of near-bed turbulence measurements have been obtained. These indicate significant variation in the intensity and 'event structure' of turbulence within and between individual tidal periods, although extreme short-term flow unsteadiness partly obscures longer-term pressure gradient effects. High rms turbulence intensities and streamwise shear stresses persist after marsh surface inundation; these may be significant in entraining sediment and in maintaining vertical concentration profiles, thus facilitating the introduction of silt-sized material to the adjacent marsh surfaces via lateral momentum exchange across channel margins. Neap tides, which do not inundate the marsh, exhibit lower turbulence intensities. These findings offer additional insight into channel - marsh interactions, and suggest new avenues of research that might lead to more sophisticated mathematical models of these environments, and to more accurate prediction of their response to natural and human-induced change.

41

Introduction

Despite published concern over the apparent imbalance, within many coastal marshes, between net vertical accretion and rising tidal levels [24], a number of modelling studies have demonstrated the potential of predominantly minerogenic marsh surfaces to track rising sea-level, subject to sediment supply constraints [10][2]. Such models necessarily involve long-term time-averaging of critical sedimentation parameters, but formulation of the underlying conceptual models is ultimately dependent upon our knowledge of fundamental hydrodynamic and sedimentary processes. Research carried out on the north Norfolk coast has demonstrated the highly unsteady nature of the tidal channel flows which dominate material exchange between the nearshore zone and vegetated marsh surfaces [3][16]. Well-defined velocity and stress transients occur close to bankfull stage on both flood and ebb phases of the tide, and result from the interaction of shallow water tide with the discontinuous prism-stage relation and large hydraulic roughness of vegetated marshes [12]. These transients are associated with i) large advective material fluxes; ii) local entrainment and suspension of channel sediments; and iii) the maintenance of vertical concentration gradients of suspended mud, which comprises the bulk of the marsh sedimentary fabric [9].

Modelling these processes using conventional sediment transport formulae which involve estimation of shear stress from the time-averaged vertical velocity profile, may result in large errors when the acceleration terms are large; previous studies within the sandy intertidal reveal significant deviations of measured current velocities from fitted logarithmic profiles close to high water and during the ebb phase of the tide [4][6]. However, electromagnetic current meters (EMCMs) provide a quasi-continuous record of velocity simultaneously in 2 or 3 dimensions, and allow the 'direct' estimation of the turbulent stress terms that are known to be important in determining the rate and mode of sediment transport.

Laboratory studies of turbulent boundary layers have demonstrated the occurrence of coherent, anisotropic, eddy structures superimposed upon the time- and space-averaged properties of the flow [19][8]. These Lagrangian structures take the form of local temporary 'ejections' of low momentum fluid away from the boundary ('bursts'; [15]), balanced by more diffuse inrushes of fluid due to 'sweeping' motions almost parallel to the boundary. The resulting 'ejection-sweep' cycles are statistically repetitive in time and space, and impart an 'event-structure' that may extend across the entire boundary layer thickness and which accounts for a large portion of turbulent kinetic energy and Reynolds stress production [7]. EMCM measurements have revealed the existence of a similar 'event structure' within a variety of marine, estuarine and fluvial boundary layers. Both laboratory and field observations show that the temporal distribution of Reynolds stress production is highly asymmetrical, with a large proportion of the total stress contributed over a small proportion of total observation time. Spatially and temporally intermittent stress generation has important implications for sediment movement [25], although relatively few data presented to date have been acquired within the context of broader geomorphological investigations [5].

The present work arises from a desire to elucidate more fully the link between marsh creek hydrodynamics and sedimentary processes on adjacent marsh surfaces. The essential characteristics of the near-bed turbulent stress field within a typical marsh channel have been described in an earlier paper [11]. This paper summarises the findings of a more recent EMCM deployment within the same channel, with particular emphasis being placed upon i) the problems of signal pre-processing and turbulence parameter estimation; and ii) the variability in turbulent stress characteristics in relation to outer flow variables, and between spring and neap tides.

Research design

Field location

Measurements were made during both flood and ebb phases of a neap tide (24 September, 1990), and a spring tide (17 March 1991), at a station located approximately 50m upstream from the mouth of Hut Creek. This is one of two large creek systems draining Hut Marsh, a 55ha back-barrier marsh at the western end of Scolt Head Island, north Norfolk. Tidal range averages 6.4m at springs, and highest astronomical tides reach 4.0m above Ordnance Datum (O.D.). Marsh surfaces lie between 2.5 and 3.1m O.D. and are flooded 200-300 times a year. The channel bed at the measurement station is at approximately 1.1m O.D. and dries out either side of low water on every tide. Bed sediment is quartz sand (median diameter 340μm) and locally abundant shell debris. Banks are comprised of cohesive marsh mud (median diameter 10-20μm), with bankfull width and thalweg depth being 30.0 and 2.6m respectively. Hut Marsh experiences essentially marine salinities, and no significant longitudinal density gradients exist within Hut Creek. A more comprehensive site description is given in reference [12].

Instrumentation

Water level was recorded at 5 minute intervals against a vertical stage in the channel thalweg. Time-averaged velocity, U, was measured at the same intervals (30s averaging) using a Braystoke flowmeter (12.7cm impellor) 0.25m above the thalweg. 'Instantaneous' streamwise ($U = \bar{u} + u'$) and vertical ($W = \bar{w} + w'$) velocities were measured using a 2-component electromagnetic current meter (Colnbrook Instrument Developments 5.5cm diameter discoidal sensor; Valeport Marine Ltd modified electronics) deployed on a rigid mounting at 0.18m above the bed. The analogue signal from the sensor was low-pass filtered at 10 or 2.5Hz and digitized at 20 or 5Hz, for spring and neap tides respectively, using a Campbell Scientific logger aboard instrument platform on the marsh surface. Series duration was typically 3-7 minutes, which, together with the digitization rate, represent a compromise between the conflicting requirements of sampling variability, low-frequency signal retention and exclusion of tidal nonstationarity. Following Soulsby [23], the effective digitization interval, f_D, depends upon sensor dimension, L, and the mean flow rate, such that

$$f_D \approx 1.4\,U\,/\,L \tag{1}$$

The 5.5cm sensor used here has an electrode spacing of 4.0cm, and over the range of U encountered here, f_D varies in the range $0.35 < f_D < 4.9$ Hz (neap tides) and $0.35 < f_D < 30.1$ Hz (spring tides). Given the constraint imposed by the fixed sensor dimension, the actual digitization rates thus represent a compromise, minimising the loss of potentially measurable high frequency contributions but at the expense of data redundancy in some series. The electronics used here exhibit significantly better signal:noise ratios than earlier Colnbrook designs: rms noise values, measured in still water, were typically 1.7×10^{-3}m s^{-1} for the U component, and 1.9×10^{-3}m s^{-1} for the W component.

EMCM signal processing

Low frequency tidal variation was identified by visual inspection of time-series plots and was removed by taking residuals from low order polynomials fitted using ordinary least squares procedures. The resulting series were examined for any remaining nonstationarity by reference to the series autocorrelation function (ACF), prior to computation of additional turbulence parameters. In line with previous work, and given the lack of data

43

TABLE 1 Classification of 'event' types using the $u'w'$ quadrant method.

Quadrant	'Event' type	Interactions		
		u'	w'	$u'w'$
$u'w'_1$	Outward interaction	>0	>0	>0
$u'w'_2$	Ejection (burst)	<0	>0	<0
$u'w'_3$	Inward interaction	<0	<0	>0
$u'w'_4$	Inrush (sweep)	>0	<0	<0

TABLE 2 Series-averaged outer flow parameters for Series N-A to N-K (neap tide), and Series S-A to S-I (spring tide), where υ is the kinematic viscosity of seawater. Series timings refer to hours-minutes relative to HW; marsh inundation occurs when h > 1.6m.

Series (timing)	h (m)	z/h	$U_{0.2}$ (m s^{-1})	Re ($U_{0.2}h/\upsilon$)	dU/dt (m s^{-2})
N-A (- 1.20; Flood)	0.85	0.24	0.12	6.8×10^4	≈ 0.0
N-B (- 1.09)	0.95	0.22	0.15	9.5×10^4	8.1×10^{-5}
N-C (- 0.54)	1.15	0.17	0.090	6.9×10^4	-1.1×10^{-4}
N-D (- 0.35)	1.30	0.15	0.062	5.4×10^4	≈ 0.0
N-E (- 0.09)	1.38	0.14	0.030	2.8×10^4	≈ 0.0
N-F (0.09; Ebb)	1.35	0.15	0.010	9.0×10^4	≈ 0.0
N-G (0.35)	1.25	0.16	0.035	2.9×10^4	4.3×10^{-5}
N-H (0.55)	1.06	0.19	0.046	3.3×10^4	≈ 0.0
N-I (1.15)	0.88	0.23	0.063	3.7×10^4	≈ 0.0
N-J (1.35)	0.65	0.31	0.090	3.9×10^4	≈ 0.0
N-K (1.50)	0.47	0.43	0.090	2.8×10^4	9.0×10^{-5}
S-A (- 2.12; Flood)	0.44	0.46	0.32	9.0×10^4	$- 4.0 \times 10^{-4}$
S-B (- 1.55)	0.93	0.22	0.28	1.8×10^5	1.6×10^{-4}
S-C (- 1.35)	1.38	0.14	0.29	2.6×10^5	$- 4.1 \times 10^{-5}$
S-D (- 1.15)	1.65	0.12	0.26	2.8×10^5	2.9×10^{-4}
S-E (- 1.00)	1.83	0.11	0.42	5.2×10^5	1.7×10^{-4}
S-F (- 0.48)	1.93	0.10	0.50	6.2×10^5	7.9×10^{-5}
S-G (- 0.10)	2.19	0.09	0.13	1.8×10^5	$- 3.4 \times 10^{-5}$
S-H (1-10; Ebb)	1.60	0.12	0.80	9.5×10^5	2.6×10^{-4}
S-I (1-26)	1.31	0.15	0.67	6.1×10^5	3.2×10^{-4}

relating to the lateral velocity field, the following analysis is restricted u'^2, w'^2 and the streamwise turbulent shear stress, $-\rho \overline{u'w'}$. The correlation, $(\overline{u'w'}/u'w')$, and the normalized horizontal and vertical turbulence intensities

$$(u'^2)^{0.5}/u_* \quad \text{and} \quad (w'^2)^{0.5}/u_* \qquad \text{where } u_* = (\overline{u'w'})^{0.5}$$

describe the variation in turbulence structure with changing outer flow variables, and allow comparison with previous field and laboratory results. The ACFs were also used to estimate an integral length scale, Λ, of the turbulence, using

$$\Lambda = \bar{u} \int_0^\phi R(\gamma) \, d\gamma \qquad (2)$$

where $R(\gamma) =$ is the ACF for the series, integrated up to the point, ϕ, where $R(\gamma)$ becomes statistically insignificant. This is given by

$$R(\phi) < 2/\sqrt{N} \qquad (3)$$

where N is the number of velocity observations.

The detailed turbulence 'event structure' was investigated by conditionally sampling the u' and w' series following the procedures outlined by Lu and Willmarth [20]. This 'quadrant analysis' involves quantitative discrimination of characteristic fluid motions ('events') on the basis of contributions within the four quadrants of the u'w' plane (Table 1), and is useful in that it allows observations made within geophysical flows to be placed within the conceptual framework provided by laboratory flow visualizations. With this technique, it is usual to apply a threshold criterion to avoid the problem of low magnitude stress contributions being assigned the status of discrete events purely on the basis of the sign of their b' and w' components. The threshold parameter is difficult to quantify on purely *a priori* grounds, and has been a source of considerable debate and inconsistency within published work. Quadrant analyses were performed for each of the series considered here, using amplitude thresholds of between 0 and 5 standard deviations about the mean stress, as well as that corresponding to the 90% stress level. The mean interval between discrete ejections, t_{ej}, was compared with the burst interval, t_b, predicted from the empirical relation devised by Rao *et al.* [21], wherein the dimensionless burst period, T_b, is given by

$$T_b = U \, \delta / t_b \approx 5 \qquad (4)$$

Boundary layer thickness, δ, is here assumed to be equal to flow depth.

Results

Outer flow nonstationarity

Time-averaged outer flow variables for the 20 series analysed here are given in Table 2. Acceleration terms, computed from linear trends in $\bar{u}$ over the duration of the series in question, are all statistically significant at $p=0.10$. Both spring and neap tides exhibit high peak accelerations, although the former are characterized by a more systematic pattern of i) acceleration during marsh inundation (Series S-D to S-F); ii) sudden deceleration close to high water (HW) (S-G); and iii) rapid acceleration on the ebb, after the flow has fallen below bankfull stage (S-H and S-I). Peak accelerations are greater than those reported for an

45

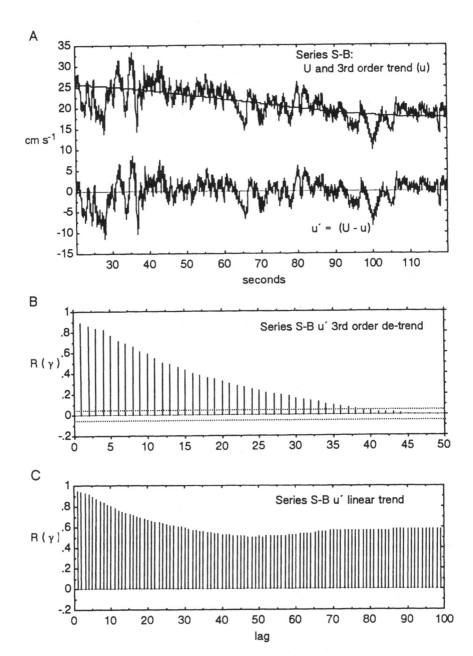

FIGURE 1 a) Portion of raw U series, with fitted 3rd order trend and resulting u′ residual; b) ACF of stationary u′ series. Horizontal dashed lines represent significance levels for R(γ); c) ACF of nonstationary series. Significance levels are the same as in b), but are here omitted for clarity.

46

TABLE 3 Mean velocities, ū and w̄, and selected turbulence parameters for series N-A to N-K (September, 1990; neap tide), and S-A to S-I (March 1991; spring tide).

Series	ū m s⁻¹	w̄ m s⁻¹	$(\overline{u'^2})^{0.5}$	$(\overline{w'^2})^{0.5}$	$-\rho\overline{u'w'}$	$\overline{u'w'}/u'w'$	$(\overline{u'^2})^{0.5}/u_*$	$(\overline{w'^2})^{0.5}/u_*$	Ψ
N-A	0.13	0.015	0.011	0.0068	0.0321	-0.416	1.99	1.22	1.64
N-B	0.14	0.015	0.013	0.0070	0.0353	-0.372	2.26	1.20	1.89
N-C	0.11	0.015	0.0097	0.0056	0.0223	-0.392	2.09	1.20	1.74
N-D	0.081	0.014	0.0076	0.0046	0.0165	-0.466	1.90	1.14	1.66
N-E	0.038	0.011	0.0040	0.0029	≈0.0	-0.0025	*	*	*
N-F	0.010	0.0091	0.0041	0.0026	0.0140	-0.184	2.94	1.85	1.59
N-G	0.087	0.0083	0.010	0.0058	0.0183	-0.293	2.46	1.39	1.78
N-H	0.077	0.0087	0.011	0.0059	0.0245	-0.381	2.19	1.20	1.82
N-I	0.072	0.0055	0.012	0.0059	0.0299	-0.407	2.24	1.10	2.04
N-J	0.100	0.0047	0.014	0.0068	0.0354	-0.348	2.46	1.17	2.11
N-K	0.120	0.0055	0.012	0.0063	0.0213	-0.274	2.62	1.39	1.88
S-A	0.22	0.031	0.029	0.016	0.136	-0.284	2.52	1.40	1.80
S-B	0.17	0.015	0.030	0.015	0.143	-0.317	2.56	1.24	2.07
S-C	0.19	0.010	0.025	0.013	0.069	-0.211	3.01	1.58	1.91
S-D	0.15	0.0054	0.030	0.014	0.114	-0.268	2.81	1.32	2.13
S-E	0.30	0.010	0.058	0.022	0.480	-0.376	2.66	1.00	2.67
S-F	0.40	0.0066	0.065	0.030	0.570	-0.280	2.78	1.29	2.16
S-G	0.061	0.0	0.021	0.012	0.082	-0.332	2.30	1.31	1.75
S-H	0.86	-0.052	0.080	0.039	1.031	-0.324	2.52	1.23	2.05
S-I	0.81	-0.049	0.075	0.037	1.246	-0.438	2.16	1.06	2.04

where $u_* = (\overline{u'w'})^{0.5}$ and $\Psi = \dfrac{(\overline{u'^2})^{0.5}/u_*}{(\overline{w'^2})^{0.5}/u_*}$ * indicates signal/noise ratio too low to allow meaningful parameter estimation estuarine flows [13][22][26], and one might reasonably expect turbulence characteristics measured here to differ from those of steady uniform laboratory boundary layers.

Obtaining meaningful turbulence data is clearly problematic where measurements have to be made at a fixed elevation within tidal flows. Laboratory work [20] has shown, however, that little variation in the mean 'bursting' rate occurs within $0.10 < z/h < 0.70$; all the series described here have z/h values which fall roughly within this range.

Examination of the U and W series revealed the former to exhibit far more complex nonstationarity, often necessitating the removal of a 3rd or 4th order polynomial trend (Figure 1a). Subtraction of the mean, or of a linear trend was usually sufficient to ensure a stationary w' series. The effect of any nonstationarity will depend upon the parameter estimated, and may lead to a misleading impression of turbulence structure. Examination of longer time-series (to be described elsewhere) indicate that series durations in excess of 6-7 minutes may introduce more complex nonstationarity in U that would be difficult to remove objectively using the above procedure, leading to inevitable nonstationarity in derived turbulence parameters. The ACFs resulting from this analysis (Figure 1b) were all indicative of successful trend removal. It should be noted that in addition to introducing errors into the estimation of individual stress terms, severe nonstationarity, as revealed by the ACF (Figure 1c), will preclude any meaningful calculation of Λ using equation (2).

Turbulence parameters

The variation in $\bar{u}$, $\bar{w}$ and selected turbulence parameters over the two sampled tides is summarized in Table 3. As expected, rms vertical and horizontal turbulence intensities, and the mean streamwise shear stress are markedly higher during spring tides. Values of the $(\overline{u'w'}/u'w')$, and the normalized horizontal and vertical intensities are generally close to those reported from previous field and laboratory studies [18][20][25]; anomalously low correlations for series N-E and N-F may be the result of turbulence intensities which are close to the sensor noise levels. The spring tide data indicate the maintenance of high near-bed velocities and turbulence intensities following the inundation of the marsh surface (S-E to S-F) - these are presumably important in maintaining silt-sized material in suspension high enough within the water column to allow advective and diffuse material transfer across the channel margins. Immediately prior to velocity reversal at HW (S-G), there is a rapid decline in turbulent stress production, with no clear evidence in favour of a lagged turbulence decay over the HW period, as suggested in our previous work [11]. In fact, such a rapid decay in both horizontal and vertical turbulence intensity is suggested by suspended sediment measurements within the same channel [9][12], which show a sudden steepening of the vertical concentration gradient as material settles through the water column at HW. This finding illustrates the need for caution in the integration of point turbulence measurements acquired within a sparse temporal sampling scheme [11] with more continuous data obtained via conventional monitoring techniques.

Several studies have demonstrated the importance of longitudinal pressure gradient effects in modifying both the vertical distribution of turbulent stress production and the magnitude and frequency of near-bed 'bursting' motions [13][19]. Surprisingly perhaps, in view of the large range of dU/dt, there is no evidence of systematic variation in the ratio, ψ. This may simply be the result of extreme short-term unsteadiness interacting with, and partly obscuring, any long-term effects.

Quadrant analysis shows the near-bed 'event structure' of tidal channel flows to be essentially similar to that reported for laboratory boundary layers [20]. In almost all cases, a more rapid fall off from quadrants 1 and 3 (diffuse inward and outward interactions) is evident, such that, for a threshold of 2 standard deviations about the mean stress [cf. 14], both account for only about 5% of the total shear stress (Figure 2). Significantly, this threshold corresponds to a stress level of 15-45%, and thus preserves far fewer high magnitude events than does use of the 90% stress level [17]. Inconsistency in the selection

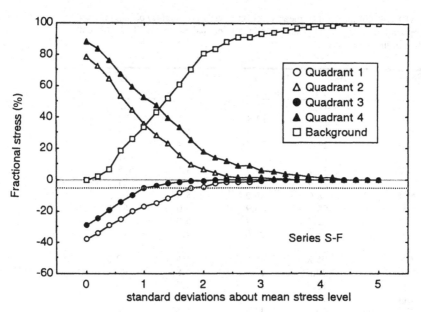

FIGURE 2 Quadrant analysis performed on series S-F. Vertical axis shows positive or negative contributions to the total shear stress from each quadrant of the u´w´ plane. The cumulative curve represents the total stress consigned to the background ('noise'). Note more rapid attenuation of stress contributions from quadrants 1 and 3, such that both contribute < 5% of the total stress at a threshold of 2 s.d.

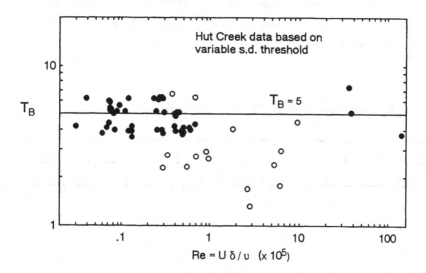

FIGURE 3 Variation in dimensionless burst interval, T_B, with outer flow Re. Open circles represent Hut Creek data; solid circles refer to data reviewed in Allen [1].

TABLE 4 Comparison of mean ejection intervals (seconds) obtained from application of equation (4), and from quadrant analysis using different event discrimination criteria.

Series	t_b (2 σ)[†]	t_b (variable σ)[†]	t_b (90% stress)[†]	t_b[††]
NA	41.5	41.5	5.4	32.7
NB	17.8	17.8	7.3	33.9
NC	28.0	28.2	4.9	52.3
ND	27.2	37.5	4.4	80.2
NE	*	*	*	181.6
NF	33.1	79.7	3.1	675.0
NG	32.9	32.9	4.5	71.8
NH	38.0	38.0	5.8	68.8
NI	80.9	80.9	5.0	61.1
NJ	32.7	32.7	5.8	32.5
NK	25.0	25.0	4.8	19.6
SA	5.8	5.8	1.6	10.0
SB	22.0	22.0	1.8	27.4
SC	12.3	12.3	1.3	36.3
SD	14.7	14.7	2.4	55.0
SE	14.8	14.8	1.5	30.5
SF	14.3	14.3	1.3	24.1
SG	10.1	7.7	1.2	179.5
SH	7.9	8.3	1.8	9.3
SI	5.3	2.9	1.2	8.1

[†] actual ejection interval (s) defined as mean period between ejection-type events

[††] theoretical burst period (s) computed from $\dfrac{t_b U}{\delta} \approx 5$ where $\delta \approx h$

* analysis not performed on this series

TABLE 5 Integral length scale, Λ, for spring tide u′ series.

Series	S-A	S-B	S-C	S-D	S-E	S-F	S-G	S-H	S-I
Λ (m)	0.44	0.39	0.35	0.53	0.84	1.42	0.04	6.02	4.70

of the event threshold has important implications for the scaling of near-bed stress events with outer flow variables. Table 4 includes mean ejection intervals (defined as the mean time between discrete quadrant 2 events) obtained using the following thresholds: i) constant 2 standard deviation about the mean stress; ii) variable, based upon the level at which quadrants 1 + 3 contribute < 5% of the total stress (i.e. between 1.6 and 2.2 deviations); and iii) the 90% stress level (i.e. lowest 10% of stress events consigned to the background). Also included are the intervals predicted by equation (4). The standard deviation-based thresholds clearly discriminate events which correspond more closely to those investigated in the laboratory [21]. Also evident from Table 4, is a tendency for the bursting rate to be maintained even at low Re, giving rise to considerable scatter in the dimensionless burst interval, T_B (Figure 3).

However, consideration of the characteristic length scales for the turbulence reveals a marked reduction in eddy length prior to HW. Table 5 shows the variation in Λ during the spring tide. Computed eddy lengths appear to scale on outer flow Re, but also exhibit some variation with z/h as reported for the Great Ouse estuary by West et al. [26]. These data illustrate the difficulty of applying simple magnitude-based criteria to discriminate near-bed stress events in time-varying flows. Use of a variable threshold produces a marginally closer fit to the predicted burst intervals, however, which suggests that the time-varying form of the fractional stress curves (Figure 2) might merit further attention.

Summary

1. Given careful attention to the problems of nonstationarity induced at non-turbulent scales, EMCM measurements provide a useful insight into the nature and magnitude of the near-bed turbulent fluid motions that are known to be important in determining the entrainment of channel bed sediments and the maintenance of vertical suspension profiles.
2. Neap tides exhibit turbulence intensities that are close to the resolution of the sensors used here, and a more comprehensive error analysis will be presented elsewhere. During spring tides, high rates of turbulent stress production are associated with the outer flow transients which occur close to bankfull stage. Although the acceleration terms are large, their short duration means that the turbulence - mean flow hysteresis observed within larger-scale marine boundary layers is less evident here.
3. Further comparison with published laboratory and field studies is complicated by the inconsistency in the criteria used to define near-bed stress 'events'. The use of a 2 standard deviation event threshold appears to provide a reasonable fit between observed ejection intervals and the 'burst' periods predicted from outer flow variables. However, the ejection rate is maintained during brief periods of low Re via the progressive decay of larger eddies.

ACKNOWLEDGEMENTS

Data were collected as part of a wider study of marsh hydrodynamics and sedimentation, in collaboration with Dr T. Spencer (University of Cambridge). The authors also acknowledge receipt of a British Geomorphological Research Group Small Grant.

REFERENCES

[1] Allen, J.R.L. (1985) Principles of Physical Sedimentology. London, Allen & Unwin.
[2] Allen, J.R.L. (1990) Salt Marsh Growth and Stratification: a Numerical Model with special reference to the Severn Estuary, southwest Britain. Mar. Geol. 95, 77-96

[3] Bayliss-Smith, T.P., Healey, R., Lailey, R., Spencer, T., & Stoddart, D.R. (1979) Tidal flows in Salt Marsh Creeks. *Est. Coastal Mar. Sci.* 9, 235-255

[4] Carling, P.A. (1981) Sediment Transport by Tidal Currents and Waves: observations from a sandy intertidal zone (Burry Inlet, South Wales). *Spec. Publs. Int Ass. Sediment.* 5, 65-80

[5] Clifford, N.J., McClatchey, J., & French, J.R. (1990) Discussion: Measurements of Turbulence in the Benthic Boundary Layer over a Gravel Bed, and comparison between Acoustic Measurements and Predictions of the Bedload Transport of Marine Gravels. *Sedimentology* 37, 161-171

[6] Collins, M.B. (1981) Observations of some Sediment Transport Processes over Intertidal Flats. *Spec. Publs. Int Ass. Sediment.* 5, 81-98

[7] Corino, E.R., & Brodkey, R.S. (1969) A visual investigation of the wall region in turbulent flow. *J. Fluid Mech.* 37, 1-30

[8] Falco, R.E. (1977) Coherent motions in the outer regions of turbulent boundary layers. *Phys. Fluids* 20, 124-132

[9] French, J.R. (1989) Hydrodynamics and Sedimentation in a Macro-tidal Salt Marsh, Norfolk, England. Unpubl. PhD thesis, University of Cambridge

[10] French, J.R. (1991) Eustatic and Neotectonic controls on Salt Marsh Sedimentation. In: N.C. Kraus, K.J. Gingerich, and D.L. Kriebel (eds.) *Coastal Sediments '91: Proceedings of a speciality conference on quantitative approaches to coastal sediment processes*, Seattle, Washington. Amer. Soc. of Civ. Engrs., 1223-1236.

[11] French, J.R., and Clifford, N.J. (1992) Characteristics and 'event-structure' of Near-bed Turbulence in a Macro-tidal Saltmarsh Channel. *Est. Cstl. Shelf Sci.* 34, 49-69

[12] French, J.R., and Stoddart, D.R. (1992) Hydrodynamics of Salt Marsh Creek Systems: Implications for Marsh Morphological Development and Material Exchange. *Earth Surface Processes and Landforms 17.*

[13] Gordon, C.M., & Dohne, C.F. (1973) Some observations of Turbulent Flow in a tidal Estuary. *J. Geophys. Res.* 78, 1971-1978

[14] Gordon, C.M. (1974) Intermittent Momentum Transport in a Geophysical Boundary. *Nature* 248, 392-394

[15] Grass, A.J. (1971) Structural features of Turbulent Flow over Smooth and Rough Boundaries. *J. Fluid. Mech.* 50, 233-255

[16] Green, H.M., Stoddart, D.R., Reed, D.J., & Bayliss-Smith, T.P. (1986) Saltmarsh Tidal Creek Hydrodynamics, Scolt Head Island, Norfolk, England. In G. Sigbjamarson (ed.) *Iceland Coastal and River Symposium Proc.*, 93-103

[17] Heathershaw, A.D. (1974) 'Bursting' phenomena in the sea. *Nature* 248, 394-395

[18] Heathershaw, A.D. (1979) The Turbulent Structure of the bottom Boundary Layer in a Tidal Current. *Geophys. J. Roy. Astr. Soc.* 58, 395-430

[19] Kline, S.J., Reynolds, W.C., Schraub, R.A., & Runstadler, P.W. (1967) The Structure of Turbulent Boundary Layers. *J. Fluid. Mech.* 30, 741-773

[20] Lu, S.S., & Willmarth, W.W. (1973) Measurements of the Structure of the Reynolds Stress in a Turbulent Boundary Layer. *J. Fluid Mech.* 60, 481-511

[21] Rao, K.N., Narashimha, R., & Badri Narayanan, M.A. (1971) The 'bursting' phenomenon in the Turbulent Boundary Layer. *J. Fluid. Mech.* 48, 339-352

[22] Shiono, L., & West, J.R. (1987) Turbulent Perturbations of Velocity in the Conwy estuary. *Est. Coast. Shelf Sci.* 25, 533-553

[23] Soulsby, R.L. (1980) Selecting Record Length and Digitization Rate for near-bed Turbulence Measurements. *J. Phys. Oceanogr.* 14, 208-219

[24] Stevenson, J.C., Ward, L.G., and Kearney, M.S. (1988) Sediment Transport and Trapping in Marsh Systems: implications of tidal flux studies. *Mar. Geol.* 80, 37-59

[25] West, J.R., Knight, D.W., & Shiono, K. (1986) Turbulence Measurements in the Great Ouse estuary. *J. Hydraul. Div. Am. Soc. Civ. Eng.* 112, 167-180

[26] Williams, J.J., Thorne, P.D., & Heathershaw, A.D. (1989) Comparisons between Acoustic Measurements and predictions of the Bedload Transport of Marine Gravels. *Sedimentology* 36, 973-979

5 The effect of interfacial friction factor on the structure of salt wedge

M. O. Lee and A. Murota

ABSTRACT

Experiments in a rectangular open channel were carried out to evaluate the interfacial friction factor for salt wedges, under steady and unsteady conditions. From the experimental results obtained under steady conditions, reverse operations by a one dimensional two-layer model were performed to obtain values of interfacial friction factor, fi, and the results were then compared with published data. Considering the entrainment mixing and a side-wall friction in an open channel, fi compared favorably with Kaneko's empirical formula. However, the results were smaller than those obtained from the curves of Dermissis *et al.*[1][2]. In particular, entrainment coefficients were generally smaller, compared with interfacial mixing rates, E_g. obtained by Grubert, when the interfacial transition layer is in a supercritical state. Therefore, if we apply $\overline{fi}$, obtained by Dermissis *et al.* or E_g. obtained by Grubert[6] for computations of salt wedges, wedge lengths become remarkably short, due to excessive evaluation of fi. On the other hand, computed results of unsteady salt wedges showed much better reappearance toward variations of flow fields as well as wedge lengths when a tidal period is 600 seconds and using Kaneko's empirical formula as fi, compared with using fi proposed by Suga[11].

1. Introduction

In order to predict accurately the shapes and lengths of salt wedges which occur in weakly mixed estuaries, firstly, an appropriate evaluation of interfacial friction factor, fi, is required. In the past, fi has been mainly expressed as a function of the Keulegan number. However, the results have shown extensive scattering. Dermissis et al.[1][2] found that the average interfacial friction factor, $\overline{fi}$, is closely related to $ReFr^2$ (where Re is a Reynolds number and Fr is a regular Froude number) with relative density differences as independent parameters.

In this study, comparisons are made between experimental results and the empirical or theoretical models proposed by various researchers for the interfacial friction factor. In particular, an emphasis is placed upon the importance of the scale of the experimental channel.

2. Theoretical Background

For open channel flow, with a constant width, assuming that entrainment mixing takes place through the interface toward the upper layer from the lower layer, and that longitudinal variations in the density of the lower layer can be ignored, then the equations of conservation of volume, mass and momentum for one dimensional two-layered flow can be expressed as follows:

the upper layer:

$$\frac{\partial h_1}{\partial t} + \frac{1}{B}\frac{\partial Q_1}{\partial x} = We \tag{1}$$

$$\frac{\partial h_2}{\partial t} + \frac{1}{B}\frac{\partial Q_2}{\partial x} = -We \tag{2}$$

$$\frac{\partial(\rho_1 h_1)}{\partial t} + \frac{1}{B}\frac{\partial}{\partial x}(\rho_1 Q_1) = \rho_2 We \tag{3}$$

the lower layer:

$$\frac{1}{g}\frac{\partial u_1}{\partial t} + \frac{u_1}{g}\frac{\partial u_1}{\partial x} + \frac{\partial h_1}{\partial x} + \frac{\partial h_2}{\partial x} + \frac{\rho_2 u_1}{\rho_1 g h_1}We + \frac{h_1}{2\rho_1}\frac{\partial \rho_1}{\partial x} - i_0 + if_1 = 0 \tag{4}$$

$$\frac{1}{g}\frac{\partial u_2}{\partial t} + \frac{u_2}{g}\frac{\partial u_2}{\partial x} - \frac{u_2}{gh_2}We + \frac{\rho_1}{\rho_2}\frac{\partial \rho_1}{\partial x} + \frac{\rho_1}{\rho_2}\frac{\partial h_1}{\partial x} + \frac{\partial h_2}{\partial x} - i_0 + if_2 = 0 \tag{5}$$

where W_e: entrainment velocity (= $E|u_1-u_2|$, E: entrainment coefficient), i_0: bottom slope, if_1, if_2: friction slopes in the upper and lower layers are given by:

$$if_1 = \frac{fi}{2gh_1}(u_1-u_2)|u_1-u_2| + \frac{fw}{4gB}|u_1|u_1 \tag{6}$$

$$if_2 = \frac{fb}{2gh_2}u_2|u_2| - \frac{fi}{2gh_2}(1-\epsilon)(u_1-u_2)|u_1-u_2| + \frac{fw}{4gB}|u_2|u_2 \tag{7}$$

54

where *fi,fw* and *fb*: the interfacial friction factor,the side-wall friction factor and the bottom friction factor, respectively, ε: relative density $(=1-\rho_1/\rho_2)$,g:the acceleration of gravity,B:channel width, $|u_1-u_2|$:relative velocity

Substituting equations (1) and (4) into equations (3) and (5), and then rearranging yields fundamental equations for the salt wedge. Substituting equation (5) from equation (3), and then considering the relationships between equation (1) and (2) (assume $u_2 = 0$),the equation for the interfacial friction factor, *fi* (including the influence of side-wall friction), can be also obtained. Blasius's resistance law was applied for a side-wall friction factor, that is, $fw = 0.224Re^{-0.25}$, $Re = u_1h_1/\nu$ $(B/(B+2h_1))$.

3.   Experimental Apparatus and Methods

The experimental channel used was made from plexiglass. It had an inclination of 1/22.5 and was 8 m long, 0.1 m wide and 0.25 m deep. It was connected to a steel tank 2.74 m long, 1.83 m wide and 0.91 m deep. Flowing freshwater into a salt water tank with a constant concentration out of the upper stream of a channel and, at the same time, supplying salt water from near the bottom of a tank, the arrested salt wedge formed. In the unsteady state experiments, a tidal motion was given by the movement of the end weir, located at the downstream edge of the salt water tank, so that the salt water supply was regulated in proportion to a tidal stage. The measurements taken included interfacial depths, salt wedge lengths, velocities and densities in the upper and lower layers, interfacial waves and wedge tip vortices. Flow visualization techniques used were hydrogen bubbles or fluorescent dyes introduced to observe slow velocities, interfacial waves and flow structures near to the wedge tip. Densities(i.e. salinities) were measured by a conductivity meter. Tidal water levels and interfacial depths were measured simultaneously by two servo water gauges and recorded by pen recorders. The experimental conditions cover Reynolds number $Re_1(= u_1h_1/\nu$) between 1564 and 3984, relative density differences $\varepsilon(=(\rho_2-\rho_1)/\rho_2)$ between 0.0031 and 0.0266 and densimetric Froude number $Fd_1(=u_1/\sqrt{\varepsilon gh_1})$ between 0.49 and 1.72.

4.   Results and Discussions

4.1 *Contribution of Each Resistance   to Interfacial Friction Factor*

Interfacial resistance for two-layered stratified flow can be caused by three factors, namely: friction caused by a molecular viscosity in laminar flow,*fiv*, friction caused by short wavelength interfacial waves, *fiw* and additional friction caused by entrainment mixing, *fim* (Egashira[3]; Tamai[12]). Therefore, the interfacial friction factor, *fi*,for salt wedge can be expressed by the linear total of resistance caused by each one,i.e.

$$fi = fiv + fiw + fim \qquad (8)$$

where $fiv=6/Re_1/(1+(3/4)(h_2/h_1))$, $fiw=0.090(Re_1Fd_1^2)$, $fim=2E$, Re_1:Reynolds number in the upper layer, h_1,h_2:the upper and the lower depth,respectively,Fd_1:densimetric Froude number in the upper layer, E:entrainment coefficient($=0.002Fd_1^3$)

The interfacial friction factor,*fi*,was calculated using equations (1),(2),(3) and (5) as described previously and the interfacial inclination obtained by experiments was then compared with each resistance cause. Results showed that the resistance induced by interfacial waves played an important part as interfacial resistance for arrested salt wedges(Figure 1). However,the resistance induced by a molecular viscosity or entrainment mixing was an order of magnitude less than the actual interfacial resistance.

4.2 Interfacial Mixing Rates

Grubert[6] found that two distinct types of interfacial mixing takes place depending on the state of the interfacial transition layer. He noted that when the interfacial transition layer is in a subcritical state, entrainment mixing takes place, but if the layer is in a supercritical(or critical) state, then the turbulent diffusion mixing takes place. The two interfacial mixing satisfy the following relationships, repectively.
 i) entrainment mixing:

$$E_g = u_e/V = 2.40(\overline{fi}/8)^{-1.5}Ri^{-1.0} \qquad (3.5<Ri<20) \qquad (9)$$

 ii) turbulent diffusion mixing:

$$E_g = q/V = 0.50(\overline{fi}/8)Ri^{-2.0} \qquad (0.5<Ri<3.5) \qquad (10)$$

where u_e:entrainment velocity, E_g: Grubert's interfacial mixing rate, V: $|u_1-u_2|$, $\overline{fi}$: average interfacial friction factor, Ri: overall Richardson number(= $g'R_1/V^2$, $g'= g(1-\rho_1/\rho_2)$, $R_1=bh_1/(b+2h_1)$)

In Figure 2, a comparison was carried out between the interfacial mixing rates,E_g, computed by equation (9) or (10), and entrainment coefficients,E, measured by the visualization techniques. The results showed that when the interfacial transition layer is in a supercritical state,E_g were mostly larger than E. The differences seem to arise from the difference in scale of the laboratory open channel used in experiments. Grubert[6] employed a comparatively large open channel, 15 m long and 0.31 m wide, to approach the actual mixing rates in salt wedge estuaries, while in the present study,(with a channel 8 m long and 0.1 m wide) dynamical similarity was probably not accomplished. i.e. if the interfacial transition layer is in a complete turbulent state in Grubert's open channel then, a quasi-turbulent state occurs in this study. Re_1 in this study ranges from 2650 to 4530, however, in Grubert's experiments, it is not clear what the range of values was.

However, when the interfacial transition layer is in a subcritical state, E_G were clearly smaller than E,i.e. entrainment coefficients. Consequently, in the present study it seems that interfacial waves, which interfere with each other and then generate the turbulent mixing, act predominently as a cause of interfacial resistance.

4.3 Relationships between Interfacial Friction Factor and Keulegan Number

The interfacial friction factor, fi, for two-layered stratified flows has been frequently expressed as a function of Keulegan number $\theta\ (=(\epsilon \nu g)^{1/3}/V)$, V:local average velocity in the moving layer) (Keulegan[9]; Iwasaki[7]; Shi-Igai[10];Kaneko[8];Suga[11]). However, Dermissis et al.[1][2] presented a new interpretation based on a systematical investigation of experimental results from arrested salt wedges. They stated that the average interfacial friction factor, $\overline{fi}$, is closely related to $ReFr^2$ (where Fr is a regular Froude number), holding the relative density difference as an independent parameter. The results showed that the values of the average interfacial friction factor, $\overline{fi}$, lay on well-defined mean lines with comparatively small scattering. However, they still contain flow variables and density parameters, similarly to Keulegan number. Figure 3 shows the relationships between Keulegan number (or Iwasaki number $\psi = ReFd_1^2 =\theta^{-3}$) and fi (where side-wall friction caused by the channel is still considered along with entrainment mixing). Also plotted are the values of $\overline{fi}$, obtained from the curves of Dermissis et al.[1]. The present results show good agreement with Kaneko's empirical formula[8] but $\overline{fi}$, obtained from the curves of Dermissis et al.[1] are larger than present results. These differences are likely to result from the differences of scale in laboratory open channels, as described previously. In Figure 4, an example of the computed shape of an arrested salt wedge using the interfacial mixing rates (or entrainment coefficients) and the interfacial friction factor is shown. The computed results shows remarkably good agreement with the experimental results, especially in the interfacial configuration near to the river mouth and the wedge length, when using the interfacial friction factor proposed by Kaneko[8] and the entrainment coefficient obtained by Suga[11]. However,if the average interfacial friction factor, $\overline{fi}$, obtained from the curves of Dermissis et al.[1] or the interfacial mixing rate, obtained by Grubert[6], is used to compute the wedge length, then the result would be an under estimate due to the excessive evaluation of fi.

4.4 The Flow and Density Fields of Salt Wedge in an Unsteady State

In Figures 5 and 6 are shown the computed results of flow fields for an unsteady salt wedge, where a side-wall friction is considered, but entrainment mixing is not considered. Firstly, in Figure 5 (T=400 sec), the upper layer mean velocity, u_1, indicates its maximum in a low water and its minimum a high water, respectively. The lower layer mean velocity, u_2, changes in direction in accordance with tidal stages so that the upstream flow comes to be most strong during the flood tide and the downstream flow during the ebb tide. In Figure 6 (T=600 sec), the variation of u_2, is rather more regular than in Figure 5. This suggests that when the tidal period is 600 seconds, a velocity field for the unsteady salt wedge was well defined, compared to when the tidal period is 400 seconds. Figure 7 shows the computed results for the upper layer mean density, ρ_1 and wedge length, l, where the symbols (o) and (t) denote values when t = 0 and t= t, respectively. The upper layer mean density, ρ_1, indicates maximum at low water and minimum at high water, similarly to u_1. However, the wedge length is maximum during the ebb tide and minimum during flood tide. In particular, it should be noted that when ρ_1 is at a maximum, the relative velocity $|u_1-u_2|$ is a minimum (therefore interfacial shear stress is much smaller than elsewhere), as shown in Figure 6. Gardner et al.[4] pointed out that when the shear stress acting on the interface becomes minimum during the ebb tide, a shear instability takes place and this causes a strong vertical mixing. On the other hand, using Kaneko's empirical formula or Suga's formula for the interfacial friction factor, they both gave identical velocities but resulted in difference in densities and in a wedge length. The computed results also showed a good agreement with the experimental results when using Kaneko's empirical formula for the interfacial friction factor, but a poorer agreement when using Suga's proposed formula.

4.5 The Shapes of Salt Wedge in an Unsteady State

In Figure 8, the shapes of computed salt wedge in an unsteady state is shown, using the interfacial friction factor proposed by Kaneko. The shapes of salt wedge vary with a tidal stage but retain similar shapes those obtained for a steady state. The depth variation at the river mouth and tidal excursion at the wedge tip were in good agreement with the experimental results.

5. Conclusions

The conclusions obtained through experiments carried out in order to evaluate the interfacial friction factor for salt wedge are as follows:
1) Considering entrainment mixings and a side-wall friction and then applying Kaneko's empirical formula for

the interfacial friction factor,the wedge shape and flow
field is able to be predicted satisfactorily by one
dimensional two-layered model. However, the effect of
entrainment mixings on interfacial resistance are
relatively quite small, compared to a side-wall friction
or interfacial waves resistance.
2) The average interfacial friction factors by Dermissis
et al.[1][2] or interfacial mixing rates (especially in a
supercritical state) by Grubert[6] are much larger than
those obtained by experiments,repectively. Therefore,the
computed wedge lengths result in considerably small
values, compared to experimental results. These results
come from not being established dynamical similarities
fully or incomplete turbulences because a relatively
small-scaled channel has been used in the experiments,
compared to those used by them.
3) Kaneko's empirical formula is still applicable for the
computation of an unsteady salt wedge,similarly in a
steady state.
In the future,through the experiments in a large-scaled
channel or field observations,it is hoped that functional
relationships between the interfacial friction factor and
variations of velocity or density fields should be
resolved.

References

[1] Dermissis,V. and Partheniades,E.,(1984)," Interfacial
    Resistance in Stratified Flows", *Journal of Waterway,
    Port,Coastal and Ocean Engineering*,ASCE,Vol.110,No.2,
    pp.231-250.
[2] Dermissis, V. and Partheniades, E.,(1985)," Dominent
    Shear Stresses in Arrested Saline Wedges", *Journal of
    Waterway,Port,Coastal and Ocean Engineering*,ASCE,Vol.
    111,No.4,pp.733-752.
[3] Egashira,S.,(1986)," Side-Wall Effects in the Open
    Channel Experimental Data",*Report for Application and
    Unification of Entrainment Concepts in the Density
    Flows*,Ministry of Education, Japan, pp.30-40.
[4] Gardner, G. B. and Smith, J. D., (1978), " Turbulent
    Mixing in a Salt Wedge Estuary", *Hydrodynamics of
    Estuaries and Fjords*, Elsevier Scientific Publishing
    Company,New York,pp.79-106.
[5] Grubert,J.P.,(1989),"Interfacial Mixing in Stratified
    Channel Flows",*Journal of Hydraulic Engineering*,ASCE,
    Vol.115,No.7,pp.887-905.
[6] Grubert,J.P.,(1990)," Interfacial Mixing in Estuaries
    and Fjords", *Journal of Hydraulic Engineering*, ASCE,
    Vol.116,No.2,pp.176-195.
[7] Iwasaki, T., Gishida, S. and Tomioka, R.,(1962),"On a
    Density Mixing in Two-Layered Stratified Flows",
    *Colletion of Annual Scientific Lectures in Japan
    Society of Civil Engineers*,Vol.2,pp.5-6.
[8] Kaneko,Y., (1966)," Exemples of Interfacial Friction
    Factor in Two-Layered Flows", *Proceeding of the 13th
    Japanese Conference on Coastal Engineering*,Vol.13,pp.
    263-267.

[9]   Keulegan, G. H.,(1966)," The Mechanism of Arrested Saline Water",*Estuarine and Coastline Hydrodynamics,* McGraw-Hill Book Company,New York,pp.546-574.

[10] Shi-Igai, H., (1966)," Studies on the Salt Wedge", *Report of the Department of Civil Engineering,* Tokyo Institute of Technology,No.1,pp.19-62.

[11] Suga,K.,(1979)," Studies on Hydraulics of Salt Wedge in the Tidal River", *Report of Institution of Civil Engineering Research,*Ministry of Construction,Japan, No.1537,pp.1-255.

[12] Tamai, N.,(1980), *Hydraulics on Density Flows,*Gibodo Publishing Company,Tokyo,pp.1-260.

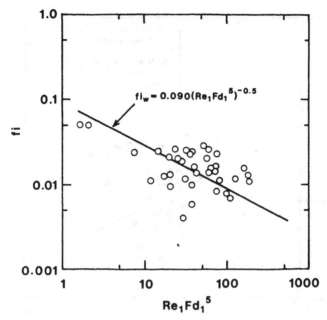

Figure 1 *Relationship between Friction caused by Short Internal Waves,fiw, and fi*

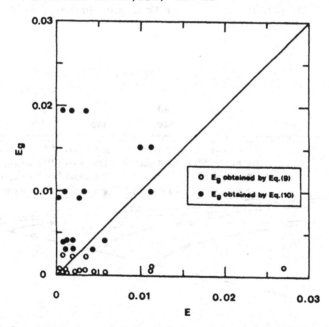

FIGURE 2 *Comparison between Entrainment Coefficients, E, and the Interfacial Mixing Rates,Eg,obtained by Grubert's Equation (9) and (10)*

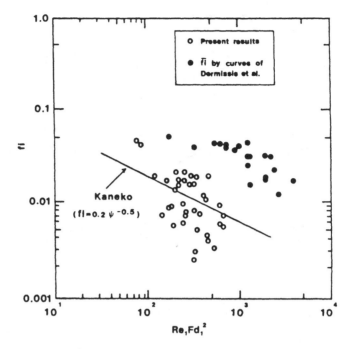

FIGURE 3 *Relationship between Keulegan Number and fi*

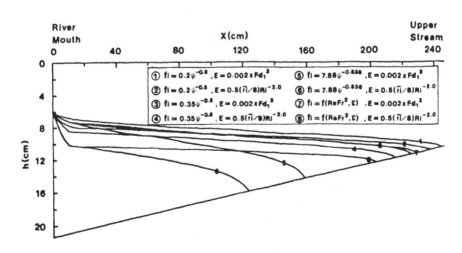

FIGURE 4 *The Example of the Computed Shape for Arrested Salt Wedges*

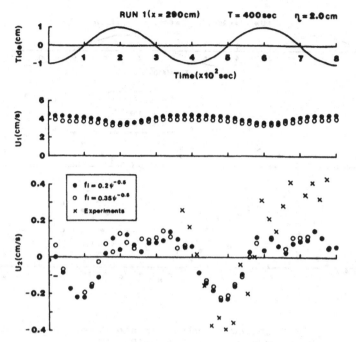

FIGURE 5   *The Computed Results of Flow Fields for an Unsteady Salt Wedge(Run 1)*

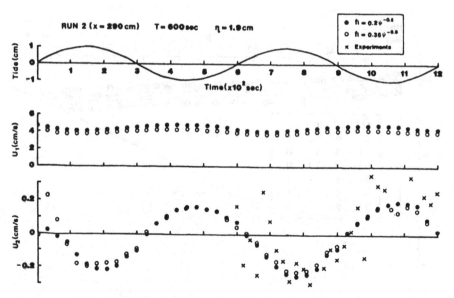

FIGURE 6   *The Computed Results of Flow Fields for an Unsteady Salt Wedge(Run 2)*

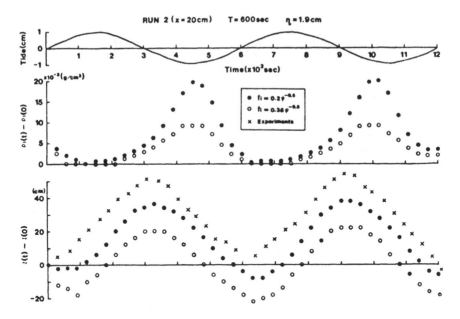

FIGURE 7 *The Computed Results for the Upper Layer Mean Density and Wedge Length(Run 2)*

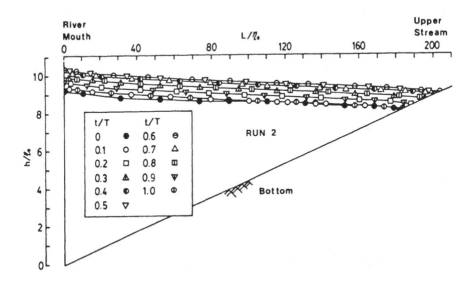

Figure 8 *The Shapes of Computed Salt Wedge in an Unsteady State(Run 2)*

6 Longitudinal dispersion in rectangular turbulent open channel flow

K. Fujisaki, H. Ohki and Y. Awaya

ABSTRACT

This paper describes the effect of channel geometry and density gradient on longitudinal dispersion coefficient D_L in turbulent open channel flow. The vertical distribution of flow properties is approximated by those of infinite width flow and horizontal variation of flow is determined by the horizontal variation of shear velocity. It is shown that D_L increases with aspect ratio and longitudinal density gradient.

1 Introduction

The longitudinal dispersion phenomena in a turbulent open channel is studied at first by Elder[1] who presented the dispersion coefficient in the 2-D open channel flow by applying Taylor's model[2] and employing the logarithmic velocity distribution. Fisher[3] also employed this modeling to the wide open channel flow and showed much large coefficients than those given by Elder's results.
Ever since Fisher discussed the dispersion in natural streams, numerous reports on the longitudinal dispersion in 3-D turbulent flow have been presented. However, there are few theoretical works which discussed the effect of channel geometry on longitudinal dispersion. Attila and Sooky[4] demonstrated that channel geometry gives serious effect dispersion in turbulent flow with triangular or parabolic cross section. Chatwin and

65

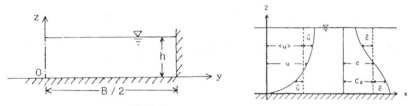

FIGURE 1   Coordinate system

Sullivan[5] also proposed turbulent dispersion coefficient in rectangular turbulent open channel flow as a modification of laminer flow. Smith[6] also discussed relationship between longitudinal dispersion and longitudinal density gradient in a shallow channel.

So far as writers are aware, there has been few reports which give sufficient informations on the relationship between longitudinal dispersion and 3-D flow properties. In the present paper, we investigate the longitudinal dispersion coefficient in turbulent open channel of rectangular cross section. The objective of this work is to describe longitudinal dispersion coefficients as a function of aspect ratio, and longitudinal density gradient, by using simple method.

2 Basic Equation

2.1 Analytical method for 3-D flow

For steady uniform turbulent flow the equation of motion can be written as

$$0 = \rho g i - \frac{\partial p}{\partial x} + \frac{\partial}{\partial y} \tau_{xy} + \frac{\partial}{\partial z} \tau_{xz} \tag{1}$$

$$0 = \rho g - \frac{\partial p}{\partial z} \tag{2}$$

in which x,y,z = the cartesian coordinate as shown in Fig.1, i = the slope of energy gradients, p = the pressure, τ_{xy}, τ_{xz} = the Reynolds shear stress, g = the acceleration due to gravity, ρ = the total density of mixed water.

The conservation of matter is given by

$$u \frac{\partial c}{\partial x} = \frac{\partial}{\partial y} \left(v_t' \frac{\partial c}{\partial y} \right) + \frac{\partial}{\partial z} \left(v_t' \frac{\partial c}{\partial z} \right) \tag{3}$$

in which u = the flow velocity in the x direction, c = is concentration of solute or injected matter, and v_t' = the turbulent eddy diffusivity.

The density of fluid is determined as

$$\rho = \rho_0 + c(\rho_s - \rho_0)$$ (4)

in which ρ_s and ρ_0 = density of solute and water respectively. Integrating Eq.1 over flow depth h, we have

$$0 = \rho ghi - \frac{1}{2} gh \frac{d}{dx} \int_0^h \rho \, dz + \frac{d}{dy} \int \tau_{xy} \, dz + \tau_{xz} \Big|_0^h$$ (5)

Turbulent shear force τ_{xy} is given by using kinematic eddy viscosity ν_t, as

$$\tau_{xy} = \rho \nu_t \frac{\partial u}{\partial y}$$ (6)

It is also assumed that the flow properties in the x-z plane in Fig.1 is similar to those of infinite width flow and lateral variation of the flow is given by the lateral variation of shear velocity $u_*(y)$, thus

$$u = u_*(y)\bar{u}(z)$$

$$\nu_t = u_*(y)\bar{\nu}_t(z)$$ (7)

Substituting Eq.7 to Eq.5, we obtain

$$0 = 1 - \sigma + \left(\frac{h}{B/2}\right)^2 \frac{\partial}{\partial y^+} \int_0^1 \{ \bar{u}_*(y)\bar{\nu}_t(z)$$

$$+ \frac{\partial}{\partial y^+} (\bar{u}_*(y)\bar{u}(z)) \} \, d\bar{z} - \bar{u}_*(y)^2$$ (8)

in which

$$y^+ = y/(B/2), \ \bar{z} = z/h, \ \bar{u}_*(y) = u_*(y)/u_{*c}, \ \left(u_{*c}^2 = ghi\right)$$ (9)

where σ is a nondimensional density gradient parameter given in Eq.12. The above Eq.8 gives shear velocity distribution in a channel and total flow field can be specified by Eq.7, provided u and ν_t are given as a function of z.

The boundary condition of Eq.8 are the symmmetry condition at y=0 and the shear force acting along the total wetted perimeter is equal to motive force of flow $\rho ghiB$. It is also assumed that the total shear force acting on the side wall is equal to the shear force acting on the bottom zone, which has the lengh of h taken from the channel corner

2.2 Effect of density gradient on 2-D open channel flow

To solve Eq.8, we must determine u and v_t as a function of z for any given value of σ. The basic equation of 2-D turbulent flow can be derived from Eq.1 and 3 by setting $\partial/\partial y = 0$

$$0 = 1 - 2\sigma(1 - \bar{z}) + \frac{d}{d\bar{z}}\left(\bar{v_t}\frac{d\bar{u}}{d\bar{z}}\right) \tag{10}$$

$$\bar{u}\frac{\partial c}{\partial \bar{x}} = \frac{\partial}{\partial \bar{z}}\left(\bar{v_t}\frac{\partial c}{\partial \bar{z}}\right), \quad \bar{x} = x/h \tag{11}$$

where

$$\sigma = \frac{gh}{2u_*}\,\Delta\bar{\rho}\,\frac{dC_{mh}}{d\bar{x}} \tag{12}$$

$$C_{mh} = \int_0^1 C\,d\bar{z}, \quad \Delta\bar{\rho} = (\rho_s - \rho_0)/\rho_0 \tag{13}$$

$$\bar{v_t}'(= \bar{v_t}) = v_t'/(hu_*) \tag{14}$$

parameter σ is made by longitudinal density gradient and shear force of flow, and from now on we write $v_t' = v_t$. To the case of $\sigma = 0$, that is the case of no density gradient, we employed the mixing length model and logarithmic flow velocity profile. In addition, the effect of density gradient on eddy viscosity is given by Monin-Obukhov length theory[7] thus,

$$\frac{u}{u_*} = \frac{1}{\kappa}\ln\frac{u_* z}{v} + A = \frac{1}{\kappa}\ln\left(\bar{z}\frac{u_*}{U_{mh}}\frac{U_{mh}h}{v}\right) + A \tag{15}$$

$$\left.\begin{array}{l} \dfrac{u_*}{U_{mh}} = \sqrt{\dfrac{f}{8}}\,, \quad f = 0.3164\,(2\,R_e)^{-1/4} \\[2em] R_e = \dfrac{U_{mh}h}{v}\,, \\[2em] U_{mh} = \displaystyle\int_0^1 u\,d\bar{z} \end{array}\right\} \tag{16}$$

$$\frac{v_t}{hu_*} = \kappa\bar{z}\left\{ 1 - \bar{z} - \sigma(1-\bar{z})^2 \right\}\left(1 - \beta R_i \right) \qquad\qquad \beta = 5.0 \tag{17}$$

$$R_i = -g\left(\frac{d\rho}{dz}\right)\Big/\left\{ \rho\left(\frac{du}{dz}\right)^2 \right\} \tag{18}$$

where Umh = the depth averaged value of flow velocity.
Concentration of matter can be approximated as

$$c = c_0(1 + Kx_1) + \hat{c}^+ \tag{19}$$

$$\bar{x}_1 = \bar{x} - \bar{U}_{mh}\bar{t} \tag{20}$$

where K = a constant which may correspond to the cross sectional mean value of longitudinal concentration gradient and x_1
= a nondimensional distance taken from the point moving with
mean flow velocity.
Substituting Eqs.19 and 20 into Eq.11, we have

$$\frac{u - U_{mh}}{u_*} = \frac{d}{d\bar{z}}\left(\bar{v}_t \frac{d\hat{c}^+}{d\bar{z}} \right) \tag{21}$$

Flow properties in 2-D turbulent open channel flow with longitudinal density gradient is determined by Eqs.10 and 11.
Eq.21 has been used to obtain longitudinal dispersion coefficient in 2-D density effected turbulent flow and will be used
again later in the 3-D case. If we put $\sigma=0$ in above analysis
then we have Elder's solution.
To get solutions of these equations, following regular perturbation method is used

$$\begin{aligned}
\bar{u} &= u / u_* = \bar{u}_0 + \bar{u}_1\sigma + \bar{u}_2\sigma^2 + \cdots \\
\bar{v}_t &= v_t /(hu_*) = \bar{v}_{t0} + \bar{v}_{t1}\sigma + \bar{v}_{t2}\sigma^2 + \cdots \\
\hat{c}^+ &= c_{10} + c_{11}\sigma + c_{12}\sigma^2 + \cdots
\end{aligned} \tag{22}$$

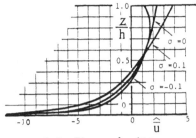

2-1 Flow velocity

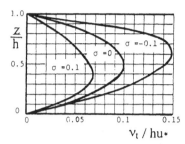

2-2 Eddy viscosity

FIGURE 2

Effect of longitudinal
density gradient on
flow properties

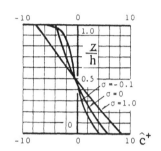

2-3 Additive concentration

The effects of σ on u, ν_t and $\hat{c}^+$ are illustrated in Fig.2. Detailed solution is given in our previous paper[8]. When $\sigma < 0$ flow becomes unstable, as a result, the vertical mixing increase and when $\sigma > 0$ turbulence is dumped and flow velocity give so called log-linear profile.

2.3 Longitudinal dispersion coefficient in 3-D flow

Making use of $u(z)$ and $\nu_t(z)$, we can determine the shear velocity distribution in the 3-D channel flow, therefore total flow velocity distribution is specified for given value of σ, by Eq.7.
Here we assume again that the concentration of matter can be approximated as

$$c(x, y, z) = c_0(1 + Kx_1) + \hat{c}(y, z) \qquad (23)$$

Substituting Eq.23 to Eq.3, we obtain

$$\left(\bar{\bar{u}} - \bar{\bar{U}}_{ma}\right) \frac{\partial c}{\partial \bar{\bar{x}}} = \frac{\partial}{\partial \bar{y}}\left(\bar{\bar{\nu}}_t \frac{\partial \hat{c}}{\partial \bar{y}}\right) + \frac{\partial}{\partial \bar{z}}\left(\bar{\bar{\nu}}_t \frac{\partial \hat{c}}{\partial \bar{z}}\right) \qquad (24)$$

70

where

$$\bar{\bar{U}}_{ma} = \frac{1}{Bh} \int \int \bar{\bar{u}} \, d\bar{y} \, d\bar{z} \, , \ \ \bar{\bar{u}} = u / u_{*c} \, , \ \ \bar{v}_t = v / u_{*c} \, , \ \ \bar{y} = y / h \qquad (25)$$

To seek solutions of Eq.24, we assume that c can be written as,

$$\hat{c} = \hat{c}_1 (\bar{z}) + \hat{c}_2 (\bar{y}) \qquad (26)$$

Substituting Eq.26 to Eq.24, we have

$$\left\{ \left(\bar{\bar{u}} (y, z) - \bar{\bar{U}}_{mh}(y) \right) + \left(' \bar{\bar{U}}_{mh}(y) - \bar{\bar{U}}_{ma} \right) \right\} \frac{\partial c}{\partial \bar{x}}$$

$$= \frac{\partial}{\partial \bar{z}} \left(\bar{v}_t \frac{\partial \hat{c}_1 (z)}{\partial \bar{z}} \right) + \frac{\partial}{\partial \bar{y}} \left(\bar{v}_t \frac{\partial \hat{c}_2 (y)}{\partial \bar{y}} \right) \qquad (27)$$

where

$$\bar{\bar{U}}_{mh} = \int_0^1 \bar{\bar{u}} \, d\bar{z} \qquad (28)$$

Taking the first term of both sides of Eq.27, we obtain the same equation to Eq.21, so that

$$\hat{c}_1 (\bar{z}) = \hat{C}^+(\bar{z}) \qquad (29)$$

When we integrate Eq.27 over flow depth, the first terms of both side of Eq.27 disappear, Then c can be obtained as

$$\hat{c}_2 (\bar{y}) = \int_0^y \frac{1}{\bar{v}_{th}} \int_0^y \left(\bar{\bar{U}}_{mh}(\bar{y}) - \bar{\bar{U}}_{ma} \right) d\bar{y} \, d\bar{y} \qquad (30)$$

where

$$\bar{v}_{th} = \frac{1}{h} \int_0^1 \bar{v}_t \, d\bar{z} \qquad (31)$$

Finally the dispersion coefficient in this case is given by

$$\frac{D_L}{hu_{*c}} = \int_A \int \left(\bar{\bar{u}} - \bar{\bar{U}}_{ma} \right) \hat{c} \, d\bar{y} \, d\bar{z} \qquad (32)$$

71

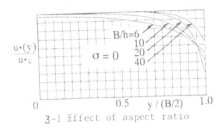

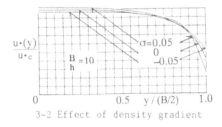

3-1 Effect of aspect ratio   3-2 Effect of density gradient

FIGURE 3   Lateral distribution of shear velocity

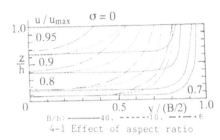

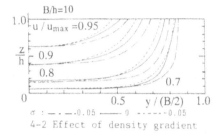

4-1 Effect of aspect ratio   4-2 Effect of density gradient

FIGURE 4   Flow velocity distribution (isovel)

3 Results and discussion

3.1 Numerical solution

Numerical solutions are obtained by using ordinary finite difference method. Vertical mesh size is 0.05*h and horizontal mesh size is varied 0.1*B–0.01*B according to the aspect ratio. Examples of lateral distribution of shear velocity are given in Fig.3. Effect of aspect ratio and longitudinal density gradient

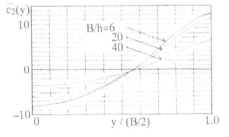

FIGURE 5   Additive concentration

are shown in Figs.3-1 and 3-2 respectively. Horizontal axis is normalized by channel width B. These figures show that with increase of aspect ratio the effect of side shear becomes negligible near the center zone. On the other hand, the effect of σ can be seen even at the center of flow, for if we put $\frac{\partial}{\partial y}=0$ in Eq.8, $u_*(y)$ is still affected by σ .
Flow velocity profiles in channel cross section are illustrated in Figs.4-1 and 4-2. The values in these figures are normalized by maximum velocity. These figures also present the effect of aspect ratio and density gradient.
Fig.5 shows the additive concentration obtained by Eq.30. This concentration is also affected by the lateral distribution of shear velocity.

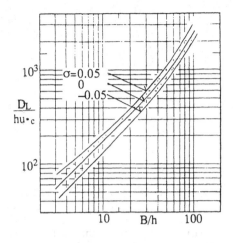

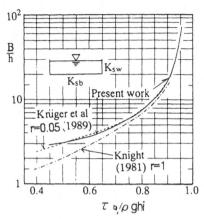

FIGURE 6 Dispersion coefficient

FIGURE 7 Bed shear stress τ_b
roughness ratio: $r = K_{sb}/K_{sw}$

Relationship between D_L, B/h and σ are given in Fig.6. With increase of B/h dispersion coefficient increases as reported so far. When $\sigma < 0$ or $\sigma > 0$ Dl gives smaller or larger value compared with the case of $\sigma = 0$ and this is a reasonable result. To the effect of density gradient, there exist the case where the secondary current due to local density difference in a channel cross section gives serious effect on dispersion[9,10]. In our study, these phenomena are not taken into account, further detailed research must be needed to discuss on the effect of density.
Relationship between bottom shear stress and aspect ratio in rectangular turbulent open channel is given in Fig.7 with previous work.[11,12] This figure shows that our analysis gives close values to those of previously reported.

3.2 Comparison with experiment

Fig.8 presents a comparison of our results with experiment. Experimental result and field measurement data are referred from Iwasa and Aya's paper[13]. Iwasa and Aya reviewed 67 laboratory data and 79 field data and derived the best fit relation as shown in Fig.8. The theoretical prediction of the present study gives rather higher value of D_L. More than half of the data plotted in Fig.8 are obtained from field measurement, therefore the flow conditions that we employed in the analysis may not be satisfied. In addition, we used simple and rough method to get the relationship between the dispersion coefficient and the aspect ratio. Considering these matter, the result of the present work can be said to be reasonable.

FIGURE 8

Comparison of theoretical
prediction with laboratory
data● and river measurement○

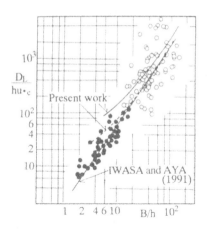

In relation to the dispersion in turbulent open channel flow,
it is reported that to get best fit to the field measurement,
the mean value of horizontal turbulent viscosity is taken as
3.4-3.5 times greater than vertical one[1,3,4]. A numerical
solution is also presented by broken line in Fig.8, which is
obtained by using the horizontal eddy viscosity multiplied by
3.45. and no other modification in our theory . A better agreement is obtained in these figures, but this method does not
necessary mean improvement, for the ratio of vertical eddy
viscosity to horizontal one cannot be taken as constant numerical value but may vary with the distance from side wall.
There are several factors that must be considered in the analysis of these phenomena, such as secondary flow due to both side
shear and density difference, turbulence decay at the water
surface, local difference of roughness along the wetted perimeter etc. Further elaborated study must be needed for more
detailed discussion.

ACKNOWLEDGMENT

We would like to acknowledge graduate student Kounosuke Taka-
hara for his computational assistance.

REFERENCES

[1] Elder,J.W.,(1959),The Dispersion of Marked Fluid In Turbu
    lent Shear Flow, Journal of Fluid Mechanics, Vol.5, pp.554
    -560
[2] Taylor,G.I.,(1954), The Dispersion of Matter in Turbulent
    Flow Through a Pipe, Proceedings, Royal Society of London,
    Vol.223, Series A, pp.446-468
[3] Fisher,H.B.,(1967),Longitudinal Dispersion in Natural
    Streams,Journal of the Hydraulics Division,ASCE,Vol.93,
    No.HY6,pp.187-216

[4] Attila,A.S.,(1969), Longitudinal Dispersion in Open Channels, Journal of the Hydraulics Division,ASCE, Vol.95,No.
HY4, pp.1327-1346

[5] Chatwin,P.C.,(1982),The Effect of Aspect Ratio on Longitudinal Diffusivity in Rectangular Channels,Journal of Fluid
Mechanics, Vol.120,pp.347-358

[6] Smith,R.,(1976), Longitudinal Dispersion of a Buoyant Contaminant in a Shallow Channel, Journal of Fluid Mechanics,
Vol.78,pp.677-688

[7] Turner,J.S.,(1979),Buoyancy Effects in Fluids, Cambridge
University Press,p.130

[8] Fujisaki,K.,Minami,Y.and Awaya,Y,(1987), Effect of Density
on Longitudinal Dispersion,Proceedings of the Specialized
Conference on Coastal and Estuarine Pollution,pp.237-244

[9] Smith, R.,(1979), Buoyancy Effects upon Lateral Dispersion
in Open Channel Flow,Journal of Fluid Mechanics, Vol.90,
pp.761-779

[10] Fisher, H.,B.,(1972), Mass Transport Mechanisms in Partially Stratified Estuaries,Journal of Fluid Mechanics,
Vol.53,pp.671-687

[11] Kruger,F., and Bollrich,G.,(1989), Boundary Shear Distribution in Rectangular and Trapezoidal Channels with Uniform and non Uniform Bed and Wall Roughness, Proceedings
of International Congress of IAHR, pp.B91-B98

[12] Knight,D.,W.,(1981),Boundary Shear in Smooth and Rough
Channels,Journal of the Hydraulics Division,ASCE, Vol.107,
No.HY7, pp.839-852

[13] Iwasa,Y.,and Aya,S.,(1991),Predicting Longitudinal Disper
sion in Open-Channel Flows,Environmental Hydraulics,pp.
505-510

Part 2
FLOW MODELLING

7   A backwater method for trans-critical flows

P. G. Samuels and K. S. Chawdhary

Abstract

Accepted practice for the computation of steady water surface profiles is that it is necessary to alter the direction of the calculation when the flow becomes super-critical. The origin of this limitation appears to be in the common practice of writing the flow equations as an explicit ordinary differential equation. This paper demonstrates by some test cases, however, that retaining the flow equations in their implicit form allows trans-critical flows to be computed by a backwater method. Finally the implications of these results for practical river modelling are discussed.

1   Introduction

Backwater calculations are one of the most common hydraulic calculations carried out in the river engineering practice and these are often formalised into computational models of varying degrees of complexity. Despite many years of research and application there is still much in hydraulic modelling that depends upon the art and experience of the modeller, whether in the development of new model code or in the application of existing methods. Part of the received wisdom passed from one generation of modellers to the next is that steady flow in an open channel must be calculated from downstream to upstream for sub-critical flow (ie a *backwater*) and in the reverse direction for super-critical flow. The material presented in this paper, however, challenges that received wisdom in that a method has been developed which will compute water surface profiles in the upstream direction for both flow regimes and will pass through the critical transitions. This effect was observed in 1989 during an investigation at HR Wallingford into the weighting of friction and area terms in the flow equations which has been presented in part elsewhere, (Samuels, 1990). It was so unexpected that it was dismissed as a numerical quirk and not pursued further. However, following an MSc project at the University

79

of Reading (Chawdhary, 1991) it appears that the result was due to solving the open channel flow equations as an implicit ordinary differential equation (ODE) rather than the more familiar formulation as an explicit ODE. Unfortunately, by the nature of the topic of this paper, some mathematics is unavoidable but it is hoped that the complexity is kept to a minimum.

2   Explicit and implicit ODEs

For ease of analysis and approximate or exact analytic solution, the differential equation(s) resulting from a mathematical model of a physical process is usually manipulated, if possible, into an explicit form. An explicit first order ODE may in general be written as:

$$y' = f(x,y) \qquad\qquad (1)$$

where y is the unknown quantity (in our case water level or depth), x is the independent variable (distance along the centre line of the channel) and f(x,y) is some relationship which does not involve the derivative y'. There are many methods available to solve explicit ODEs (Lambert, 1973) and for the open channel flow equation either the trapezium rule or a Runge-Kutta method is often employed. Each of these methods has its range of stability which determines the maximum step size and the direction of numerical integration that can be used.

In many mathematical models of physical systems, for example steady flow in an open channel, the model equations do not occur naturally as explicit ODEs but rather as implicit ODEs. Implicit ODEs involve the highest order derivative on both sides of the equation, the general form of the first order equation is:

$$y' = F(x,y,y') \qquad\qquad (2)$$

where F may be a non-linear function of its arguments. This change in the form of the differential equation precludes the direct application of many common types of numerical method which require an explicit formula for the derivative, y'. That is why the model equations are often manipulated into an explicit form before numerical methods are employed to obtain approximate solutions. There are advantages, however, in retaining the implicit form of the ODE and using a numerical approximation to estimate the value of the coefficient y' in the function F in equation (2). Two important advantages are that improved stability and accuracy properties can be obtained (Chawdhary, 1991) but the theoretical analysis demonstrating this will not be presented here. The penalty is that it almost certainly will be necessary to iterate to obtain the discrete solution at each computational step but if, in equation (1), f is non-linear and involves y, then iteration cannot be avoided with most methods. Unfortunately, the mathematical theory of numerical solution of implicit ODEs appears to be sparse.

3   A simple illustration

Before considering the model equations for steady flow in an open channel, a more simple illustration of the use of implicit ODEs will be given. Although this case is not derived directly from an analogy in hydraulics, it contains a singularity in some ways similar to the transition between sub- and super-critical flow. We consider the linear ODE:

$$2y' = |x|^{-1/2} \qquad (3)$$

$$y(-1) = -1$$

this has the analytic solution (see Fig 1):

$$y = x |x|^{-1/2} \qquad (4)$$

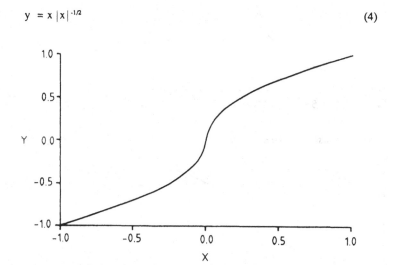

FIGURE 1   *Analytic solution*

Obviously as any ODE solution method approaches $x = 0$ and evaluates the unbounded derivative from equation (3) the approximate solution will break down. Rewriting the original ODE in implicit form as:

$$y' = F(x,y,y')$$

$$F = \{1/2 + (p - |x|^{1/2})y'\}/p \qquad (5)$$

where p is a free parameter, removes the singularity from F at $x = 0$. The trapezium rule, for example, can now be applied to equation (5) with a suitable discrete replacement for y' in F and a numerical solution obtained by iteration. If the trapezium rule is used on the original equation (3), it fails at $x = 0$. Figs 2 & 3 show the numerical solutions to equation (5) for $p = 1$ using space steps of 0.05 & 0.005 respectively.

4   Steady flow in a open channel

The usual model for unsteady flow in an open channel is the St Venant equations. These are formed from applying the principles of mass and momentum conservation to the free-surface flow in a straight open channel. The steady flow equations are then obtained in the limit as the time derivatives vanish. Written in terms of discharge, Q, and water level, h, they are:

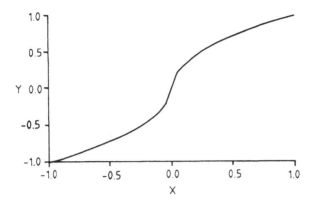

FIGURE 2   *Numerical solution (step size 0.05)*

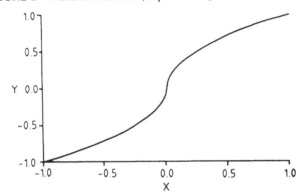

FIGURE 3   *Numerical solution (step size 0,005)*

$$Q = \text{constant} \tag{6}$$

$$\frac{d}{dx}\left\{\frac{Q^2}{A}\right\} + gA\left\{\frac{dh}{dx} + S_f\right\} = 0 \tag{7}$$

where x is the distance along the channel in the direction of the flow, A is the flow area, S_f is the friction slope and g is acceleration due to gravity. Often the implicit form of dynamic equation (5) is rearranged to give an explicit backwater equation for the water depth, y, in gradually varied flow, see Henderson (1966) or French (1986).

$$\frac{dy}{dx} = \frac{S_0 - S_f}{1 - Fr^2} \tag{8}$$

Here Fr is the Froude number of the flow and S_0 is the bed slope of the channel. [Note that, in usual engineering practice, the direction of x in the backwater equation runs against the flow but that here in equation (8) x has the opposite direction.] This form of the equation can be used to develop approximate analytic solution to the water surface profile for sub-critical flow

82

(Samuels, 1989). Obviously as the Froude number passes through the critical condition (Fr = 1) a numerical algorithm for the solution of equation (8) may become ill-defined. Equation (8), however, forms the basis of many backwater programs in engineering practice and education. For small variations onf water level around normal depth, the first order perturbation equation derived from (8) is given by Samuels (1989) as:

$$\frac{dh}{dx} = Gh \tag{9}$$

where

$$G = -\frac{I}{I - Fr^2} \frac{dS_f}{dy} \tag{10}$$

The origin of the choice of the direction of integration lies in the need for the method to be stable to growth of rounding error. The condition for the trapezium rule to be *zero-stable* is that:

$$G \Delta x < 0 \tag{11}$$

that is, the calculation must proceed against the flow in subcritical conditions and with the flow in supercritical conditions. Lambert's (1973) discussion of convergence on pp 33-34 implies that zero-stability is a prerequisite for a linear multistep method to be of practical value, precisely because of the growth or decay of rounding error as the solution progresses. These comments are made in the context of solving explicit ODEs.

The open channel flow equations (6) & (7) occur naturally in implicit form and there is no over-riding reason why the rearrangement into explicit form need be done in order to achieve approximate numerical solutions. Hence Chawdhary (1991) considered the following implicit formulation of the dynamic equation (7) for flow in a rectangular channel of gradually varying width.

$$\frac{dy}{dx} = \frac{Q^2}{gB^3 y^2} \frac{dB}{dx} + \frac{Q^2}{gB^2 y^3} \frac{dy}{dx} + S_o - S_f \tag{12}$$

Here the convective acceleration term has been expanded in symbolic form and written using the breadth, B, as the main geometric parameter. Furthermore, Chawdhary used Manning's equation to determine the friction slope:

$$S_f = \frac{Q|Q|n^2 (2y + B)^{4/3}}{(By)^{10/3}} \tag{13}$$

It is clear that the dynamic equation (12) is implicit and its solution may be approximated using the trapezium rule iteratively. The value of dy/dx on the right-hand-side must be estimated by some convenient rule, in the results presented below this was approximated as the slope of the chord joining the water depth at the two ends of the trapezium rule calculation.

5 A test case

The computation of the steady flow profile for water level and Froude number have been carried out for some test cases examined by Priestley (1990). The geometry represents a rectangular channel 10 km long with width varying smoothly from 10m at either end to 5m in the centre, see Fig 4. The bed slope is uniform except for the central 1 km where the bed gradient is doubled, see Fig 5, this forces transitions between sub- and super-critical flow for some bed gradients. In each test the value of Manning's n was 0.03, the discharge was 20 m³/ s and a step size of 25 m was used. Figs 6, 7 & 8 show the water depth and Froude number for the bed slopes of magnitude 0.002, 0.01 and 0.02 respectively. The results show no oscillation and capture the transitions in depth cleanly. In all cases the calculations proceeded from the downstream end of the channel and passed through the transitions (Fig 7) without any special

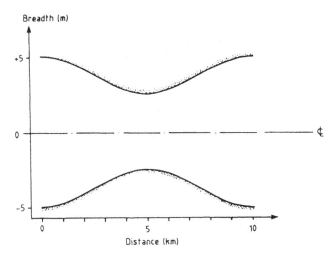

FIGURE 4   *Channel breadth*

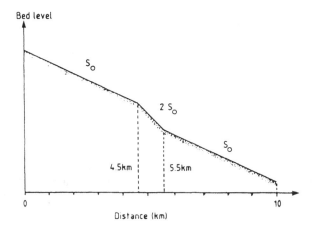

FIGURE 5 *Bed profile*

action to locate these points. The results are nearly indistinguishable from the corresponding plots given by Priestley (1990). Priestley studied the conservative form of the unsteady flow equations using the Riemann invariants of this hyperbolic set of PDEs and produced steady state solutions by applying steady boundary conditions.

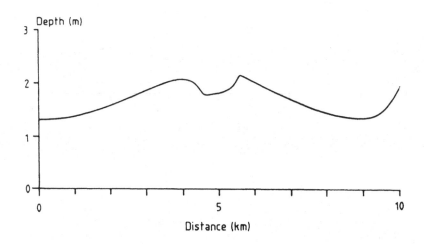

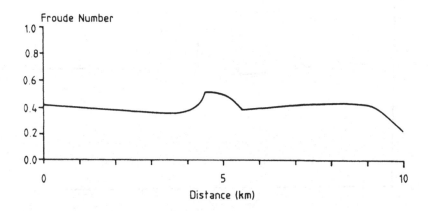

FIGURE 6   *Depth and Froude number (slope = 0.002)*

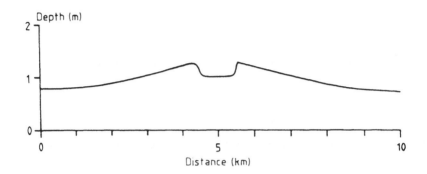

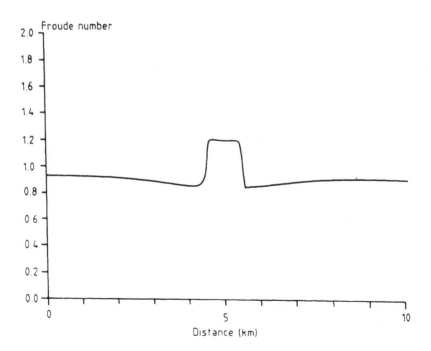

FIGURE 7   *Depth and Froude number (slope = 0.01)*

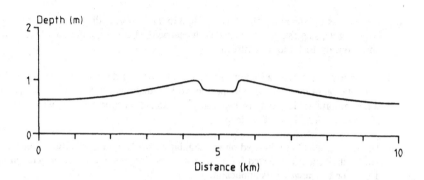

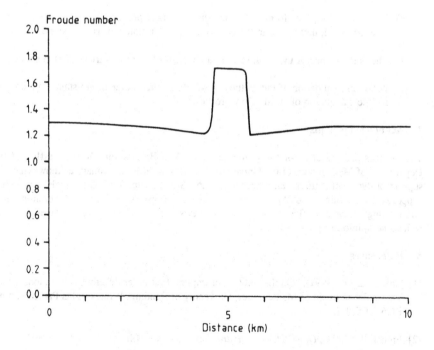

FIGURE 8 *Depth and Froude number (slope = 0.02)*

6 Discussion and future work

The results given above were obtained in the course of an MSc project and necessarily this had a restricted scope. There are several further tests and extensions that could be considered, some of these are in hand.

(a) The success of integrating the flow equation in the "reverse" direction for supercritical flow (Fig 8), suggests that tests should be carried out for computation of subcritical flow from upstream to downstream.

(b) The ability to compute a steady surface profile in either direction will simplify the computation of the flows in a looped network in that the ordering of the computation is not important. This may be especially significant for pipe systems where the slope of the invert may be locally steep.

(c) The tests to date have focused on rectangular channels with gradually varying width. The method should be tested for more severe artificial and natural river geometries to determine the range of its applicability.

(d) Extensions of the method should be sought for unsteady flow. In this context the scheme that corresponds to the trapezium rule is the Preissmann box scheme. Samuels and Skeels (1989) have shown that the range of formal linear stability of the Preissmann scheme extends into the super-critical regime.

(e) The most appropriate form of treating the non-linear products in the equations should be established, these arise in the convective acceleration and friction slope terms.

(f) The performance of the method should be established for a variety of step sizes.

(g) Following validation of the method, it should be incorporated into standard packages for the calculation of steady flow profiles.

7 Acknowledgements

The authors are grateful for the guidance of Dr M J Baines and Dr A Priestley of the Department of Mathematics of the University of Reading. K S Chawdhary acknowledges the support of the SERC through an Advanced Course Studentship. Dr P G Samuels' research is supported by the Ministry of Agriculture, Fisheries and Food (MAFF) research commission to HR Wallingford on river flood protection. The views expressed in this paper, however, do not reflect the opinion or policy of MAFF

8 References

[1] Chawdhary K (1991), "On the solution of implicit first-order differential equations", MSc Thesis, Department of Mathematics, University of Reading, Whiteknights Park, Reading, RG6 2AX, UK

[2] French R, (1986), *Open-Channel Hydraulics*, McGraw Hill

[3] Henderson F (1966), *Open Channel Flow*, MacMillan

[4] Lambert J (1973), *Computational Methods in Ordinary Differential Equations*, John Wiley & Son

[5] Priestley A (1990), "A quasi-Riemann method for the solution of one-dimensional shallow water flow", Numerical Analysis report 5/90, Department of Mathematics, University of Reading, Whiteknights Park, READING, RG6 2AX.

[6] Samuels P G (1989), "Backwater lengths in rivers", *Proc Instn Civ Engrs*, Pt 2, vol 87, Dec, pp 571-582.

[7] Samuels P G (1990), "Cross section location in 1-d models", Paper K1 at the International Conference on *River Flood Hydraulics*, Wallingford, Sept 1990, Proceedings ed WR White, John Wiley & Sons.

[8] Samuels P G & Skeels C P (1989), "Stability limits for Preissmann's scheme", *Journal of Hydraulic Engineering*, Proc ASCE, Vol 116, No 8, Aug, pp 997-1012

8 Comparison of four simulation tools used in river engineering

J. Vodslon

ABSTRACT

This paper is devoted to testing the ability of various models to simulate particular hydrodynamic processes in rivers and flood plains in their complexity. Four simulation tools were applied in the reach of the Ohře river close to the village of Pisty in Northern Bohemia.
Two independent 1D morphological models, NIMMPH (Dekker, 1989) and MIKE-11 (DHI, 1989), were applied for morphological investigation. A 2D hydrodynamic model FLUVIUS (Hydroinform, 1989) and 1+D model VEGA (CVUT 1992) were used to simulate velocity fields and stages in the dominion of interest.
The objective of the study was mainly to investigate the applicable limits of simpler tools for projects within the framework of revitalization. This paper presents results achieved by applied models and gives conclusions concerning their applicability under the given conditions and adopted aims.

1. Introduction

Man-made structures and changes in rivers and adjacent flood plains have always caused feedback effects on hydrological and hydrodynamic characteristics in a basin. The history of the basin of interest involves the construction of dams, hydraulic structures, buildings, water intakes, along with dredging (sand

91

or gravel mining) activities. These actions have always caused a certain impact on the basin environment.

The main attempt of water management today is to return river channels to their natural status - refered to under the heading **"revitalization"**. However designers are facing problems how to get accurate information about velocity fields, discharges, and stages in a particular river channel with a natural shape. The highest uncertainties are related to estimates of velocities close to banks and in flood plains adjacent to the channel because of the thick cover of vegetation. The values of velocity fields and stages are very important for design activities such as bank protection, sediment transport sensitivity and stability of a channel bed and/or flood plain close to the river channel itself. A common approach of engineers today is to utilize some group of simulation tools to provide a technical solution within the framework of **revitalization**. However, there does not exist any general recommendation or methodology concerning quality and type of models or tools to be applied. There are still several areas for discussion left, whichmay be specified as follows:

* how accurately various models describe the velocity field under specific conditions (such as alluvial roughness, sediment transport ratio, crossings of river channel and flood plain flow streams).
* the ability of models to perform as tools for the design of river channels and revitalization of rivers, with higher emphasis on shape variety providing a more natural type of channel using vegetation cover for bank protection and river improvement.
* evaluation of roughness characteristics with respect to varieties of vegetation, locatoin and time of year.

2. Reach of the Ohře river under the investigation-pilot study

Since reach on the Ohře river has been used several times for various studies and investigations, there are available complete data sets, which were utilized for the current study. The reach is situated downstream from the Nechranice dam located close to the Hostěnice weir which is situated approximately 12 km upstream from the confluence with the Elbe river fig.[1]. The experimental reach is approximately 2,5 km long, the river width is about 65 m, and the river bed consists of gravel and sand. The reach represents a quite natural river course without river improvements, with four severe bends. The river channel is situated in the middle of a 2 km wide flood plain which is partly covered by forest.

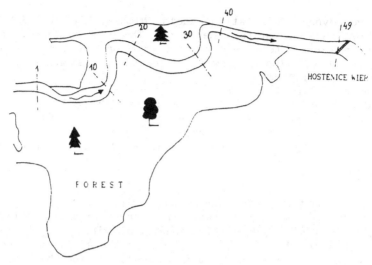

FIGURE 1   *Situation of pilot study.*

2.1.Data availability

A unique data set has been collected during research activity over the last two decades.Two detailed surveys of the river channel were taken during this time:
- 1974/75 - 57 cross-sections carefully investigated, average
                  distance from 30 m up to 50 m
- 1989 - 54 cross-sections with a distance from 35 m up to 55 m
The comparison of the two surveys has shown quite severe displacement of banks and bed levels in the reach over time.
Grain size distributions were available from two independent sampling surveys; required grain size distribution curves wereprepared for the previous study. [3]
Hydrology data in the form of time series for a numerical experiment were available in two forms:
- record of daily discharges in the Louny gauging station for
   the period 1975 - 1989
- record of selected flood events (recorded hourly)
Roughness characteristics of flood plains (partly forested) for 2D simulation were estimated by various experts in hydraulics. (manning coefficient n veried in interval 0.25 - 0.03) The forest was divided into more or less homogeneous sections, in order to provide unified data for simulation purposes. The delineation of a section for this study was dependent upon the density of the trees in a given area, and the trunk diameter of the individual trees.

3. Steady flow simulation using four hydrodynamic models in the reach

Four hydrodynamic simulation models, NIMMPH, MIKE 11, FLUVIUS, and VEGA were chosen for the study. These models were chosen for their ability to give some sort an answer to the objectives outlined above. The following characteristics were compared among the applied models:

* dimensions (1D, 1$^+$D, 2D)
* condition of the flow(steady - unsteady, Fr, Re)
* hydrodynamic parameters and characteristics (R,A,P)
* resistance parameters (C,n,ks,λ ..)
* fixed bed, alluvial roughness, sediment transport processes

3.1.MIKE 11

simulates unsteady one-dimensional flows, transports and biological-chemical reactions in one layer (vertically homogeneous fluids). MIKE 11 includes a rainfall-runoff model which simulates the catchment runoff to the river composed of overland flow, inter flow and base flow. The hydrodynamic description is based upon the Saint Venant equations of conservation of mass and momentum. For sediment transport description, two formulations were implemented. The Engelund-Hansen formula, which gives the total bed material transport and Engelund - Fredsoe formula, which gives the bed load and suspend load separately. The water quality module is coupled to the transport-dispersion module and used to simulate the reaction processes of multicompound systems. The model is able to respect alluvial roughness.

3.2.NIMMPH

The NIMMPH model is based on 1-D quasi-steady approach and de Vries (1987) theory valid for subcritical flow, i.e.(Fr<0.6). For NIMMPH morphological calculation rectangular schematization of real cross sections is expected. There is the VIPP supporting programme unit allowing to a user schematize cross sections quite easily. The separation of discharges related to river channel and flood plain had to be done beforehand, since the NIMMPH may handle only a channel part of the flow. For sediment transport calculation has been employed 5 standard methods: Meyer-Peter-Müller, Engelund and Hansen, Acker and White standard method, Ackers and White wide distribution method and van Rijn method. Alluvial roughness should be applied. Since maximum 500 time steps can't be exceeded, there is a restart option in the programme menu solving problem with longer time series. Boundary conditions necessary for computations are as following: discharge Q upstream, sediment transport or bed level z upstream and stage h downstream.

MIKE 11 and NIMMPH simulations
The author of this study relied upon previous work done by Jůza
[3] and Zeman [6], which were mainly focused on the
morphological changes of bed levels and displacement of banks.
The author used the known results of their simulations to
determinate discharges for the beginning of sediment transport
and to estimate conditions when alluvial roughness appears.
These two threshold conditions give limits for implementation
of the following models.
Simulation by these models provided several conclusions:
- the reach is in balance from point of view of sediment
  transport
- alluvial roughness has no important influence for discharges
  less then 100 m^3/s (valid for the cross sections of the river
  channel only)
- the flow in certain parts of the reach are typically three
  dimensional, such as the crossing of the river stream with
  the flood plain stream which is responsible for banks
  displacement of banks.

3.3. Fluvius

The mathematical model Fluvius, based on a numerical
integration (FDM) of the complete non-linear equation for 2D
unsteady flow in the horizontal plane is presented. In the flow
equations hydrostatic pressure distribution is assumed, the
algorithm used in the mathematical model is based on a
fractional step method in which wave propagation is evaluated
using an Alternating Direction Implicit Algorithm, and momentum
advection terms are calculated using the modified method of
characteristics. The algorithm is completed by diffusive
interface, enabling the user to control the rate of numerical
diffusion during calculation. The algorithm has been
incorporated in the computational module of the Fluvius
package. Flooding and drying of the flood plain, wind effect
and various methods for bed resistance calculation have been
taken into account.
From the point of view of this study, the model was used for a
more accurate description of the flow field in the channel and
adjacent flood plain, with the possibility to affect the
crossing of the river channel and flood plain flow stream and
space variety of flow.

Fluvius simulation
Simulation was done for flood event Q_{10}, on the part of the
reach which included forest and several severe curves. Fig.
[1]. The reach was described by 50 x 69 points and the grid
space were 15 x 15 m. The size of grids was a compromise
between demanding areas of interest, required accuracy of
result, proper description of flow in cross sections and

ability of the model. Resistance of flood plains was estimated using the Manning roughness coefficient. The result of simulation Q10 is shown in Fig. [2].

3.4.VEGA

A 1⁺D mathematical model for steady non uniform flow in vegetatively roughened rivers and flood plains covered by vegetation. The approach Pasche [5] suggested is based on a mathematical algorithm for computation of turbulent flow. The cross sections are divided into the several parts with homogeneous resistance. Resistance phenomena are described by four parameters: space density of trees and bushes (two parameters), the thickness of trees and bushes, and resistance of the bed - using the Manning-Strikler coefficient. The velocity profile is computed individually for each part of the cross section.

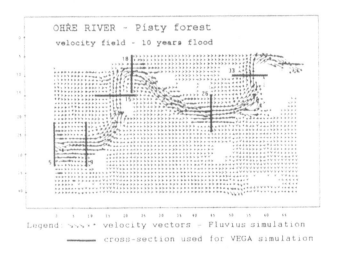

FIGURE 2   *Description of flow field by Fluvius model.*

Vega simulation
For simulation was used the same reach as the previous simulation. The Model was calibrated for bank-full discharge and other discharges less than bank full. (especially 50, 60, 80, 100, ...,180 and 200 m³/s). Fig.[3]. Results of calibrations were compared with the results produced by MIKE-11 and FLUVIUS for the same conditions. Differences in water level were: max. 0.25 m average 0.12 m.
A secondary simulation was for the flood event Q10. A flow of approximately 300 - 350 m³/s is required for VEGA's

schematization of cross sections and a flow of approximately 120 m^3/s for simulation by morphological models. This discharge(300-350) was chosen because of its low rate of

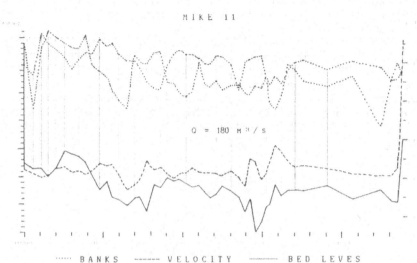

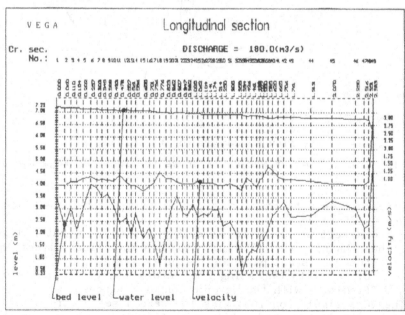

FIGURE 3   *Result of Vega model calibration - longitudinal profile.*

sediment transport, and under these conditions it is not
necessary to evaluate alluvial roughness. Simulation can be
provided using a fixed bed.
Two different sets of cross sections were used for these
simulations. The first set focused only on channel flow. The
second set was selected in such a way that part of the flood
plain in the vicinity of the river channel was also involved
(100 meters on each side).
Results for the ten year flood simulation (example of cross
section show Fig.[5].), were compared with the results from the
2 D model FLUVIUS. The Vega simulations involving the first set
of cross sections corresponded well with the 2D simulation.
Velocities and water levels were comparable, but some
differences occurred in the sharp bends. Fig.[6] shows
velocity profiles. These differences caused 1^+D schematization
Fig.[4].

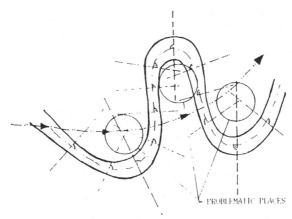

Legend: —·—·—·— center line
        – – – – cross sections
        —·—·—·— flow stream during the flood

FIGURE 4   *Schematizacion of cross sections for Vega model*

Difficulties with the proper description of the cross sections
were as follows :
- cross sections have to be orthogonal to the center line of
  the channel, but during flood events the direction of
  velocities is not parallel to the center line;
- problems with crossing of cross sections cause some
  difficulties with the description of roughness.
The Vega simulation involving the second set of cross sections
show in some locations shows significant differences which are
probably due to the problems with schematization which were
mentioned above.

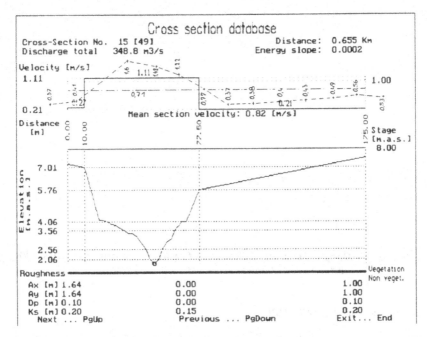

Legend: — · — · — MIKE 11 simulation
         — — — — FLUVIUS simulation
         ————— VEGA simulation

FIGURE 5   *Result of Vega simulation in cross section and -
            comparison of velocities among Fluvius, Vega, Mike11*

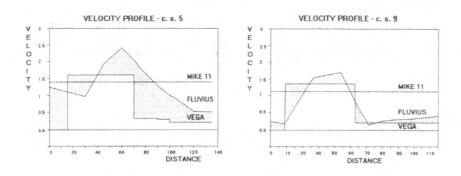

FIGURE 6   *Comparison of velocity fields - among Fluvius, Vega
            and Mike11 (part 1)*

99

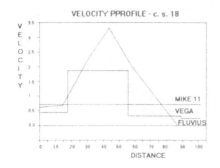
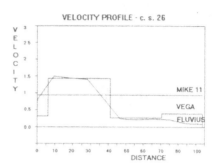

FIGURE 6   *Comparison of velocity fields - among Fluvius, Vega and Mike11 (part 2)*

4. RESULTS AND DISCUSSION

- the reach is in balance from the point of view of sediment transport
- alluvial roughness is not an important influence for discharges less then 100 m^3/s (valid for the cross sections of the river channel only)
- flow in certain locations of the reach is typically 3-D (crossing of river stream with flood plain stream), which is responsible for bank displacement and sediment extraction from the river channel. It is impossible to simulate these sections using 1-D models
- six cross sections were used to compare the differences between the velocity fields described by the VEGA model and by the FLUVIUS model for discharges of 350 [300] m^3/s :

$$\Delta = v_{fl} - v_{veg} \tag{1}$$

$$\Delta_{avrg} = \frac{1}{n} \cdot \Sigma \Delta_i \tag{2}$$

$$\sigma_\Delta = \frac{1}{n-1} \cdot \Sigma(\Delta_{avrg} - \Delta_i)^2 \tag{3}$$

where: v_{veg} velocity obtain by Vega simulation

v_{fl} velocity obtain by Fluvius simulation

Δ velocity differences

max. $\Delta = 2.04$ [2.22] $\Delta_{avrg} = 0.131$ [0.203]

standard deviation of differences for all six profiles:

$$\sigma_\Delta = 0.07 \quad [0.315]$$

for individual cross section:

$$max \ \sigma_\Delta = 0.407 \quad [0.418]$$

$$min \ \sigma_\Delta = 0.046 \quad [0.047]$$

There is also presented the differences between profile velocities $v = Q/A$ (obtain by MIKE 11) and 2D modelling by FLUVIUS for $Q = 350 \ m^3/s$:

max. $\Delta = 3.19$ $\Delta_{avrg} = -0.139$

standard deviation for all six profiles $\sigma_\Delta = 0.099$

for individual cross section:

$$max \ \sigma_\Delta = 0.616 \qquad min \ \sigma_\Delta = 0.145$$

5.CONCLUSION

The morphological models NIMMPH and MIKE-11 have prove their ability to give a reasonable morphological forecast for bed levels of river channels from the qualitative point of view. The above models can also be used for the prediction of influence of planned man-made structures in the river environment. The impact of the Hostěnický weir on observed changes in the river reach were considered minor. The influence of Nechranice dam (app. 90 km upstream from the reach) on morphological changes in the lower part of the basin of the Ohře river can't be judged correctly from the pilot study.
There are certain limits in the application of 1-D models to typically 3-D flow characteristics, such as the extraction of sediment from a river channel to a flood plain. These morphological models can provide limits of implementation for other models which have computational algorithms based on a fixed bed approach (for Ohře river were judged the limit as discharge $100 \ m/s^3$), and they can evaluate differences if the limits are exceeded.
The hydrodynamics model FLUVIUS produces a 2D description of a flow field, but requires huge data. In cases where the flood plain is covered by vegetation, the evaluation of roughness characteristics is difficult. The model defines places where streams crossing appear which help to clearly understand discontinuity and differences between results of simulations by morphological models and observed state. In this study we were not able to judge accuracy of model simulation due to the fact that we did not have observed real data set.

The VEGA model proved to be a good tool for designing channels and for revitalization of rivers when evaluation of vegetation resistance is required. It is also useful for the estimation of discharge of a given cross section. The Vega model, in comparison to other 1 D models, predicts with greater accuracy velocities near the bank. The comparison with result of 2D simulation shown good agreement. The flow simulations of large flood plains used by this model produced results that were unsatisfactory. Flow simulations of severely meandering rivers using the Vega model, therefore, cannot be recommended. In cases where alluvial roughness was present, the Vega simulation were inaccurate and it was necessary to proportionally increase the resistance of bed. This study compares the advantages and disadvantages of specialized simulation tools, Vega and Nimmph, which focus only on specific hydrodynamic phenomena.

Using these specialized tools in cooperation with the more general application such as MIKE 11, is shown to give greater accuracy for the cases we studied. This demonstrates the importance of choices made in the context of model usage and site demands. We hope that this work will help to more clearly stress upon engineers that it is necessary to take into account the specific needs of simulations together with the abilities of certain models.

6. ACKNOWLEDGEMENT

The author would like to thank P. Ingedult, B. Jůza, S. Vaněček and E. Zeman for their help and support.

7. REFERENCES

[1] Dekker J.,Kolař V.,(1989), Mathematical modeling of morph. processes in river channel and flood plain, *Vodohosp. Čas.*,**Vol.37.No.3.**,pp.303-327.

[2] Ingeduld P.,(1990), A study on boundary conditions related to two dimensional mathematical model for unsteady flow in rivers and adjacent flood plains. Diploma thesis, TU Prague, Deptm. of Civil Engineering.

[3] Jůza B.,(1990), The application of the NIMMPH and MIKE 11 simulation packages on the Ohře river; a case study. Diploma thesis, TU Prague, Deptm. of Civil Engineering.

[4] Mike - 11 Manual I and II. Danish Hydraulic Institute

[5] Pasche E.- Rouvé G.,(1985), Overbank flow with vegetatively roughened flood plain. *Journal of Hydraulic Engineering,* **Vol.111,No.9.**,pp.1262-1277.

[6] Zeman E.,(1989), 2D mathematical model for unsteady flow in river channel and flood plains. Ph.D. thesis, TU Prague.

9 A reclamation study through method of characteristics

L. K. Ghosh and C. B. Singh

ABSTRACT

In this paper a 1D network system model based on three point method of characteristics has been developed to study the effect of proposed reclamation of low lying area along Mahim Creek, situated on the west coast of India. The creek is characterised by presence of very shallow channels, tidal flats on both sides and rock outcrop which acts as a sill. These features were schematised in the model. The entire Mahim creek system has been considered to consist of 11 channels, 12 nodes and five tidal flats. Mathematical model results obtained for the existing condition indicate that the model simulates the prototype behaviour satisfactorily. Thereafter the model was effectively used to predict the impact of reclamation and channelisation.

1 Introduction

The Mahim creek which debouches into the Arabian sea acts as one of the principal drainage channel for south western Bombay during monsoon season (Figure 1).

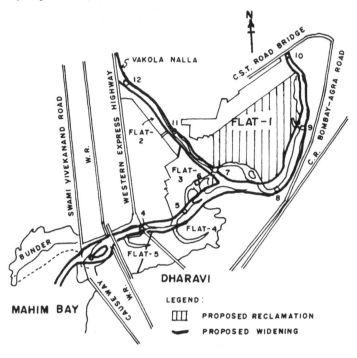

Figure 1 : Plan view of Mahim creek system

It is a 13 km long tidal creek, the upper part of which is known as Mithi river. The channel is very shallow, with average bed elevation lying at about M.S.L. Further, at the entrance to the creek, at Mahim causeway there existis a natural rock sill with its elevation at +2.1 m above chart datum (C.D.) which restricts the flow both during the flood and ebb. A small natural channel known as Vakola Nalla, about 5 km in length is the only stream which joins the Mithi river, bringing in some drainage flow of the order of 185 cumecs during the monsoon period. The Mithi river receives the upland flow of the order of 405 cumecs during monsoon period from its catchment including the over flows from two lakes which are important soucres of water supply to the city of Bombay.

The Mithi river is very shallow with an average bed level at about mean sea level. The maximum and minumum widths of the channel are 300 m and 30 m respectively. The tides in the Mahim creek are of the semi diurnal type and the maximum tidal range is of the order of 5 m. Due to the presence of the rock sill at Mahim causeway, the tidal flow cannot enter the creek until the tide has risen to mean tide level. Similarly during the ebb, the flow from the creek gets cut off as soon as the tidal level reaches the sill elevation.

It has been proposed by the Bombay Metropolitan Region Development Authority to reclaim low lying area (approximately 136 hectares as shown in the Figure 1) along the creek as part of a development plan and provide suitable measures to improve the drainage and flushing capacities of the creek.

2 Mathematical Modelling Technique

The mathematical model of 1D unsteady flow basically requires consideration of continuity equation and equation of motion (Cunge, Holly and Verway 1980).

Continuity Equation :

$$\partial Q/\partial x + B. \ \partial Y/\partial T = 0 \tag{1}$$

Equation of Motion :

$$\partial V/\partial t + V.\partial V/\partial x + g.\partial Y/\partial x + g(Se - So) = 0 \tag{2}$$

where, x is distance along channel axis, t is time, Q is discharge, B is width, V is Velocity, Y is depth of water, So is bed slope, Se is friction slope given by, $Se = n^2 \ V^2/ \ R^{4/3}$, where, n is Manning's friction factor, R is Hydraulic radius.

The usual method of characteristics (Fisher, 1970) transforms the basic set of equations (1) and (2) into the following set of four equations.

$$dx/dt = V \pm \sqrt{gh} \tag{3}$$

105

dv/dt +L . dY/dt = g (So - Se)   (4)

where, h = A/B = mean depth, A = area of cross
section and L = $\pm \sqrt{g/h}$
The equations are solved by explicit finite
difference method satisfying the Courant condition
i.e. (C+V).$\Delta t / \Delta x$ < 1 for stability. Necessary
modifications have been made in the numerical
scheme which enables to arrive at the distance step
as a function of the time step depending upon depth
of the channel.

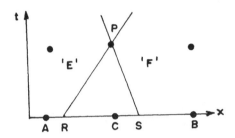

Figure 2   Mesh point notation for method of
             characteristics

Referring to Figure 2, the finite difference
scheme derived by means of the fixed lattice point
method in the x-t plane makes it feasible to
transform equations (3) & (4) as following.

Vp - Vr + Lc(Yp - Yr) = Δt . g(So - Se)   (5)

Which is true along the forward characteristic line
RP

where, Lc = $\sqrt{g/hc}$ = g/Cc with Cc = $\sqrt{g.hc}$

Vp - Vs + Lc (Yp - Ys) = Δt .g(So - Se)   (6)

which is true along the backward characteristic
line SP

where, Lc = - $\sqrt{g/hc}$ = -gCc

Rearranging equation (5) and (6)

$$Vp+g.Yp/Cc = Vr + g.Yr/Cc+ \Delta t.g(So-Se) = E \quad (7)$$

$$Vp-g.Yp/Cc = Vs + g.Ys/Cc+ \Delta t.g(So-Se) = F \quad (8)$$

The values of the unknown quantities at the point P are obtained by solving equations (7) and (8) simultaneously to yield.

$$Vp = (E + F) / 2 \tag{9}$$

$$Yp = (E - F)/2 . Cc/g \tag{10}$$

Equations (9) and (10) are sufficient to solve for velocity and depth at all interior grid points where two characteristic lines are available. Either the velocity or the elevation must be specified at the boundary and the quantity not specified can be computed from one of the characteristic equations.

The equations given above apply only to prismatic channel i.e. one method of representing a non-prismatic natural channel is to divide it into a series of prismatic segments. If one defines a node as function of any two or more segments the equations which must be satisfied at the node are the equations of continuity (flow out = flow in) and that the elevations of the water surface by the same in all segments at the node. These equations along with one characteristic equation from each segment forming the node, provide enough equations to solve for the water surface elevation at the node and the velocities at junction end of all segments.

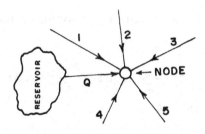

Figure 3. Schematic diagram for junction of channels

Supposing the conditions are imposed as shown in Figure 3. Then channels 1, 2 and 3 combine and flow into channels 4 and 5, the downstream grid points of the first three channels and the upstream grid points of the last two channels are considered to form the node.

Since the grid points are coincident, the unknowns are the five velocities at the end section of each segment and the depth at the node. The five characteristic equations and the continuity equation may be written as follows :

$$E1 \ = V1 + g \ (d1 + Z) \ / \ C1 \qquad\qquad (11)$$

$$E2 \ = V2 + g \ (d2 + Z) \ / \ C2 \qquad\qquad (12)$$

$$E3 \ = V3 + g \ (d3 + Z) \ / \ C3 \qquad\qquad (13)$$

$$F4 \ = V4 - g \ (d4 + Z) \ / \ C4 \qquad\qquad (14)$$

$$F5 \ = V5 - g \ (d5 + Z) \ / \ C5 \qquad\qquad (15)$$

$$\sum_{i=1}^{3} AiVi = \sum_{i=4}^{5} AiVi - Qn \qquad\qquad (16)$$

where, $Ai = Bi \ Yi$ and $Yi = di + Z$. Z is fluctuation due to tide w.r.t. m.s.l. and d is depth of channel below m.s.l. Qn is nodal discharge from tidal flat considered as reservoir.

Replacing velocity terms in equation (16) by equations (11) to (15) and solving for the stage at node yields

$$Z = \{- S/R + \sqrt{(S/R)^2 - 4T/R} \ \} \ / \ 2 \qquad\qquad (17)$$

where R,S and T are given by expressions

$$R = g \sum_{i=1}^{5} Bi/Ci$$

$$S = 2g \sum_{i=1}^{5} Bi \ . \ di/Ci - \sum_{i=1}^{3} Ei.bi + \sum_{i=4}^{5} Fi.Bi$$

$$T = -R \ di^2 + S \ di - Qn$$

The velocity in each segment can be computed by substituting the result of equation (17) into equations (11) to (15). Once the elevations at the nodal points are known the computation can proceed to determine velocity and water levels at all interior sections in all the channels independently.

3   Schematisation of the model region

The entire Mahim creek system has been considered to consist of 11 channels, 12 nodes and 5 tidal flats.
As input data, channel dimensions such as width, depth and length and area of tidal flats were provided to the model. There is a provision in the model to vary the area of tidal flats as a function of the phase of tide. Upland discharges can also be simulated appropriately at nodal points.

4   Numerical Experiments

From the observations in the prototype, it is evident that the flow at Mahim causeway section has three distinct phases during a tidal cycle as described below and shown in Figure 4.

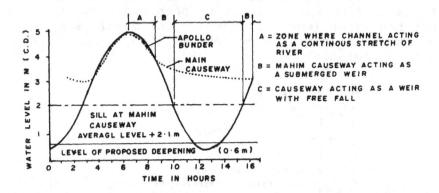

Figure 4 : Sill acting as continuous stretch
            of river, submerged weir and weir
            with free fall

When the water level down stream of the Mahim

causeway is higher than +3.65 m CD the channels upstream and downstream of the sill behave as a continuous stretch of Mithi river. The sill acts as a submerged weir when the water levels down stream of Mahim causeway are between +3.65 m and +2.0 m CD. The sill acts as a weir with a free fall when the water level goes below +2.0 m CD.

The above aspects have been modelled by adding a channel downstream of the causeway and defining the law governing the flow over broad crested weir by

$$Q = CD\ B\ \sqrt{2g}\ H^{3/2}$$

where CD is coefficient discharge, B is width of the sill, H is the effective head. Cd and H both changes depending upon the submergence and emergence of sill.

A spring tide having a range of 4.7 m (Figure 4) was provided as downstream boundary condition. The model was run with dry season conditions for proving purposes i.e. zero discharge at the upstream boundary. The friction coefficient was tuned to obtain the model water surface profile equivalent to that observed in nature. The coefficient of weir was required to be adjusted when the tidal level was lower than the bed level at Mahim causeway. It may be seen from Figure 5 that the water level obtained in the model is close to those observed in prototype with a marginal phase difference.

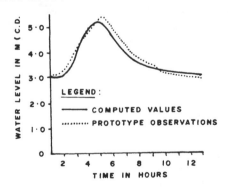

Figure 5 : Observed and computed tidal level at Mahim causeway

The final proposal for reclamation and channelisation (deepening and widening) has been indicated in Figure 1. The improved channel will have bed level of +0.6 m(CD) at Mahim causeway which will gradually rise to +2.0(CD) at CST road bridge thus giving a bed slope of 1:3000 in the improved reach. The velocities at Mahim causeway during dry season without and with the improvement have been compared in Figure 6. It may be seen that the peak velocities during flood and ebb have increased due to combined effect of reclamation and channelisation. The tidal prism at Mahim causeway was computed using the results of mathematical model and it was found that the tidal influx in improved condition would increase by about 2.6 times when compared with the existing condition.

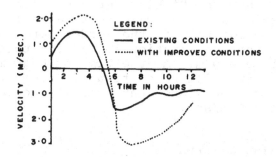

Figure 6 Comparison of velocities in existing and improved condition (dry season) at Mahim causeway.

The impact of improvement on water level has been compared in Figure 7 which shows the water profile along the creek under different experimental conditions. Under the existing condition the tidal amplitude reduces drastically upstream of causeway. The tidal range which is 5 m at the sea reduces to 2.5 m upstream of Mahim causeway , 0.9 m at Dharwar bridge and 0.5 m at CST road bridge. With the proposed improvement of the channel the tidal propagation in the creek would considerably improve all along the creek which is evident from the figure.

111

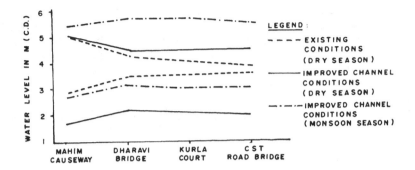

Figure 7 : Water profile along the creek in
existing and improved condition

Water surface profile with monsoon discharges i.e.
405 cumecs through Mithi river and 185 cumecs
through Vakola Nalla indicate that water level in
this area would be of the order of +6.0 m.
Accordingly the reclamation level in this area
should be above+6.0 m. At the time of high monsoon
discharge, there will be continuous flow towards
sea at Mahim causeway and peak velocity would be of
the order of 3.5 m/sec.

5   Concluding Remarks

     The   1D   model   based   on   method   of
characteristics which is long being used for the
study of single nonprismatic channel has been
extended to deal with network of natural channel
system. The model is found to be reasonably
efficient to simulate the highly irregular flows
that exist in the Mahim creek system.   Though
Courant criteria restricts the time step, the total
computer time required for 1D model is usualy very
small and time requirement does not prohibit the
use of explicit characteristic model for larger
system. Studies indicate that with the improvement
of the channel the reclamation level should be kept
above +6.0 m (CD).

ACKNOWLEDGEMENTS

Authors express their deep gratitude to Director, Central Water and Power Research Station, Pune, for his encouragement and support during the studies and for the permission to publish this paper.

REFERENCES

[1]   Cunge, J.A., Holly, F.M., and Verway, A., (1980), Practical Aspects of Computational River Hydraulics, Pitman Advanced Publishing Programme.

[2]   Fisher, H.B., (1970), A method for predicting pollutant transport in tidal waters, water resources centre contribution No.132, College of Engineering, University of California, Berkeley, USA.

10 An investigation of the effect of different discretizations in river models and a comparison of non-conservative and conservative formulation of the de St Venant equations

C. D. Wnitlow and D. W. Knight

Abstract

The purpose of this paper is to discuss new developments in mathematical modelling of hydraulic networks from both a theoretical and a practical perspective. Particularly important are the steady flow problem where the governing hydraulic equations can be reduced to an ordinary differential equation and the hydraulic jump problem where the hyperbolic nature of the St. Venant equations in conservation or integral form can be exploited.

Two methods of discretising the convective acceleration term of the St. Venant Conservation of Momentum Equation using the Preissmann 4 point scheme were examined and the results compared for a subcritical test problem at various spatial and temporal resolutions. These results were further compared against those from Halcrow's Direct ODE Solver which uses a highly accurate formulation of the steady flow equations, does not use the Preissmann Scheme and thus can be considered as a test of the accuracy of the Preissmann Scheme itself.

The applicability of the Conservative or Integral form of the St. Venant Equations is discussed and results presented for two dambreak problems using this formulation. In addition results are presented from both a conservative scheme and the Preissmann Scheme for the non-conservative (differential) form of the equations for two problems involving hydraulic jumps.

1. Introduction

The equations describing one dimensional free surface flow in an open channel were first formulated by de St Venant [14] and first solved by computer in the nineteen fifties. Various numerical schemes have been proposed since then and several are discussed in Cunge et al [3]. The most popular schemes in use in modern river models for discretising functions and their derivatives in the St Venant Equations are the Preissmann 4 point scheme and the Abbott Ionescu Scheme. It must be emphasised that there are many ways of formulating the equations prior to the discretization process even if the same scheme is chosen. This and the choice of weighting factors or simplifications of the equations can greatly affect the accuracy and robustness of the corresponding mathematical model. The linear stability of the numerical methods has been reviewed recently by Skeels and Samuels [13] among others and it is not proposed to discuss them further here. However it is worth emphasising that linear stability analysis by the Fourier expansion associated with von Neumann can at best yield criteria for permissible ratios for space step and timestep. Experience at Sir William Halcrow and Partners Ltd suggests that the stability problem has its roots in the nonlinear differential equations themselves and the inherent scale of the physical processes being modelled. A program has been written at Halcrow to determine appropriate space and time scales for a given problem which employs the nonlinear relationships in the St. Venant equations. The results agree well with the scalings obtained by the Direct Solver discussed in the following section.

The notion of what constitutes hydraulic research has been discussed recently by Abbott et al [1] who incorporate software engineering into this overall umbrella. It is the view of the authors that the development of proprietary software packages in terms of menu systems, database manipulation and a graphical options does not constitute hydraulic research at all since it can often be undertaken by software engineers who know nothing at all about hydraulics or mathematical modelling. The purpose of this paper is to discuss new developments in mathematical modelling of hydraulic networks from both a theoretical and a practical perspective. Considered here are the steady flow problem where the governing hydraulic equations can be reduced to an ordinary differential equation and the hydraulic jump problem where the hyperbolic nature of the St Venant equations in conservation or integral form is important.

2. The Steady Flow Problem

Recent research in the UK (Evans and Whitlow [4]), and subsequently Chawdhary [2], has addressed this issue by considering the ordinary differential equation (ODE) formulation of the equation:

$$\frac{dH}{dx} - \frac{Q^2}{gA^3}\frac{\partial A}{\partial x} - \frac{dz}{dx} + \frac{Q|Q|}{K^2} = 0 \qquad (1)$$

(See Section 4 for nomenclature)

This equation can either be solved implicitly or explicitly according to how the $(\partial A/\partial x)$ term is used to rearrange the equation. The Halcrow River model, ONDA has since 1988 incorporated a fourth order Runge Kutta ODE Direct solver which is used in conjunction with appropriate boundary conditions to solve steady state problems.

Steady state solutions are very important. After survey data has become available, the modeller assembles the information into a data file to represent an often looped or branched network including appropriate boundary conditions. At this point, the modeller would have no idea, except for a rough guess, what would be the appropriate initial conditions for a proposed unsteady run with an input flow hydrograph. In our experience this problem of obtaining a steady state initial condition compatible with the raw input survey data is one of the most difficult to overcome. The traditional backwater method does not work in looped systems, where the split of flows is not known.

Conventionally [3] it has been done via a pseudo time stepping strategy. The computation is begun at a relatively low timestep and run for a long time until all the waves generated by the inaccuracies in the initial conditions have been propagated or dissipated out of the system. The length of the pseudo timestep is increased gradually until the computation can proceed at large timesteps until the change in the solution is negligible as time increases and the equation residuals have dropped below a threshold value. Experience indicates that this process is something of a black art since great care is required to avoid computational breakdown. It can also be extremely time consuming and tedious for the modeller. In addition an ODE solver does not rely on the Preissmann scheme but still solves the full steady St Venant equations thus providing an independent check on the solution which is extremely important to any Quality Assurance requirement.

It also pinpoints data problems much quicker than the implicit matrix method and informs the user where supercritical flow has been encountered. It is perceived that the identification of local problem areas automatically is a great advantage over global rules of thumb.

The method deals accurately and consistently with the problem of cross section spacing. During the computation, the method checks whether the solution is grid dependent, and if necessary will add extra interpolated nodes implicitly. The user is informed where this has been done so that extra surveyed sections can be added to the model if

117

available, or extra nodes interpolated between the existing sections. If a large number of interpolated sections have been added, this indicates either that the channel properties of the two sections are significantly different (alerting the modeller to potential data discrepancies) or there is large surface curvature in the channel eg as the Froude Number approaches unity. The steady solver has performed well with Froude Numbers as close to unity as 0.98. An example of a $20m^3/s$ flow in a triangular channel with a Mannings' n of 0.035 and a slope of 0.0133 is shown in Figure 1. This corresponds to a Froude Number in excess of 0.95 and many interpolated sections were added in the region of high surface slope curvatures.

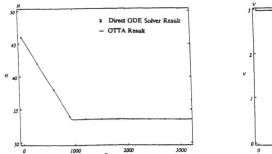 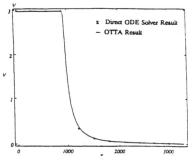

FIGURE 1. Water surface elevation and velocity for a triangular channel
-Subcritical flow at upstream (Froude Number = 0.98 at transition)

Several existing models which appeared unstable using the 4-point implicit scheme have been run steady using the new method which suggested areas where additional spatial resolution was required. When this was added, the models became stable and solutions agreed with the steady solver whereas previously they had not, sometimes differing by as much as hundreds of millimetres.

Grid dependent solutions are meaningless where local model refinements are tested, eg to investigate the effect of new river engineering works. In Computational Fluid Dynamics in general, a grid independent situation is always required.

The new method is much faster, and more robust than pseudo - timestepping sometimes by orders of magnitude. It is also more accurate. It has been used on many projects now, both in the UK and overseas.

Apart from initial flow estimates at confluences, it requires <u>no initial conditions</u>.

The method is capable of accurately modelling mixed super and sub-critical flows.

118

3. A comparison between the Direct Method and Two Discretisations using the Preissmann Scheme

This section is intended to illustrate the practical difference in model results to be expected in steady and unsteady state computations between the differential form of the convective acceleration term

$$\frac{\partial}{\partial x}\left(\frac{Q^2}{A}\right) = \frac{Q}{A}\frac{\partial Q}{\partial x} - \frac{Q^2}{A^2}\frac{\partial A}{\partial x} \qquad (2)$$

discretised as in Cunge et al [3] and a directly discretised form

$$\frac{\partial}{\partial x}\left(\frac{Q^2}{A}\right) = \frac{1}{\Delta x}\left(\theta\left(\left(\frac{Q^2}{A}\right)_{j+1}^{n+1} - \left(\frac{Q^2}{A}\right)_{j}^{n+1}\right) + (1-\theta)\left(\left(\frac{Q^2}{A}\right)_{j+1}^{n} - \left(\frac{Q^2}{A}\right)_{j}^{n}\right)\right) \qquad (3)$$

The two options for discretising the convective acceleration terms were written into the source code of Halcrow's ONDA Program [8] and their effect on model results investigated by comparison with results from Halcrow's Direct Solver. This not only provided a baseline for comparison of the two discretisations but also of the Preissmann Scheme itself.

An example of the models tested is shown in Figure 2. The channel has a bed slope of 1 in 500 and a discharge of $10m^3/s$. A summary of the stages produced by the different model runs, using models of varying spatial resolution is shown in Table 1.

By comparing the two Preissmann model runs it can be seen that the directly discretised version produces stages closer to the Direct Solver stages than those produced by the differentiated version. As the spatial resolution is progressively decreased, the rate of increase in the error with respect to the Direct Solver stages is smaller for the directly discretised version then for the differential version. When the severity of the channel constriction is increased, the errors produced by the differentiated discretisation model runs increase while they remain virtually unchanged for the directly discretised model runs.

The results show that the differences in stage are remarkably small even for a relatively severe channel constriction at a lower resolution of 80m and that the errors diminish as expected as the spatial resolution increases. What can be clearly seen is that the Direct ODE solver is insensitive to changes in spatial resolution, as discussed in the

119

Spatial Resolution (m)	Model Version	Stage at Sections						
		1	5	9	13	17	21	25
10	Direct Method	2.408	2.399	2.391	2.388	2.272	2.193	2.113
10	Cunge Version	2.410 (1)	2.400 (1)	2.392 (1)	2.387 (1)	2.272	2.193	2.113
10	Direct Discretization	2.408	2.399	2.391	2.388	2.272	2.193	2.113
20	Direct Method	2.408	2.399	2.391	2.388	2.272	2.193	2.113
20	Cunge Version	2.412 (3) (2)	2.403 (4) (3)	2.394 (3) (2)	2.370 (4) (3)	2.272	2.193	2.113
20	Direct Discretization	2.411 (2) (2)	2.401 (2) (2)	2.393 (2) (2)	2.388 (2) (2)	2.272	2.193	2.113
40	Direct Method	2.408	2.399	2.391	2.388	2.272	2.193	2.113
40	Cunge Version	2.420 (11) (8)	2.411 (12) (8)	2.403 (12) (9)	2.378 (12) (8)	2.272	2.193	2.113
40	Direct Discretization	2.416 (7) (5)	2.405 (7) (5)	2.398 (7) (5)	2.373 (7) (5)	2.272	2.193	2.113
80	Direct Method	2.408		2.390		2.272		2.113
80	Cunge Version	2.446 (37) (25)		2.429 (39) (28)		2.272		2.113
80	Direct Discretization	2.433 (25) (17)		2.415 (25) (17)		2.272		2.113

where 2.433

(25) is the difference in stage in mm from the Direct Method result

(17) is the difference in stage in mm from the stage at the next higher spatial resolution

TABLE 1. Comparison of results for different discretisations and spatial resolutions

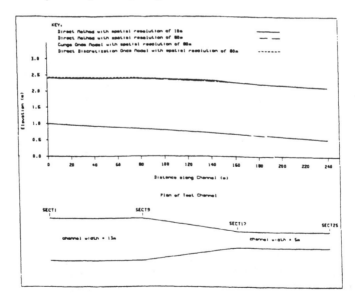

FIGURE 2. Water surface elevations for different discretisations and a plan of the test channel

120

previous section. However as the spatial resolution decreases, the two Preissmann model runs produce slightly higher water levels than the ODE solver.

The effect of timestep variation on unsteady state model runs was also tested for the simple test channel. The differences in the results produced by the two Preissmann models was negligible so long as the timestep was not so large as to lose definition from the input hydrograph.

A curious feature of the results was that the individually computed components of the momentum equations do not always appear to sum to zero in the Cunge discretisation. Two possible reasons for this are that the iterative scheme is not converging properly or that the difference equations themselves are not a consistent representation of the differential equations.

4. Conservative Schemes

In order to model accurately problems of hydraulic jumps, it is necessary to use the conservation or integral form of the St Venant Equations.

$$\frac{\partial A}{\partial t} + \frac{\partial Q}{\partial x} = q \tag{4}$$

$$\frac{\partial Q}{\partial t} + \frac{\partial}{\partial x}(p + Av^2) = p_w + gA(i - \frac{Q|Q|}{K^2}) - qv\cos\alpha \tag{5}$$

where

t	is time,
x	is the coordinate along the channel
q	is the rate of inflow
$\propto$	is the angle of the inflow
g	is the acceleration due to gravity
i	is the bottom slope
y	is the depth
w(x,y)	is the width at depth y
v	is the mean streamwise velocity
Q	is the flow
A	is the cross sectional area

121

$k^2 = AR^{4/3}/n^2$ is the conveyance

n   is Manning's n

$R = A/P$ is the Hydraulic Radius

P   is the length of the wetted perimeter

$$p = g\int_0^y (y - y')\, w\,(x,y')\, dy' \qquad \textit{is the effective presssure}$$

$$p_w = g\int_0^y (y - y')\, \frac{\partial w}{\partial x} dy' \qquad \textit{is the wall presssure}$$

In order to derive the usual differential form of the St. Venant Equations the flow variables such as stage, area and flow must be continuous differentiable functions. Clearly this is not the case across a hydraulic jump.

These conservation equations resemble closely the Euler equations from compressible gas dynamics and it is expected that schemes which perform well for those problems will perform similarly for the St. Venant equations where strong shocks and contact discontinuities are not usually encountered. The pseudo viscosity method whereby a hyperbolic problem is simplified locally to a parabolic one [1] is rarely used in modern research in gas dynamics and for the St. Venant equations has been criticised by Vasiliev et al [15] and Samuels [12] for distorting the physics for the phenomena. The DAMBRK UK program [5] for modelling of dam breaks does not use the correct integral form of the St. Venant equations, requires the flow type in river reaches (subcritical, supercritical or mixed) to be known 'a priori' and makes no serious attempt to mathematically model the correct jump conditions for the propagation of a shock. These limitations may not be a problem for the steady collapse of a typical UK earthen dam, where a shock wave would not form but would be important for the modelling of the catastrophic failure of a concrete dam for example.

Halcrow's OTTA program developed in 1990 uses a second order Riemann problem based shock capturing scheme which is conservative, accurate, robust and provides high resolution of sharp flow features such as hydraulic jumps. These methods constitute a new generation of numerical techniques in Computational Fluid Dynamics and recent developments have simplified the algorithms and increased the efficiency such that they are now comparable in speed with traditional techniques such as artificial viscosity methods. Essentially the discontinuities are ignored in OTTA by applying the same numerical scheme at all points in the flow so no 'a priori' knowledge is required by the user. This approach was pioneered by Roe [10, 11] for the Euler Equations and used by Glaister [6] to model a dam-break in an infinitely wide frictionless rectangular channel.

Three test problems were solved using OTTA. Figure 3 shows the evolution of a classical dambreak problem in a flat 20m wide channel. Figure 4 shows the solution of a dambreak problem with real cross section and bridge data. The problem considered was the catastrophic failure of the dam at Tubb's Bottom Reservoir near Bristol which forms part of the flood alleviation works on the River Frome.

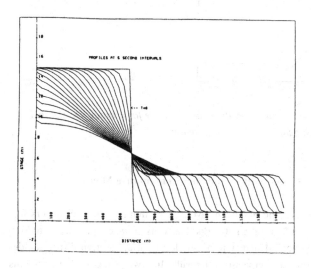

FIGURE 3. OTTA results for the evolution of a classical dambreak

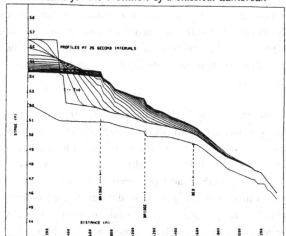

FIGURE 4. OTTA results for the evolution of a simulated dam break at Tubb's
Bottom, Bristol

Figure 5 shows results from the same channel as Figure 1 but for supercritical flow of 1083m^3/sec at the upstream end.

123

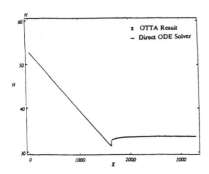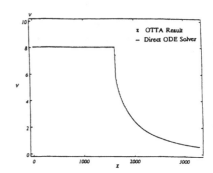

*FIGURE 5. Water surface elevations and velocity for a triangular channel
- Supercritical flow at upstream*

5. Comparison of Conservative and Non Conservative Models

The Preissmann scheme has been favoured by mathematical modellers for generally subcritical flows in hydraulic networks due to its simplicity, accuracy and implicitness. It is known however [9] that it performs less well at solution discontinuities. In this section we present a comparison of results for two stringent test problems between OTTA and the Preissmann Scheme written into ONDA by Merrick [8]. No additional effort was made to optimise the performance of the Preissmann scheme either for steep fronts in general or specifically for the two problems considered. The approach of Havno et al [7] to phase out the convective acceleration in the neighbourhood of a discontinuity was also used in conjunction with the Preissmann Scheme and the Direct Solver for comparative purposes.

The programs were run on two models the results of which are shown if Figure 6 and 7. What is immediately apparent is that OTTA produces hydraulic jumps in the water profiles. The non-conservative models produce a smoothed out form of the conservative model profile with lower peaks and higher troughs. It is observed that the conservative model can predict higher depths than those produced by the non-conservative models. Therefore, although as can be seen in Figure 6, the different programs converge on the normal depth upstream of the hydraulic jump, for localised investigations a conservative formulation may be necessary to accurately determine water profiles.

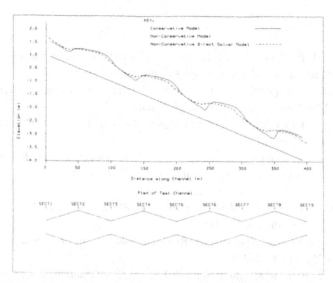

FIGURE 6. *A comparison of results using the conservative (OTTA) and non-conservative (ONDA) formulations for a channel of variable width*

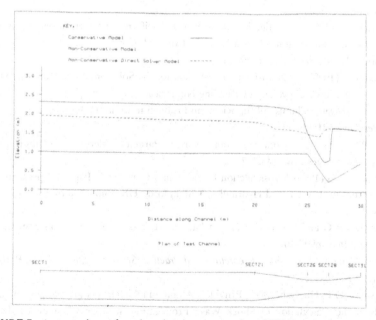

FIGURE 7. *A comparison of results using the conservative (OTTA) and non-conservative (ONDA) formulations for the high resolution modelling of a flume*

References

[1] Abbott, M B, Havno K and Lindberg S (1991) "The Fourth Generation of Numerical Modelling in Hydraulics" *J Hyd Res* 29,5,581.

[2] Chawdhary K S (1991) "On the solution of Implicit First Order Differential Equations". MSc Dissertation, University of Reading.

[3] Cunge J A, Holly F M and Verwey A (1980) *"Practical Aspects of Computational River Hydraulics,"* Pitman, London.

[4] Evans E P and Whitlow C D (1991) "Recent Developments in 1-D Modelling of Open Channel Networks", Paper presented at Scottish Hydraulic Study Group Seminar on River and Flood Plain Management, 22nd March.

[5] Fread D L (1991) "DAMBRK UK *Dam Break Flood Simulation Program, User Documentation*". Published by UK DoE.

[6] Glaister, P (1988) "Approximate Riemann Solutions of the Shallow Water Equations", *J Hyd Res* **26**, 3, 293.

[7] Havno K, Brorsen M and Refsgaard J C (1985), "Generalised mathematical modelling system for flood analysis and flood control design". Paper F3 presented at 2nd Int Conf on Hydraulics of Floods and Flood Control Cambridge UK, published by BHRA Cranfield Beds UK.

[8] Merrick A J (1991) "Investigation of the Effect of Different Discretisations in River Models and a Comparison of non-Conservative and Conservative Models". MSc Dissertation, University of Birmingham.

[9] Priestley A (1990) "A Quasi-Riemann Method for the Solution of one Dimensional Shallow Water Flow". University of Reading Numerical Analysis Report.

[10] Roe P L (1980) "The use of the Riemann Problem in Finite Difference Schemes". *Lect Notes Phys* **141**:354-59

[11] Roe P L (1981) "Approximate Riemann Solvers, Parameter Vectors and Difference Schemes". *Lect Notes Phys* **143**:357-72.

[12] Samuels P G (1990) "Cross Section Location in 1-D models". Paper K1 presented at the Int Conf on River Flood Hydraulics, Wallingford, UK, Published by John Wiley and Sons Ltd.

[13] Samuels P G and Skeels C P (1990) "Stability Limits for Preissmann's Scheme". *J Hyd. Eng.* **116**, 8 p997-1012

[14] de St. Venant A J C (1848) *Theoretical and Practical Studies of Stream Flow*. Paris, France (in French).

[15] Vasiliev O F, Gladyshev MT, Pritvits N A and Sudobicher VG (1989) "Numerical Methods of the Calculation of Shock Wave Propagation in Open Channels". Paper presented at XI IAHR Congress.

126

11 Implicit and explicit TVD methods for discontinuous open channel flows

P. Garcia-Navarro and F. Alcrudo

Abstract

Two high resolution methods for the simulation of open channel or river flow are presented. They are specially suited for flows including phenomena such as hydraulic jumps and bores. The methods are based on the theory of Total Variation Diminishing (TVD) schemes applied to the shallow water equations via flux difference splitting. The addition of a TVD conservative dissipation step to the widely used McCormack numerical scheme provides an interesting time stepping procedure for unsteady simulations, while an efficient implicit time integration method is used in cases where only the steady state solution is sought. Several computations are shown and comparison with the analytical solution, when available, asseses the validity of the numerical treatment.

1 Introduction

The use of numerical methods to predict the water profile and discharge for unsteady as well as for stationary situations of hydraulic systems is now of common use in engineering work. Most difficult situations occur when mixed sub-supercritical regimes with hydraulic jumps are present in the system, that invalidate some of the numerical methods available for calculation. Solutions of that kind usually appear when modelling steady flows in steeply sloping channels, rapidly varied steady and unsteady flows, or in dam collapse simulations. Among the techniques that succeed in those difficult cases are the *through* or *shock capturing* methods in which the equations governing the model are solved in conservation form by means of a suitable numerical scheme. This approach can locate and propagate the discontinuities present in the solution with the physically correct speed and strength without the need for any a priori information or fitting procedure.

During the last decade, much effort has been paid to the numerical solution of systems of conservation laws, mainly driven by the need for efficient Euler solvers in aerodynamics. This has led to a new class of methods, based on the Total Variation Diminishing (TVD) concept, known as *high resolution methods* which do not suffer from the known penalties of classical ones [6], [8] and to a generation of techniques able to improve the performance of conservative second order classical schemes. In this paper, unsteady as well as stationary computations are performed with two high resolution TVD methods: A modified version of McCormack scheme and an implicit upwind scheme.

2 Governing equations

One dimensional open channel flow is usually described in terms of water depth and discharge, and the evolution of these quantities taken to be governed by the de Saint Venant equations which simply express the conservation of mass and momentum along the flow direction [2] and can be written in conservative or divergent form as:

$$\frac{\partial \underline{U}}{\partial t} + \frac{\partial \underline{F}}{\partial x} = \underline{G} \tag{1}$$

Where:

$$\underline{U} = \begin{pmatrix} A \\ Q \end{pmatrix} \qquad \underline{F} = \begin{pmatrix} Q \\ \dfrac{Q^2}{A} + gI_1 \end{pmatrix} \qquad \underline{G} = \begin{pmatrix} 0 \\ gI_2 + gA(S_o\text{-}S_f) \end{pmatrix} \tag{2}$$

$A(x,h(x,t))$ is the wetted cross section area and h the water depth. $Q(x,t)$ is the discharge, g is the acceleration of gravity and I_1 stands for the hydrostatic pressure force term that can be written:

$$I_1 = \int_0^{h\,(x,t)} [h\text{-}\eta]\,\sigma(x,\eta)\,d\eta \qquad \text{with} \qquad \sigma(x,\eta) = \frac{\partial A(x,\eta)}{\partial \eta} \tag{3}$$

The integral I_2 appearing in the source term is defined as follows:

$$I_2 = \int_0^{h\,(x,t)} [h\text{-}\eta]\,\frac{\partial \sigma(x,\eta)}{\partial x}\,d\eta \tag{4}$$

It accounts for the forces exerted by the channel walls at contractions and expansions. S_o and S_f are the bottom and friction slopes. In this work Manning's formula has been used for the latter.
It is interesting to write the jacobian matrix of the flux:

128

$$J = \frac{\partial E}{\partial U} = \begin{pmatrix} 0 & 1 \\ g\frac{A}{\sigma} - \frac{Q^2}{A} & \frac{2Q}{A} \end{pmatrix} \tag{5}$$

Which has as eigenvalues and eigenvectors:

$$a^{1,2} = u \pm c \qquad \underline{e}^{1,2} = \begin{pmatrix} 1 \\ a^{1,2} \end{pmatrix} \tag{6}$$

$$u = Q/A \qquad c = \sqrt{gA/\sigma} \tag{7}$$

The matrix P which has as columnns the eigenvectors of J, and its inverse P^{-1} are such that $P^{-1} J P = \text{diagonal}(a^{1,2})$. The eigenvalues of J are the characteristic speeds and their signs provide information about the directions of propagation of perturbations in the flow.

3 TVD McCormack scheme

In order to solve Eq.1 by means of a numerical time stepping procedure, the domain of integration is discretized as $x_j = j\Delta x$, $t^n = n\Delta t$ where Δt and Δx are the mesh spacings. The presence of a source term together with the requirement of second order accuracy in both space and time makes impossible the use of a one step algorithm without a cumbersome treatment. Hence, the presented high resolution scheme is based on McCormack's two step procedure [7] which reads:

$$\underline{U}_j^p = \underline{U}_j^n - \lambda \cdot (\underline{E}_{j+1}^n - \underline{E}_j^n) + \Delta t \cdot \underline{G}_j^n \tag{8}$$

$$\underline{U}_j^c = \underline{U}_j^n - \lambda \cdot (\underline{E}_j^p - \underline{E}_{j-1}^p) + \Delta t \cdot \underline{G}_j^p \tag{9}$$

Where superindexes p and c stand for predictor and corrector steps, n for the time level, $\lambda = \Delta t / \Delta x$, and $F_j = F(U_j)$. The solution at the next time level becomes:

$$\underline{U}_j^{n+1} = \frac{1}{2} (\underline{U}_j^p + \underline{U}_j^c) \tag{10}$$

A third step is proposed in this paper. It furnishes the scheme with total variation diminishing (TVD) dissipation capable of rendering the solution oscillation free while retaining second order accuracy in space and time almost everywhere (except at extrema points) [5], [6], [11]. This is a very remarkable property indeed when dealing with supercritical and rapidly varied flows with hydraulic jumps and bores. To achieve this, Eq.10 is replaced by:

$$\underline{U}_j^{n+1} = \frac{1}{2} (\underline{U}_j^p + \underline{U}_j^c) + \frac{\lambda}{2} \left(\overline{B}_{j+1/2}^n (\underline{U}_{j+1}^n - \underline{U}_j^n) - \overline{B}_{j-1/2}^n (\underline{U}_j^n - \underline{U}_{j-1}^n) \right) \tag{11}$$

The form of the $\overline{B}$ matrix is:

$$\overline{B}_{j+1/2} = \overline{P} \cdot \text{diag} \left[\left| \overline{a}_{j+1/2}^k \right| (1 - \lambda \left| \overline{a}_{j+1/2}^k \right|) (1 - \varphi) \right] \cdot \overline{P}^{-1} \tag{12}$$

Where $\bar{a}^k_{j+1/2}$ is a discrete average characteristic speed of the states at j and j+1 expressed as:

$$\bar{a}^{1,2}_{j+1/2} = \bar{u}_{j+1/2} \pm \bar{c}_{j+1/2} \tag{13}$$

The quantities $\bar{u}_{j+1/2}$ and $\bar{c}_{j+1/2}$ are discrete approximations to the water velocity and speed of sound between mesh points j and j+1. Formulae for these quantities together with a detailed derivation can be found in [1]. The matrices $\bar{P}$ and $\bar{P}^{-1}$ are those that diagonalise the aproximate jacobian $\bar{J}$ whose eigenvalues are the $\bar{a}^k_{j+1/2}$, in complete analogy with the P and P^{-1} matrices with respect to J of previous section.

Factor φ in Eq.12 is called a flux limiter and is a nonlinear function of the gradient of the flow variables in the neighbourhood of points j and j+1 [6], [9]. Its purpose consists of adding sufficient artificial dissipation to the scheme when there is a discontinuity or a steep gradient in the solution, while adding very little or no dissipation in regions of smooth variation. In the examples shown in this work the limiting function φ used is that proposed by Van Leer whose form can be found in the above mentioned references.

It should be stressed that, strictly speaking, the theory of high resolution TVD schemes has been rigorously developed only for homogeneous nonlinear scalar equations or linear systems. However present work evidences a very good performance of the method for the non linear open channel flow equations with inclusion of source terms.

The TVD property guarantees stability since it is a stronger requirement. A sufficient condition for the method to be TVD imposes a limit on the CFL number given by [9]:

$$CFL_{max} = \frac{2}{2+\varphi_{max}} \tag{14}$$

Where φ_{max} is the maximum value attainable by the limiting function which varies between 1 and 2. Nevertheless in this paper monotone calculations have been performed up to CFL numbers close to 1 which is the stability limit for standard McCormack scheme. The treatment of the external as well as internal boundary conditions is based on the theory of characteristics [4].

4 Implicit TVD scheme

If one defines the upwind TVD numerical flux:

$$E^{*n}_{j+1/2} = \frac{1}{2}\left[E^n_{j+1} + E^n_j - \bar{D}^n_{j+1/2}\cdot\left(U^n_{j+1} - U^n_j\right)\right] \tag{15}$$

Where matrix $\bar{D}$ can be written in a similar way as matrix $\bar{B}$:

$$\bar{D}_{j+1/2} = \bar{P}\cdot diag\left[\left|\bar{a}^k_{j+1/2}\right| (1-\varphi)\right]\cdot\bar{P}^{-1} \tag{16}$$

And all symbols as defined in previous section, using Euler implicit time integration the following implicit scheme for the solution of Eq.1 can be obtained:

$$U^{n+1}_j - U^n_j + \lambda\left(E^{*n+1}_{j+1/2} - E^{*n+1}_{j-1/2}\right) = \Delta t\, G^{n+1}_j \tag{17}$$

This is a system of nonlinear equations where j ranges from 1 to the number of mesh points, and its solution is impractical. A better approach is to seek first a linearised form by approximating:

$$\underline{E}_j^{n+1} \approx \underline{E}_j^n + J_j^n \cdot \Delta \underline{U}_j^n \qquad \Delta \underline{U}_j^n = \underline{U}_j^{n+1} - \underline{U}_j^n \tag{18}$$

$$\underline{G}_j^{n+1} \approx \underline{G}_j^n + G_U{}_j^n \cdot \Delta \underline{U}_j^n \qquad \overline{D}_{j+1/2}^{n+1} \approx \overline{D}_{j+1/2}^n \tag{19}$$

Where G_U is the jacobian matrix of the source term $\underline{G}$. After rearrangement, the linearised form is found to be a block tridiagonal system of equations:

$$AA_j \cdot \Delta \underline{U}_{j-1} + BB_j \cdot \Delta \underline{U}_j + CC_j \cdot \Delta \underline{U}_{j+1} = -\lambda \left(\underline{E}_{j+1/2}^{*n} - \underline{E}_{j-1/2}^{*n} \right) + \Delta t \, \underline{G}_j^n \tag{20}$$

The right hand side of Eq. 20 is often called the residual. The block matrices AA, BB, an CC read:

$$AA_j = -\frac{\lambda}{2} \left(J_{j-1}^n + \overline{D}_{j-1/2}^n \right) \tag{21}$$

$$BB_j = I + \frac{\lambda}{2} \left(\overline{D}_{j+1/2}^n + \overline{D}_{j-1/2}^n \right) + \Delta t \, G_U{}_j^n \tag{22}$$

$$CC_j = \frac{\lambda}{2} \left(J_{j+1}^n - \overline{D}_{j+1/2}^n \right) \tag{23}$$

From Eqs. 21, 22 and 23 one sees that the implicit operator is strictly diagonally dominant. The authors have used either the efficient Thomas procedure or relaxation methods for its solution. Better convergence can be found by setting the factor φ in the left hand side of Eq. 20 to zero.

The above scheme is only first order accurate in time, but second order in space and is linearly unconditionally stable. Moreover the stationary solution is independent of the time step used. It is thus well suited for the calculation of steady flows by solving the unsteady equations with fixed boundary conditions until the residual (right hand side of Eq. 20) drops to zero. Boundary conditions can be treated either explicitly or implicitly and are easily incorporated into the tridiagonal structure but a detailed description of the technique used falls out of the scope of this paper. The reader can find very interesting material in [10].

5 Numerical results

In this section the numerical results obtained for several test problems used to validate the two previously described algorithms are shown and compared with the exact solution, when available. All the examples consider either rectangular or trapezoidal channels with friction, bottom slope or cross section variation, so that source terms are always relevant.

5.1 Flood wave propagation on a sloping trapezoidal channel

The first example is concerned with the propagation of a surge wave over steady flow on a variable bottom slope channel 1000m long. The trapezoidal channel is 6m wide at the bottom, with lateral wall slope of 0.25. Three regions can be labelled by their different bed slopes ($S_{01}=0.001$, $S_{02}=0.009$, $S_{03}=0.001$). Manning's friction coefficient (n) is n=0.009 for all three regions.

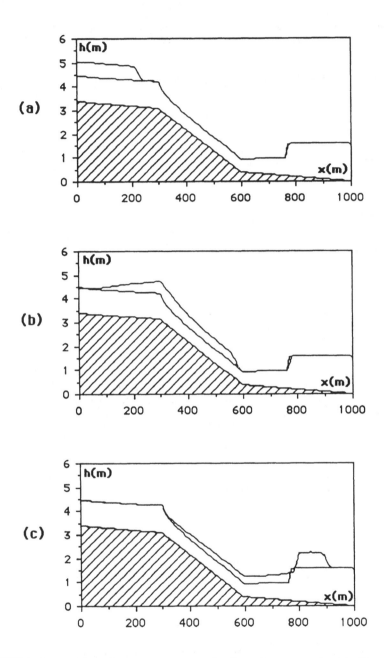

FIGURE 1 *Propagation of a surge wave over the discontinuous steady state flow on a variable bottom slope channel as calculated with McCormack TVD scheme.*

It was run at a CFL number of 0.9 by means of the TVD McCormack scheme of section 3 on a grid of 100 points and the results can be seen on Figure 1. The initial conditions are given by a discontinuous steady profile corresponding to a discharge of $Q(x,0)=20$cum/s previously calculated. A flood hydrogram ($Q=Q(t)$) is introduced at the upstream boundary by a sudden increase of the discharge to $Q=50$cum/s which persists for a time interval of 30s. This means a supercritical advancing surge and therefore requires two upstream boundary conditions.

The second condition stems from the value of the water level that satisfies the mass and momentum conservation relationships across the shock [3]. Then the discharge is linearly reduced to its initial value during another 30s. The downstream boundary is determined by a stage-discharge rating curve of the weir type with a weir height $H_w=0.25$m. The unsteady situation at three different times ($t_a=29$s, $t_b=70$s and $t_c=109.5$s) can be seen on Figure 1. Despite the abrupt changes in the bottom slope and the supercritical flow conditions, the numerical solution shows up stable and well behaved everywhere and the surge neatly propagated along the different parts of the channel.

5.2 Surge propagation through converging-diverging channel

In a 500m long rectangular channel a sinusoidal width variation is supposed to exist between x=100m and x=400m from a maximum value of 5m to a minimum of 3.6m. The plan view of the channel shape can be seen at the top right corner of Figure 2a.

A bore 9.79m deep of 1000cum/s (Froude number of 2.09) is introduced at the left hand side of the channel and propagates downstream over still water 1m deep. The situation at a time t=5s after the bore entered the channel can be seen on Figure 2a. A 2m high weir is supposed to be placed downstream hence initially closing the right hand channel end.

As the supercritical front advances through the contracting channel it increases its height and decelerates. Once surpassed the point of maximum contraction it starts lowering and accelerating again. Figure 2b displays the profile at t=15s.

The downstream end is then reached by a front similar to the initial one. It is partially reflected and partially transmitted over the weir. The reflected surge, 25.25m deep starts travelling upstream very slowly and is shown on Figure 2c (t=150s). It is worth noting here that the height of the reflected front is the one simultaneously satisfying the jump equations together with the weir condition. It propagates upstream until it becomes a stationary hydraulic jump at a certain location in the contracting region. The final steady state is shown on Figure 2d (t=600s). The plotted solutions were calculated with TVD McCormack scheme.

Although the results are not shown here for space reasons, the same problem was also computed with the original McCormack scheme (Eq.10). Due to the generation of oscillations around discontinuities associated with classical schemes, the reflected front was not accurately represented. McCormack scheme kept it standing at the weir location instead of propagating it upstream as expected, spoiling so the numerical solution. From a technical point of view, it is important to remark the serious mistakes that can be introduced by the limitations of a given numerical scheme as this example makes plain.

5.3 Flow over a bump

Finally the stationary discontinuous flow in a rectangular channel 1m wide of variable bottom slope in the form of a triangular shaped bump was computed and compared to the analytical solution. At the upstream end a constant head of 10m is imposed, while downstream water depth is held fixed at a value of 8.5m. These conditions lead to subcritical accelerating flow before the bump that reaches critical conditions at its top and then becomes supercritical downhill. A hydraulic jump must develop to connect the supercritical profile with the subcritical one imposed by the downstream boundary condition.

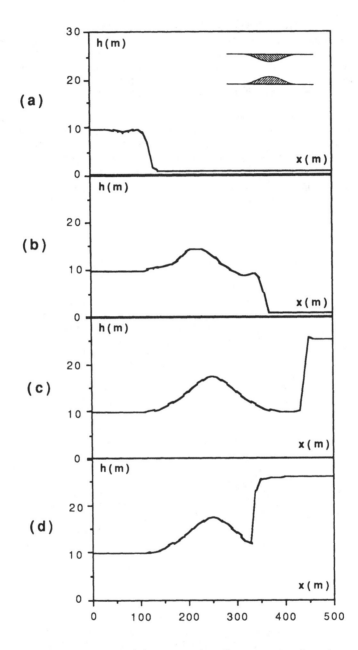

FIGURE 2 *Four sequences of the propagation of a surge wave through a contraction as calculated with McCormack TVD scheme.*

An exact solution can be calculated by solving for each point a cubic equation that derives from imposing water head conservation. The two profiles on each side of the shock are connected by the jump relations.

An initial situation of constant discharge and water stage was marched in time towards the steady state with the two numerical schemes proposed. 81 points were used in the calculation.

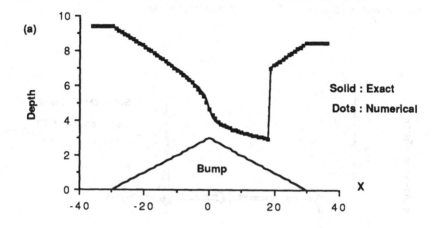

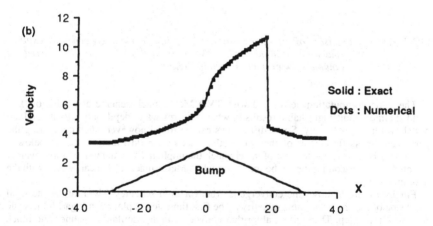

FIGURE 3 *Computation of stationary transcritical discontinuous flow in a channel with a bump by means of the implicit TVD scheme compared to the analytical solution. a) Water depth. b) Water velocity.*

Figure 3 corresponds to the results obtained with the implicit TVD method at a CFL number of 50. The linear system obtained was solved by Thomas algorithm every time step. Figure 3a represents the water depth and Figure 3b the water velocity. The numerical results (square dots) follow closely the exact solution (solid line) representing accurately all the flow regimes, including a smooth transition from subcritical to supercritical flow at the top of the bump. The shock is sharply captured at the correct position between two mesh points with no over or undershoot.

It must be pointed out here that the triangular shape was chosen for this example because it imposes very exacting conditions on the numerical scheme despite its apparent simplicity. In particular the discontinuous changes in the bottom slope at the three triangle vertices are difficult to deal with for any high order classical scheme.

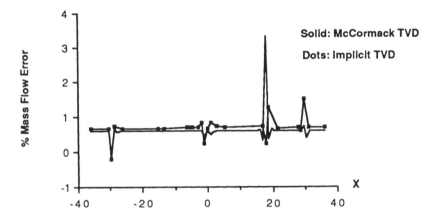

FIGURE 4 *Comparison of the percentage mass flow error between the explicit McCormack TVD and the Implicit TVD numerical schemes for the transcritical discontinuous flow example shown in Figure 3.*

The same computation performed with TVD McCormack scheme at a CFL of 0.95 yielded almost indistinguishable results in what concerns water depth and velocity and are not shown here for brevity. Some differences can be found however when comparing the percentage mass flow error of the two methods as shown on Figure 4. Being below 1 percent for both schemes in most of the channel, the implicit TVD method is more severely affected by the discontinuities in bottom slope while McCormack TVD deals worse with the hydraulic jump.

Finally Figure 5 shows the convergence rate for both schemes. The ratio of the right hand side of Eq. 20 to the same quantity at the first time step is plotted in logarithmic scale for every time step. The explicit algorithm converges in an oscillating manner and much slower than the implicit one (notice the different scaling in the number of time steps of Figure 6 a and b). Although the computational cost per time step is much cheaper for the former, TVD schemes always require the evaluation of complicated nonlinear expressions every time step and this is very expensive. Actually for this test case the overall computational cost of the implicit scheme was about one fifth that of the explicit one.

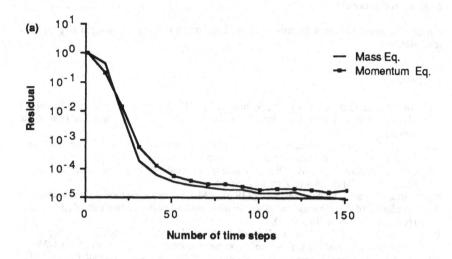

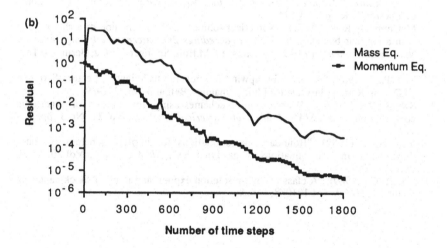

FIGURE 5 *Convergence history of the stationary discontinuous solution shown in Figure 3. The residuals of the mass and momentum conservation equations are shown for: a) Implicit TVD scheme and b) Explicit McCormack TVD scheme.*

5 Conclusions

The use of high resolution TVD methods for open channel flow calculations appears as an interesting and reliable option whenever mixed sub-supercritical or discontinuous flows are to be expected.

Acknowledgements

Financial support provided for this work by Diputación General de Aragón is gratefully acknowledged.

References

[1]   Alcrudo, F., García-Navarro P., Savirón, J.M., (1991), "Flux difference splitting for 1D open channel flow equations", *Journal of Numerical Methods in Fluids* (to appear).

[2]   Cunge, J.A., Holly, F.M., Verwey, A. (1980), *Practical Aspects of Computational River Hydraulics* , Pitman, London, U.K., pp.7-29.

[3]   García-Navarro, P., (1989), "Estudio de la propagación de ondas en cursos fluviales", PhD. Thesis, University of Zaragoza, Spain.

[4]   García-Navarro, P., Savirón, J.M., (1990), "McCormack's method for the numerical simulation of one-dimensional discontinuous unsteady open channel flow", *Journal of Hydraulic Research*, (to appear).

[5]   Harten, A., (1984), "On a class of high resolution total-variation-stable finite difference schemes", *SIAM Journal of Numerical Analysis*, Vol.21, No.1, pp.1-23.

[6]   Hirsch, Ch., (1990), *Numerical Computation of Internal and External Flows. Vol.2: Computational Methods for Inviscid and Viscous Flows*, John Wiley&Sons, Chichester, U.K., pp.493-574.

[7]   McCormack, R.W., (1971), "Numerical solution of the interaction of a shock wave with a laminar boundary layer", *in Proceedings 2nd International Conference on Numerical Methods in Fluid Dynamics*, ed. M.Holt, Springer-Verlag, Berlin, pp.151-163.

[8]   Roe, P.L., (1989) "A survey on upwind differencing techniques", Lecture Series in CFD, Von Karman Institute for Fluid Dynamics, Belgium, March 1989.

[9]   Sweby, P.K., (1984), "High resolution schemes using flux limiters for hyperbolic conservation laws", *SIAM Journal of Numerical Analysis*, Vol.21, No.5, pp.995-1011.

[10] Yee, H.C., (1982), "Boundary approximations for implicit schemes for one-dimensional inviscid equations of gasdynamics", *AIAA Journal*, Vol.20, No.9, pp.1203-1211.

[11] Yee, H.C., (1989), "A class of high-resolution explicit and implicit shock-capturing methods", NASA-TM 101088, U.S.A.

12 A numerical simulation of open-channel flows on a sinusoidal bed by using the k-ε turbulence model

N. Matsunaga, Y. Sugihara and T. Komatsu

ABSTRACT

Open-channel flows on a sinusoidal bed are classified into three types on the basis of the phase-relation between the free surface profile and bed configuration, i. e., in-phase flow, out-of-phase flow and chute & pool flow. The spatial distributions of mean velocity, Reynolds stress and turbulence energy for these three types have been analyzed numerically by using the k- ε turbulence model. The free surface profiles have been also calculated from the kinematic boundary condition. The analytical results on the mean velocity, the free surface profiles and the Reynolds stress have been compared with the experimental ones. The comparison shows good agreement for the in-phase and the out-of-phase flows. However, the chute & pool flow has not been simulated accurately because of its complexity.

1. Introduction

Shear flows past wavy boundaries have attracted considerable research attention because of the central role which they play in the generation of wind waves, the formation of sedimentary ripples and dunes in deserts and river channels, etc. The occurrence of the waves changes remarkably the turbulence structure of the flows, and interaction between the movable boundaries and the mean flows makes it more complicated. Furthermore, a high technique is necessary to the velocity measurement on the movable boundaries (see [1] and [2]). Therefore, many studies (e. g., [3], [4] and [5]) on the turbulent flows with a solid, wavy boundary have been performed as the first step.

Hsu & Kennedy [5] made an experimental study on steady, non-separated, axisymmetric air flows though circular wavy pipes. They measured radial and longitudinal distributions of mean velocity, Reynolds shear stress, three components of turbulent velocity, etc. Their measurements revealed that there is a central core region in which the turbulence quantities are constant along the pipe, and that the turbulent velocity and the Reynolds stress near the wall vary periodically along the boundary waves. Zilker & Hanratty [6] and Buckles et al. [7] investigated the turbulent flow over large-amplitude solid waves. This flow formed a large separated region in one wavelength. They discussed the characteristics about the separation region, the reattached boundary layer and the free shear layer.

On the other hand, there seems to be few studies on open-channel flows on a sinusoidal bed. Their turbulence structure is more complicated than that of the wavy pipe flows because of the existence of free surface. It is well-known that the open-channel flows are classified into three types on the basis of the phase-relation between the bed configuration and the free surface profile, i. e., in-phase flow, out-of-phase flow and chute & pool flow [8]. Matsunaga et al. [9] showed that their formation depends on both the Froude number and the rate of the mean water depth to the wavelength of the bed. Matsunaga et al. [10] investigated the turbulence structure of the three flows, and pointed out that the periodic occurrence of coherent turbulence induces the relative minimum values in the vertical distributions of the Reynolds stress.

In this paper, the turbulence properties of these three flows have been analyzed numerically by using the k- ε turbulence model, and some characteriatic quantities have been compared with the experimental results.

2. Numerical analysis

The k- ε turbulence model is used in the numerical analysis of open-channel flows on a sinusoidal bed. The bed profile η (x) is given by

$$\eta \ (\ x\ ) = a \sin \frac{2\pi}{L} x - \frac{a}{2} \ ,$$

where a is the amplitude and L is the wavelength. When the x- axis is taken in the flow direction and the y- axis in the upward direction in a Cartesian coordinate system, the continuity equation, the momentum equation for x - and y - components, turbulence energy equation and the equation of energy dissipation are expressed respectively by

$$\frac{\partial U}{\partial x} + \frac{\partial V}{\partial y} = 0 \ ,$$

$$\frac{\partial U}{\partial t} + \frac{\partial U^2}{\partial x} + \frac{\partial UV}{\partial y} = g_x \ - \frac{\partial P}{\partial x} + \frac{\partial}{\partial x}\left(-\overline{u'u'}\right) + \frac{\partial}{\partial y}\left(-\overline{u'v'}\right) + \nu\nabla^2 U \ ,$$

$$\frac{\partial V}{\partial t} + \frac{\partial UV}{\partial x} + \frac{\partial V^2}{\partial y} = g_y \ - \frac{\partial P}{\partial y} + \frac{\partial}{\partial x}\left(-\overline{u'v'}\right) + \frac{\partial}{\partial y}\left(-\overline{v'v'}\right) + \nu\nabla^2 V \ ,$$

$$\frac{\partial k}{\partial t} + \frac{\partial(kU)}{\partial x} + \frac{\partial(kV)}{\partial y} = \frac{\partial}{\partial x}\left\{\left(\nu + \frac{\nu_t}{\sigma_k}\right)\frac{\partial k}{\partial x}\right\} + \frac{\partial}{\partial y}\left\{\left(\nu + \frac{\nu_t}{\sigma_k}\right)\frac{\partial k}{\partial y}\right\}$$

$$+ \nu_t\left\{2\left(\frac{\partial U}{\partial x}\right)^2 + \left(\frac{\partial U}{\partial y}\right)^2 + 2\left(\frac{\partial V}{\partial x}\right)\left(\frac{\partial U}{\partial y}\right) + \left(\frac{\partial V}{\partial x}\right)^2 + 2\left(\frac{\partial V}{\partial y}\right)^2\right\} - \varepsilon , \quad \text{and}$$

$$\frac{\partial \varepsilon}{\partial t} + \frac{\partial(\varepsilon U)}{\partial x} + \frac{\partial(\varepsilon V)}{\partial y} = \frac{\partial}{\partial x}\left\{\left(\nu + \frac{\nu_t}{\sigma_\varepsilon}\right)\frac{\partial \varepsilon}{\partial x}\right\} + \frac{\partial}{\partial y}\left\{\left(\nu + \frac{\nu_t}{\sigma_\varepsilon}\right)\frac{\partial \varepsilon}{\partial y}\right\}$$

$$+ c_1\frac{\varepsilon}{k}\nu_t\left\{2\left(\frac{\partial U}{\partial x}\right)^2 + \left(\frac{\partial U}{\partial y}\right)^2 + 2\left(\frac{\partial V}{\partial x}\right)\left(\frac{\partial U}{\partial y}\right) + \left(\frac{\partial V}{\partial x}\right)^2 + 2\left(\frac{\partial V}{\partial y}\right)^2\right\} - c_2\frac{\varepsilon^2}{k} ,$$

where g_x and g_y are the x- and y- components of the gravitational acceleration, P the rate of the pressure to the fluid density, ν the kinematic viscosity and ν_t the eddy viscosity. The Reynolds stresses and the eddy viscosity are related to the mean quantities as follows.

$$-\overline{u'u'} = 2\nu_t\left(\frac{\partial U}{\partial x}\right) - \frac{2}{3}k, \qquad -\overline{u'v'} = \nu_t\left(\frac{\partial U}{\partial y} + \frac{\partial V}{\partial x}\right) ,$$

$$-\overline{v'v'} = 2\nu_t\left(\frac{\partial V}{\partial y}\right) - \frac{2}{3}k , \quad \text{and} \quad \nu_t = c_\mu\frac{k^2}{\varepsilon} .$$

As the values of the constants in these governing equations,

$C_1 = 1.44$, $C_2 = 1.92$, $\sigma_k = 1.0$, $\sigma_\varepsilon = 1.3$ and $C_\mu = 0.09$

are used.

The free surface profile h (x , t) can be derived from the kinematic condition

$$\frac{\partial h(x,t)}{\partial t} = V - U\frac{\partial h(x,t)}{\partial x} .$$

The mean velocities must satisfy both the no-slip condition at the bed surface and zero shear stress at the free surface. The boundary conditions for the turbulence energy k and the dissipation rate ε are

$k = u_*^2/\sqrt{c_\mu}$, $\varepsilon = u_*^3/\kappa y_0$; at $y = \eta(x)$, and

$\partial k/\partial y = 0$, $\partial \varepsilon/\partial y = 0$; at $y = h(x,t)$,

where u_* is the local friction velocity on the bed surface, which is calculated by extraporating the Reynolds stress to the bed surface, y_0 is the height at which the U- profile begins to become logarithmic and κ is von Kármán's constant. Cyclic conditions are used as the boundary conditions at both the inlet of the flow and the outlet.

The ABMAC method is used in the numerical calculation. The velocity and the pressure are analyzed by using the simultaneous relaxation method. The advancing difference is applied to the time differentials, the second-order upwind-difference to the convection terms and the central difference to the other terms. The horizontal and the vertical mesh widths are 0.55 cm and 0.1 cm, respectively. In the numerical calculation, the kinematic viscosity v is neglected, and the height of one mesh is taken as y_0. The steady solutions are obtained by iterating the calculation until the relative errors of U, V, P, k, ε and h at the present time step and before one time step become smaller than 10^{-2}, respectively.

3. Experimental methods

Experiments were carried out by using a channel illustrated in Fig.1. It was 12 m long, 0.15 m wide and 0.3 m deep. Its slope can be varied from the horizontal position to 1/50. Six blocks of flat plate were placed at the upstream side as an approach section. The plate, which was made of plaster, was 66 cm long. Eight blocks of wavy plate were set at the downstream side of the approach section. The wavy plate was also 66 cm long and consisted of three sinusoidal waves. Their wavelength and wave height were 22 cm and 1.0 cm, respectively. The three type flows were made by adjusting the channel slope and flow rate. Their hydraulic conditions are summarized in Table 1. The Froude number Fr is defined by $\hat{U}/\sqrt{g\hat{h}}$ and the Reynolds number Re by $\hat{U}\,\hat{h}/v$.

One wavelength section, in which the overall slope of water surface agreed well with the channel slope, was chosen as a data-taking section. The profile of the water surface in the section was measured by setting a static pressure tube near the surface. Dynamic pressures were also measured by pressing a pitot tube of diameter 2.0 mm along the bed surface. After local mean velocities u_w at 0.1 cm above the bed surface were obtained from the dynamic pressures, local friction velocities u_* were evaluated from

$$u_* = \frac{u_w}{U_0}\ \sqrt{gRI}\quad,$$

where Uo is the mean velocity at 0.1cm above the bed of the approach section

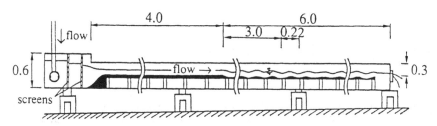

Fig. 1 Schematic diagram of experimental set-up
(dimensions in m)

Table 1. Hydraulic conditions

Run	1	2	3
Type of flow	Out-of-phase flow	Chute & pool flow	In - phase flow
Bed slope	2/1000	14/1000	8/1000
Averaged velocity $\hat{U}$ (cm/s)	52.6	27.3	68.6
Averaged friction velocity $\hat{u}_*$ (cm/s)	2.74	4.69	4.18
Averaged water depth $\hat{h}$ (cm)	7.84	2.04	3.19
Fr	0.600	0.611	1.22
Re	3.38×10^4	4.56×10^3	1.79×10^4

where uniform flow forms, and R and I are the hydraulic radius and the hydraulic gradient of the uniform flow, respectively. The friction velocities $\hat{u}_*$ given in Table 1 are averaged values of u_* over one wavelergth. Vertical profiles of the mean velocities, the turbulence intensities and the Reynolds stress were obtained by traversing a x- shaped hot-film probe vertically at nine sections. The signals were recorded by using a data recorder and were digitized under the sampling time of 8/500 s. The number of the sampled data was 8192.

4. Comparisons between analytical and experimental results

The numerical simulations have been performed under the same conditions that were shown in Table 1. Figures 2 (a) to (c) show comparisons about the free surface profiles and the vertical distributions of mean velocity. The calculated and measured surface profiles are drawn by solid and dashed lines, respectively. The phase difference of about $4\pi/11$ is seen between these profiles of the out-of-phase flow. The calculated profile for the in-phase flow has the phase lag of about $3\pi/11$ as compared with the measured one. The difference of amplitude rather than the phase difference is obvious in the chute & pool flow. The distributions of the calculated velocity are shown by solid lines and the experimental data are dotted. They are normalized by making the use of velocities averaged cross-sectionally at each section. The disagreement between the calculated and measured velocities is considerable near the bed and the free surface. It is very large in the case of the chute & pool flow. The influence of the boundaries appears through the overall depth of this flow, because it accompanies a hydraulic jump and its depth is shallow relatively. In

143

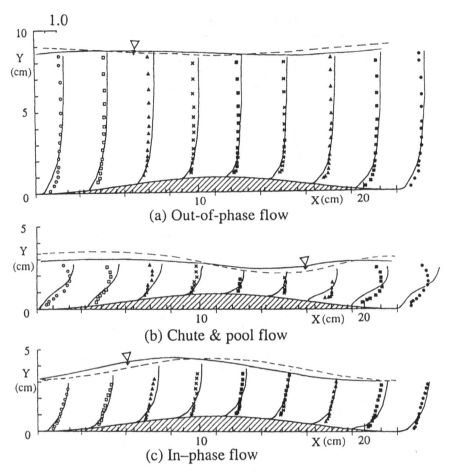

1.0

(a) Out-of-phase flow

(b) Chute & pool flow

(c) In–phase flow

Fig. 2 Comparison on profiles of mean velocity and free surface.

order to simulate chute & pool flow correctly, therefore, it may be necessary to reconsider the boundary conditions.

Figures 3 (a) to (c) show contour maps of the non-dimensional Reynolds stress $-\overline{u'v'}/\hat{u}_*^2$ which is obtained from the numerical analysis. Spatial distributions of the msasured one are shown in Figs. 4 (a) to (c). These distributions of the out-of-phase flow are similar qualitatively and quantitatively. The Reynolds stress in the region of $y > 4$ does not vary rapidly in the flow direction. This region corresponds to the constant core region, whose formation Hsu & Kennedy [5] found in wavy pipe flows. A region of strong Reynolds stress is seen near the bed surface of the lee side. From the comparison of Figs. 3 (b) and 4 (b), it is seen that the analytical results do not agree well with the measured ones. This is caused by the reason that the mean velocity profile which plays an important role on the formation of the Reynolds stress could not be simulated well in the chute &

144

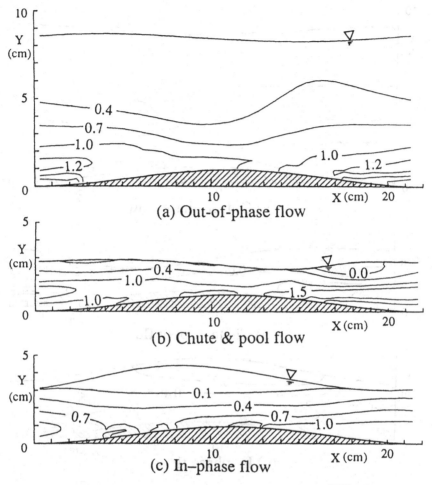

Fig. 3 Contour maps of calculated values of $-\overline{u'v'}/\hat{u}_*^2$

pool flow. It is clear from the experimental results that a layer, where the Reynolds stress vanishes, forms at a half of water depth. The Reynolds stress is negative above this layer because the mean velocity profile with adverse gradient occurs due to a hydraulic jump. On the other hand, the distribution of the in-phase flow seems to be simulated relatively well. It is seen that the constant core region is not formed in both the in-phase and the chute & pool flow.

The contour maps of turbulence energy are shown in Figs. 5 (a) to (c). These results are obtained by the calculation. The out-of-phase flow accompanies a large energy region at $y/\hat{h} < 0.2$. In the range of $y/\hat{h} > 0.4$, the turbulence energy is nearly homogeneous in the flow direction. The distribution of the chute & pool flow shows that the turbulence energy becomes large in the region of the hydraulic jump and small at the crest section. Thus,

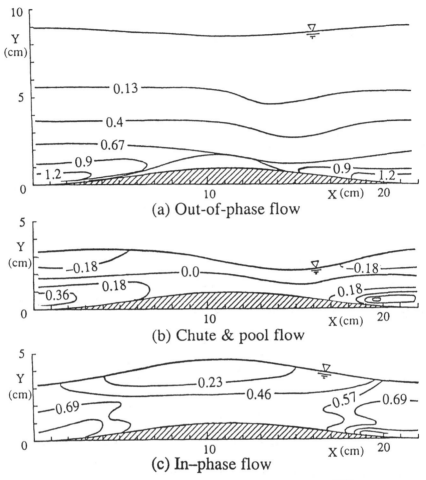

(a) Out-of-phase flow

(b) Chute & pool flow

(c) In–phase flow

Fig. 4 Contour maps of measured values of $-\overline{u'v'}/\hat{u}_*^2$

the turbulence energy of this type flow varies complicatedly in the flow direction because of the rapid acceleration and deceleration in one wavelength. The in-phase flow has a large energy region in the range of $y/\hat{h} < 0.6$. Though the pattern of this map is similar to that of the out-of-phase flow, on the whole, the non-dimensional turbulence energy of the in-phase flow is smaller than those of the other type flows.

5. Conclusions

The three types of flow, i. e., out-of phase flow, chute & pool flow and in-phase flow were analyzed numerically by using the k - ε turbulence model. Furthermore, these results were compared with the experimental results. The out-of-phase and in -phase flows were simulated relatively well in regard to the

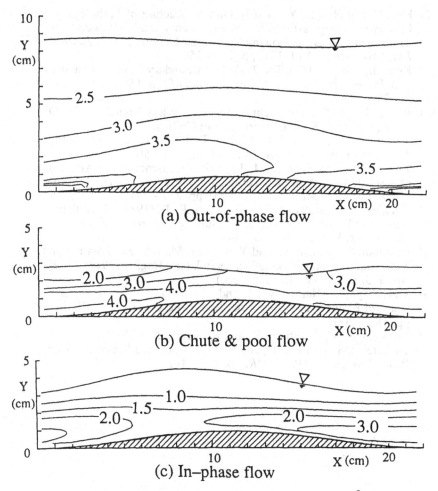

Fig. 5 Contour maps of calculated values of $k / \hat{u}_*^2$

free surface profile, the vertical profile of mean velocity and the spatial distribution of the Reynolds stress. However, these characteristic quantities of the chute & pool flow could not be analyzed accurately, because the rapid transition from subcritical flow to supercritical occurs in one wavelength. In addition of the reconsideration about the boundary conditions, it seems to be necessary for more acculate simulation to use a modified k - ε turbulence model which is also applicable to anisotropic turbulence near boundaries.

REFERENCES

[1]   Hsu, C., Hsu, E.Y., and Street, R. L., (1981), On the Structure of Turbulent Flow over a Progressive Water Wave: Theory and Experiment in a Transformed, Wave-Following Coordinate System, *Journal of Fluid Mechanics*, Vol. 105, pp. 87 - 118.

[2] Hsu, C. and Hsu. E. Y., (1983), On the Structure of Turbulent Flow over a Progressive Water Wave: Theory and Experiment in a Transformed Wave-Following Coordinate System, Part 2, *Journal of Fluid Mechanics,* Vol. 131, pp. 123 - 154.

[3] Kendall, J. M., (1970), The Turbulent Boundary Layer over a Wall with Progressive Surface Waves, *Journal of Fluid Mechanics*, Vol. 41, pp. 259 - 282.

[4] Beebe, P., (1972), Turbulent Flow over a Wavy Boundary, Ph. D. thesis, Colorado State University, Fort Collins.

[5] Hsu, S. and Kennedy, J. F., (1971), Turbulent Flow in Wavy Pipes, *Journal of Fluid Mechanics*, Vol. 47, pp. 481 - 502.

[6] Zilker, D. P. and Hanratty, T. J., (1979), Influence of the Amplitude of a Solid Wavy Wall on a Turbulent Flow. Part 2. Separated Flows, *Journal of Fluid Mechanics*, Vol. 90, pp. 257 - 271.

[7] Buckles, J., Hanratty, T. J. and Adrian, R. J., (1984), Turbulent Flow over Large-Amplitude Wavy Surfaces, *Journal of Fluid Mechanics*, Vol. 140, pp. 27 - 44.

[8] Fukuoka, S., Okutsu, K. and Yamasaka, M., (1982), Dynamic and Kinematic Features of Sand Waves in Upper Regime, *Proc. JSCE*, No. 323, pp. 77 -89. (in Japanese)

[9] Matsunaga, N., Namikawa, T. and Komatsu, T., (1986), Open-Channel Flows over a Sinusoidal Bed, *Proc. 5 th Cong. APD-IAHR*, pp. 185 - 200.

[10] Matsunaga, N., Takehara, K. and Komatsu, T., (1990), Turbulence Structure of Open-Channel Flows on a Sinusoidal Bed, *Proc. 7 th Cong. APD-IAHR*, pp. 135 - 142.

13 Hydrodynamic simulation of small-scale tidal wetlands

P. Goodwin, J. Lewandowski and R. J. Sobey

1. Introduction

Tidal wetlands are an essential ecological resource but have been reduced dramatically during the past century. In California, the area of coastal wetlands has been reduced from over 120,000 ha at the turn of the century to less than 25,000 ha in 1992 [12]. In some specific regions the loss of coastal wetlands has been more dramatic, for example, over 80% of the coastal wetlands in Southern California and 95% of historic wetlands in San Francisco Bay have been lost due to development, changes in the watershed hydrology or accelerated sedimentation. These tidal marshes and lagoons are not isolated ecosystems but can influence regional biological diversity and complex biological chains; for example, migratory bird flyways or salmon runs along rivers tributary to the tidal wetland.

Tidal wetlands represent a fine balance between sedimentation due to tidal action, sedimentation due to floods in the watershed tributary to the wetland, local subsidence, wave energy and changes in mean sea level. If the balance is disturbed and the wetland cannot respond at a rate comparable with the disturbance, there is a deterioration in the habitat values and loss of wetlands. Heerdt Marsh in San Francisco Bay [1] is an example of tidal wetland loss due to sediment depletion and San Elijo Lagoon (refer to section 6) is an example of tidal wetland loss due to accelerated sedimentation.

The sensitivity of marsh plain elevation to the productivity of the wetland has been demonstrated by Reed and Cahoon [11]. Detailed measurements in a Louisiana tidal wetlands showed that a difference in elevation of 10 cm between two test sites would make a difference between being inundated for 30% of a year or 52 % of a year. Both test sites contained *Spartina alterniflora* although the test site at the lower elevation showed plants of reduced height, density and biomass than the higher site. The reason for the vegetation differences were attributed to waterlogging and accumulation of toxic sulphides. In general, the marsh plain elevation also influences the type of vegetation within the wetland which is important if a particular habitat is to be restored.

It is therefore essential to be able to predict accurately the tidal elevation-duration relationships for marshes when developing a restoration plan for an existing wetland or designing an artificial wetland. Marsh plain slopes are typically less than one percent, so that the formation of the marsh plain at an incorrect elevation due to inaccuracies in the construction procedure or inaccurate predictions from an hydrodynamic model can make a substantial difference in areas of specific habitat types or in the productivity of the wetland. A hydrodynamic model suitable for simulating the duration, extent of tidal inundation and degree of tidal damping in tidal wetlands is described herein.

2.   Hydrodynamic Model

Most tidal marshes comprise a network of channels with overbank flow on the marsh plains. A one-dimensional model capable of simulating complex networks of channels and overbank flows is generally adequate for predicting the aerial extent and duration of inundation by tidal flows. California coastal lagoons or tidal marshes with large areas permanently inundated or subject to significant salinity variations may require more complex 2-d or 3-d models to determine the circulation patterns. A one-dimensional model solves the gradually varied unsteady flow equations for the conservation of mass and momentum, i.e.

$$\frac{\partial h}{\partial t} + \frac{\partial Q}{\partial x} = O \tag{1}$$

and

$$\frac{\partial Q}{\partial t} - \frac{2BQ}{A} \frac{\partial h}{\partial t} - \frac{Q^2}{A^2} \frac{\partial h}{\partial x} = -gA \frac{\partial h}{\partial x} - \frac{g|Q|Q}{C^2 AR} \tag{2}$$

where t is time (s), x is the longitudinal distance coordinate (m), B is the channel width (m), Q is the discharge (m^3/$_s$), A is the cross-sectional area of flow (m^2), h is the water surface elevation (m), and C is the Chezy resistance coefficient.

The solution to these equations can be obtained using a suitable numerical algorithm, for example the results presented herein are generated with ESTFLO (ESTuarine FLOw) model which uses a three point implicit finite difference scheme [14].

There are three wetland-specific difficulties to resolve before equations (1) and (2) can be solved. The first difficulty is the choice of a suitable value of the Chezy coefficient for flow in the marsh channels and on the vegetated marsh plain. The second problem

150

is to accurately simulate the wetting and drying of the marsh plain and slough channels since the propagation of the leading edge of the wetting front will determine which areas will be inundated. Thirdly, it is necessary to select a representative channel size at the tidal inlet to the wetland since a large variability can occur depending upon current, wave, tide and freshwater inflow conditions.

3.   Resistance in Tidal Wetlands

The flow resistance occurring within the tidal wetland can be classified in two categories, the resistance provided in the meandering slough channels incised in the marsh plain and the resistance occurring on the marsh plain surface. The slough channels are generally free from vegetation and contain beds of cohesive sediments and insignificant bedforms. The slough channel resistance to flow can be represented by the frictional resistance in the form of the Chezy resistance coefficient, C, or the Darcy-Weisbach friction factor, f. The resistance coefficient can be determined from the equivalent roughness of the boundary, k_s, and the Colebrook-White equation.

The flow resistance occurring on the marsh plain is due to surface boundary resistance and drag resistance caused by the vegetation throughout the depth. The combined resistance [7,16] can be represented by

$$f_T - f_V + f_b \qquad (3)$$

where the subscripts V and b denote vegetation and boundary respectively. The evaluation of f_V and f_b for California' salt marshes is the subject of a current study, but the results of Chen [6] give a useful estimate of f_V. Generally, the friction factors at low Reynolds numbers are large and have a significant influence on the propagation of the wetting or drying front.

4.   Tidal Inlet Conditions

4.1   Equilibrium inlet area

The cross-sectional area of the tidal inlet channel influences the degree of tidal damping, tidal setup in the lagoon, the circulation patterns and flushing characteristics of the lagoon. The area of the inlet channel represents a balance between the total volume of water flowing through the inlet in a tidal cycle acting to scour the inlet channel and the nearshore sediment transport acting to close off the inlet. The sediment particle size and geologic features such as rock sills constraining the lateral or vertical scour of the channel also affect the inlet area. O'Brien [10] determined that the most significant parameter controlling the inlet area is the tidal prism. Jarrett extended the finding and correlated 17 lagoon sites to derive an empirical relationship for the inlet area, A_c, with no jetties:

$$A_c - 8.95 \times 10^{-6} P^{1.1} \quad m^2 \qquad (4)$$

where P is the mean diurnal tidal prism (m^3).

151

4.2   Stability of inlet area

The simplest method to determine the stability of a tidal inlet is to consider the maximum velocity during a tidal cycle. As a rule-of-thumb; if the ebb velocity exceeds 1.0 m/s the inlet channel is stable, whereas if it is less than 1.0 m/s, the lagoon is subject to closure [5].

Johnson [9] assumed that the longshore transport acting to close the lagoon entrance is proportional to the annual deep water wave power. It is then possible to predict whether a coastal lagoon will be closed or open to tidal action under normal hydrologic conditions. Williams [17] refined Johnson's data and developed an empirical curve useful for predicting the frequency of lagoon opening (Figure 1) in the absence of more detailed site data.

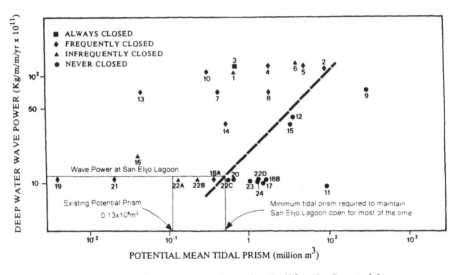

Figure 1   Entrance Closure Conditions for California Coastal Lagoons

The channel inlet area will change with the intensity of wave action, tidal range on a given day or during major flood flows into the lagoon. In small tidal inlets, such as the Tijuana River estuary, the tidal inlet area can be changed by 50% between spring and neap tide cycles, or during a diurnal tide cycle [9, 15].

A numerical model predicting the precise configuration of the inlet channel for any tide or nearshore wave and current conditions requires extensive field data for calibration and validation [15]. A simpler technique to determine whether an inlet is stable was proposed by Escoffier [8], who used the peak ebb velocity (U_{MAX}) as a stability indicator. U_{MAX} can be computed for different inlet areas using a numerical or analytical model. U_{MAX} is proportional to the scouring ability of the inlet channel

152

and reaches a maximum at $(A_c)_{CRIT}$. For any given peak velocity, U_{MAX}, there are two possible values of the inlet area (Figure 2). If the value of A_c (evaluated from field observations or predicted by an equation similar to [4]) normally exceeds $(A_c)_{CRIT}$, the inlet is likely to be stable. If storm generated waves reduce the inlet area from $(A_c)_1$ to any value of A_c greater than $(A_c)_2$, U_{MAX} is increased. U_{MAX} is proportional to the scouring ability of the inlet, and the increase in U_{MAX} leads to an increase in A_C and the original inlet area $(A_c)_1$ is restored. However, if the storm reduces the inlet area to an area less than $(A_c)_2$ or if the original area was less than $(A_c)_{CRIT}$, then U_{MAX} and the scouring ability of the channel is reduced.

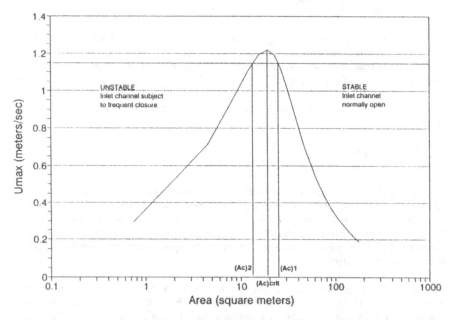

Figure 2: Escoffier Stability Curve - San Elijo Lagoon

This initiates a continuous infilling of the channel until the inlet is closed to tidal exchange. The inlet will not open again unless the sand bar is breached artificially or until the freshwater inflows into the lagoon fill up the lagoon and the sand bar is breached by overtopping.

If data on the longshore drift of sediment is known, Bruun [4] has developed an empirical estimate of the stability of the inlet channel, based upon the ratio of the neap tidal prism, $P_N(m^3)$ to the longshore sediment transport, $Q_{LS}(m^3)$. If $P_N/Q_{LS} < 50$ the inlet is closed except during significant flows from the watershed. If this ratio exceeds 125, the inlet channel will be nearly always open.

5. Wetting and Drying — a Parabolic Approximation

Equations [1] and [2] may be non-dimensionalised, rewritten in terms of the variable velocity u rather than flow discharge Q and if unit width of the marsh plain is considered, then

$$\alpha \frac{\partial h}{\partial t} + u\frac{\partial h}{\partial x} + h\frac{\partial u}{\partial x} = 0 \tag{5}$$

$$\alpha \frac{\partial u}{\partial t} + \frac{\partial h}{\partial t} + u\frac{\partial u}{\partial x} + \beta(S_o + S_f) = 0 \tag{6}$$

where $x = x \cdot W$, $t = t \cdot T$, $u = u\sqrt{gH_T}$, $h = h \cdot H_T$, $S_o = S_o(H_T/W)$, $S_f = S_f(H_T/W)$. H_T is a representative vertical length scale and taken as the tide range, T is a representative time scale, and W a representative horizontal length scale. The scaling parameters is $\alpha = W/(T\sqrt{gH_T})$.

Equations [5] and [6] represent an hyperbolic system of equations with characteristic directions

$$\frac{dx}{dt} = \frac{1}{\alpha}(u \pm \sqrt{h}) \tag{7}$$

In subcritical flows, there are two families of characteristics. The positive family emanating from the upstream boundary and propagating in the direction of flow through the solution domain; and a negative family emanating from the downstream boundary and propagating against the flow, eventually passing through the upstream boundary [2]. One point boundary data must be specified at the upstream and downstream boundaries.

Most approximations of this hyperbolic problem assumes that the α or the advection term are small. However in tidal marshes, the depth of flow becomes zero at the leading edge of a wetting front on a marsh plain or mud flat, and the problem is reduced to a parabolic problem with a single characteristic direction

$$\frac{dx}{dt} = \frac{1}{\alpha}u \tag{8}$$

and along this characteristic, h remains constant and the Riemann quasi-invariant is

$$[u + 2h]_{t1}^{t2} = -\frac{1}{\alpha}\int_{t1}^{t2}(S_o + S_f)dt \tag{9}$$

For the wetting front, this results in a solution which propagates forward across the marsh plain from a point which is always inundated and represents the limit of a conventional tidal hydrodynamic model (Figure 3). The parabolic problem requires two point boundary data at the interface between the hyperbolic and parabolic problem.

154

This can be achieved by running a negative characteristic from the parabolic domain to the interface boundary, allowing both h and q (or u) to be determined. The speed of propagation of the wetting front is sensitive to the choice of f (Section 3). Generally the vegetation resistance is low on the bed of the marsh plain but increases 2-5 cm above the bed as the density of plant material increases. f is therefore a function of h, but will be constant along each characteristic since h is also constant in the parabolic approximation.

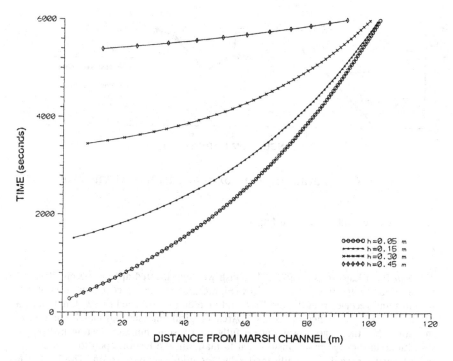

Figure 3   Characteristic Directions for Wetting Front

This algorithm prevents the generation of transients due to the addition, or removal of nodes in the solution field, and enables realistic tracking of the leading edge of the wetting front. The precise delineation of the wetted area is important when determining the hydroperiod, salinity distribution and habitat distribution within a tidal marsh. The applicability of the parabolic approximation was verified by a series of field measurements at Heerdt Marsh in San Francisco Bay (Figure 4).

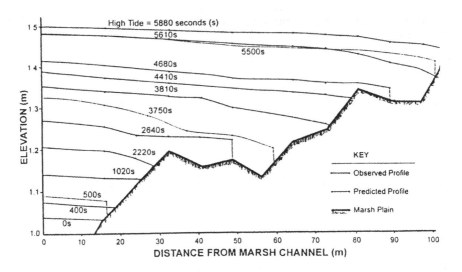

Figure 4   Parabolic Approximation for Wetting Front

6.   A case study: San Elijo Lagoon

6.1   Background

The San Elijo Lagoon is a shallow brackish wetland, located approximately 20 miles north of San Diego in California, USA (Figure 5). The lagoon covers an area of 220 hectares and receives runoff from Escondido Creek (a watershed of 190 km^2) and La Orilla Creek (a watershed of 13 km^2). Archeological analyses and marsh cores have indicated that the Lagoon was once a fully tidal system but the existing hydrologic conditions in the lagoon are insufficient to maintain the lagoon open to tidal action, except during periods of significant freshwater inflow to the lagoon. The lagoon is divided into the East, Central and West Basins by the Interstate 5 Highway, the Santa Fe Railroad and the Pacific Coast Highway connected by narrow channels at the bridges. Changes in watershed management practices, particularly the rapid urban growth in this area, and the constrictions imposed on the natural flow patterns have resulted in rapid sedimentation of the lagoon during the past few decades which has resulted in a significant loss in tidal prism. The inlet channel connecting the lagoon to the Pacific Ocean is an artificial channel confined to a sinuous alignment against the northern bluff by the railroad bridge and Pacific Coast Highway. A rock sill in the inlet channel prevents downcutting, and the inlet channel is wide and shallow when open to tidal action.

156

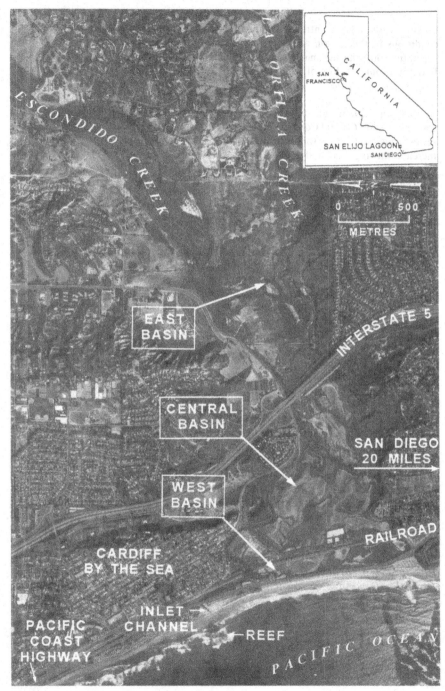

Figure 5 *San Elijo Lagoon*

157

Closure of the lagoon entrance results in extended periods of freshwater or brackish conditions in the lagoon and there has been extensive encroachment of freshwater vegetation into the salt marsh areas of the lagoon. Lagoon closure also creates local flooding problems, loss of foraging habitat for wildlife, water quality problems and disease carrying vector problems associated with the poor circulation.

In 1990, the County of San Diego, Department of Parks and Recreation and the California State Coastal Conservancy undertook a feasibility study to investigate means of restoring tidal action to San Elijo Lagoon. The primary goals of the project are to increase the regional and local wildlife habitat value of the lagoon. The project attempts to recreate the natural historic gradient of habitats from tidal saltwater marsh adjacent to the ocean, to brackish marsh, freshwater wetlands, riparian, and upland habitat at the inland end of the study area.

6.2 Results

The existing potential mean tidal prism in San Elijo Lagoon is $0.13 \times 10^6 m^3$ and field observations have shown that the lagoon inlet channel is closed frequently. As a first approximation, Figure 1 indicates that the tidal prism should be increased to approximately $0.7 \times 10^6 m^3$ to maintain the lagoon open to tidal circulation for most of the time. The equilibrium inlet area for this tidal prism is approximately $25m^2$ (Equation [4]). Reference to the Escoffier curve shows that this inlet channel will remain open unless a storm event reduces the inlet area to less than $12m^2$.

The maximum longshore transport rate at the site is $1550m^3/day$, resulting in a ratio of P_N/Q_{LS} of approximately 25. If the tidal prism is increased to $0.7 \times 10^6/m^3$ this ratio exceeds 125 implying the inlet channel will probably be self maintaining [4].

On February 25, 1991, the inlet channel was breached to allow field observations of the tidal exchange characteristics to be monitored and model calibration data to be collected. Figure 6 shows the sensitivity of the numerical simulation to choice of resistance coefficient in the main marsh channels. $k_s = 0.05$ cm was the roughness found to be the most representative value during the model validation, and 0.25m is an extreme choice of the order of the flow depth at low tide. Accurate estimates of the roughness coefficient on the marsh plain is of greater importance since the duration of inundation will govern the type of habitat which will develop in the enhancement project. The sensitivity of the frequency of inundation to the choice of roughness coefficient can be predicted by the model described herein. Preliminary findings show that a small variation in the resistance coefficient leads to significant changes in the frequency and duration of inundation on the vegetated marsh plain. For example, a 10% reduction in the Chezy coefficient results in a reduction in the predicted annual duration of inundation from 30% to less than 8% at peripheral locations in the Central Basin.

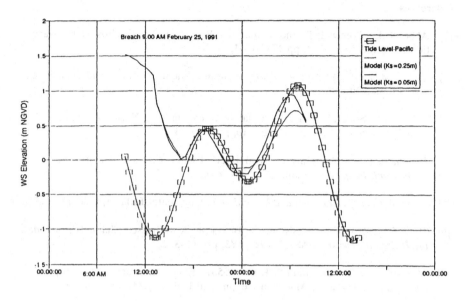

Figure 6: Simulation of Flows in San Elijo Lagoon

7. Conclusions

Accurate simulation of flow within a tidal wetland, particularly in areas subject to wetting and drying, is essential in designing a restoration or enhancement project. For the restoration of tidal wetlands in California, it is necessary to determine whether the tidal inlet channel will remain open to tidal exchange under most hydrological conditions. Numerical models can then be used to predict the tidal flows in the tidal marsh channels and across the marsh plain. The success of a tidal wetland restoration project or an artificial wetland is dependant upon drainage ability, as well as the period and frequency of tidal inundation. Elevation differences of 0.1 metres can result in significant reductions in the productivity of the wetland or result in different distributions of habitat than was intended in the original design, since many salt marsh plants are sensitive to the duration and frequency of inundation.

Acknowledgements

The field data for San Elijo Lagoon was collected as part of a current study for the County of San Diego supervised by Barbara Simmons and funded by the County of San Diego Department of Parks and Recreation and the California State Coastal Conservancy. Additional observations and historic data for San Elijo Lagoon has been compiled by Susan Welker and Robert Patten, Park Rangers for the San Elijo Natural Reserve.

References

[1] Abbe, T.A., Goodwin, P. and Williams, P.B. (1991) "Marsh Erosion by Wave Action", *Coastal Zone 91*, **2**, pp 1747-1761.

[2] Abbott, M.B. (1985) "Computational Hydraulics, Elements of the Theory of Free Surface Flows", Pitman, p. 326.

[3] Bode, L, and Sobey, R.J. (1984) "Initial Transients in Long Wave Computations", *Journal Hydraulic Engineering*, ASCE, **110**, pp 1371-1397.

[4] Bruun, P. (1978) "Stability of Tidal Inlets, Engineering and Theory", *Elsevier Scientific Publishing*, Amsterdam, The Netherlands.

[5] Bruun, P. (1968) *Tidal Inlets and Littoral Drift*, Universitetsforlaget, Oslo, Norway.

[6] Chen, C. (1976) "Flow Resistance in Broad Shallow Grassed Channels", *Journal of the Hydraulics Division*, ASCE, **102**, HY3, pp 307-322.

[7] Einstein, H.A., and Banks, R.B. (1950) "Fluid Resistance of Composite Roughness", *Transactions*, American Geophysical Union, **31**(4), pp 603-610.

[8] Escoffier, F.F. (1940) "The Stability of Tidal Inlets", *Shore and Beach*, October 1940, pp 114-115.

[9] Goodwin, P. and Williams, P.B. (1991) "Short Term Characteristics of Coastal Lagoon Entrances in California", *Coastal Sediments 91*, **1**, pp 1192-1206.

[10] O'Brien, M.P. (1969) "Equilibrium areas of inlets on sandy coasts", *Journal of Waterways and Harbors Division*, ASCE, **95**, WW1, pp 43-52.

[11] Reed, D.J. and Cahoon, D.R. (1992) "The Relationship between Marsh Surface Topography, Hydroperiod and Growth of *Spartina alterniflora* in a Deteriorating Louisiana Salt Marsh", *Journal of Coastal Research*, **8**(1), pp 77-87.

[12] Salvesen, D (1990) *Wetlands: Mitigating and Regulation Development Impacts*, UCI–The Urban Land Institute, Washington, D.C., p. 117.

[13] Skou, A. (1990) "On the Geometry of Cross-Section Areas in Tidal Inlets", *ISVA Series Paper 51*, Technical University of Denmark, p. 109.

[14] Sobey, R.J., Adil T.S. and Vidler P.F. (1980) "ESTFLO: A numerical hydrodynamic model of unsteady flow in a Well Mixed Estuarine System", *Research Bulletin No. CS22*, James Cook University, Townsville, Australia, p. 40.

[15] Webb, C.K. (1989) *Coastal Inlet Dynamics for Southern California*, Thesis submitted in partial fulfillment of the degree of Master of Arts, Department of Geography, San Diego State University, San Diego, California.

[16] Wessels, W.P.J and Strelkoff, T. (1968) "Established Surge on an Impervious Vegetated Bed", *Journal of the Irrigation and Drainage Division*, ASCE, IR1, pp 1-22.

[17] Williams, P.B. (1984) *An evaluation of the feasibility of a self maintained ocean entrance at Bolsa Chica*, Report for the State of California, Coastal Conservancy, Oakland, California, p. 26.

14 Feasibility study of a two-barrier system in the Dutch sea water defence

I. C. M. Helsloot and J. P. F. M. Janssen

Abstract

In the southwest of the Netherlands a storm surge barrier is under construction in the waterway that gives access to the port of Rotterdam. A feasibility study into the effect of an extra harbour entrance to the sea is being performed. The entrance requires a second barrier. Combined with dikes along the rivers, the barriers have to provide the required safety against flooding. The study is concentrated on the influence of the second barrier on design water levels for dikes which were already set by the first barrier. The frequency of closure for both barriers has to be minimized, because of harbour interest. With this restriction, no significant increase in design water levels is allowed. A hydraulic network model is used to calculate the effect of the barriers on water levels. To calculate design water levels an efficient probabilistic calculation procedure is developed. Reliability aspects related to the barriers are integrated into the calculations.

1   Introduction

The project area covers the lower river reaches in the southwest of the Netherlands, as shown in figures 1 and 2. Figure 2 represents the Rhine and Meuse Delta network, in which water levels will be influenced by operating the storm surge barrier to be built in the New Waterway (S_w). Combined with the dikes along the river branches this barrier will provide safety against flooding.

The main southern branch (Haringvliet and eastward) was separated from the southern delta by the Volkerak dam in 1969 and from the sea by the Haringvliet sluices in 1970. These sluices are closed during high tide to prevent sea water from entering the delta region during low tide only when the river discharge is low. In this way the water is forced to discharge through the northern branches, to stop the salt intrusion.

The conditions for control of the northern branches are different. The northern branches give access to the port of Rotterdam and therefore have to render free transit for navigation whenever possible. From this infrastructural point of view the plan was drawn up to create an extra harbour entrance to the sea. In 1990 a feasibility study was started into the effects of the project. One of the aspects incorporated was the safety against flooding, which has to remain guaranteed. Since the entrance again will connect the rivers to the sea it was obvious that another (storm surge) barrier, located somewhere in the Hartelchannel (figure 2, S_H), was needed. The feasibility study was concentrated on the influence of this Hartel

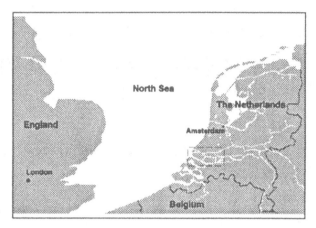

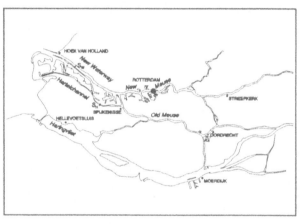

Figures 1 and 2 *The project area*

barrier on the design water levels of dikes. This level strongly determines the safety against flooding and can be affected by barrier closure. However, on economical grounds, originating from port and navigational interest, the closure of the Hartel barrier has to be minimized, in concordance with the closure of the barrier in the New Waterway. The frequency of closure of the Waterway barrier is restricted to once every ten years. For the Hartel barrier, which is not situated in the main entrance, once every one to three years, seems to be acceptable.

This paper deals with the approach, chosen to meet the requirements of both safety against flooding and minimal closure of the system of two barriers. To this end an integral safety concept has been developed, where both the combined effect of storm surges and river floods and the performance of the two-barrier system are handled probabilistically and integrated to assess safety conditions in the whole Rhine and Meuse delta region.

In section 2 the project and the hydraulic system will be described. Next the calculation procedure will be outlined (section 3). In section 4 the results of an indicative sensivity analysis are discussed. Finally some conclusions are drawn.

2 Project description and impact on the hydraulic system

2.1 Project description

Water levels in the lower river reaches in the Rhine and Meuse delta area in the southwest of the Netherlands are determined by discharges of both rivers and by sea levels. When the storm surge barrier in the New Waterway will be in full operation the extreme water levels in this area will be influenced by using it.

The western part of the area to be influenced is shown schematically in figure 3. The combined action of both river discharges and sea levels is indicated. The figure also shows the location of the storm surge barrier in the New Waterway (S_w) and the main channels through the port of Rotterdam where the extra harbour entrance is projected. It involves the removal of the Beerdam (BD), which will create a second connection with the sea. Tidal movement and storm surges will enter the delta region through both the New Waterway and the Hartelchannel. Water from the rivers Rhine and Meuse will then discharge through these two branches and at high

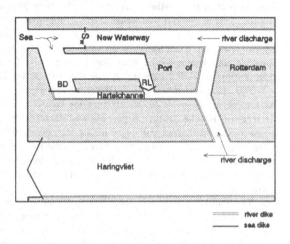

Figure 3 *Schematic view of the system*

discharges also through the Haringvliet sluices that are to be opened (figure 3).

In the project even a third connection to the sea is considered. This involves the removal of the Rozenburg lock (RL). Studying the effects of this alternative, it is obvious that the Hartel barrier has to be located in the Hartelchannel east of the Rozenburg Lock.

The strategy of control of the two-barrier system has to be settled, whereby it has to comply with two demarcations defined in the project. First it is required that the safety against flooding remains guaranteed. Secondly, the closure of the Waterway barrier is restricted to once every ten years. These requirements make demands on the performance of the Hartel barrier within the two-barrier system.

2.2 The hydraulic system

Safety against flooding will be expressed by Design Water Levels (DWLs), a local water level with a fixed frequency of exceedance. Closing the barriers very frequently would give a large reduction of the DWLs, but would not comply with the requirements. From the characteristics of the hydraulic system of the delta area it is possible to describe the effects of operating the barriers on the DWLs qualitatively, given these requirements.

The effects of operating a barrier on River Water Levels (RWLs) are twofold. By closing the barrier RWLs are reduced because storm surges are prevented from entering the system. On the other hand RWLs are increased because of the accumulation of river discharge. In this respect the delta area upstream of the barriers, where water levels are influenced by operating them, can be divided into two regions as follows:

-   The 'tidal region' where storm surges which have access to the area due to the restriction in frequency of closure are dominant.
-   The 'transition region' where water levels attained by accumulation of river discharge are dominant.

Taking into account the characteristics of the hydraulic system these regions can be indicated roughly. The exact extension of the regions is dependent on various aspects, e.g. the management of the barriers and the river regime.

The tidal region is located in the vicinity of the barriers. Roughly this covers the western part of the northern branch with the New Meuse and most part of the Old Meuse (figure 2). It is obvious that water levels in the area just upstream of the Waterway barrier will be determined by storm surges that are allowed through by not closing this barrier. A similar reasoning concerning the area upstream of the Hartel barrier is legitimate. Generally speaking the closure of the Waterway barrier has a far more pronounced effect on RWLs than closing the Hartel barrier, because the cross-section of the New Waterway is larger.

The characteristics of the tidal region can explain the fact that the Hartel barrier will have to be closed more frequently than the barrier in the New Waterway. Here DWLs are determined by the storm surges allowed through by the Waterway barrier. An extra connection with the sea will result in an increase of RWLs. Closing both barriers as often as was intended for the Waterway barrier in the present geometrical situation will lead to an increase in DWLs. Because the frequency of closure of the Waterway barrier is limited, reduction of the DWLs to the required safety standard can be accomplished by closing the Hartel barrier more frequently.

The transition region is located at sufficient distance from the barriers. It covers the river branches on the eastern part of the northern side and the whole southern branch (figure 2). The upstream boundary can be drawn where the barriers have virtually no effect on DWLs. From the characteristics of the transition region some conclusions can be drawn concerning effects on DWLs. Since accumulation of river discharge is dominant in this area RWLs have to be minimized when the barriers are to be closed during periods of high discharge. Therefore, at high discharges, it is important to:
-   drain water through the barriers (during closure) whenever possible,
-   drain water through the Haringvliet sluices whenever possible,
-   minimize inflow into the area during closure of the barriers.

It can also be concluded that:
-   phenomena such as changes in storm surge or river regime (due to climatic changes), resulting in an increase in water levels at sea and discharges on the rivers cannot be fully reduced by a change in the frequency of closure of the barrier(s),
-   reduction of the now large storage capacity in the area (e.g. Haringvliet estuary) will directly lead to an increase in DWL in this region.

3   Safety system; calculations

First the system and the safety calculations will be described for a one-barrier system. This will enable a good understanding of the calculation procedures followed. Then the procedure will be extended to the two-barrier system.

3.1   The one-barrier system in general

A schematic representation of the one-barrier system is given in figure 3. The system consists of two major parts: the barrier (S_w) that is used to prevent the sea from coming up the river, and the dikes to protect the land against the water. Since it is not necessary to close the barrier for all storm surges a criterion for closure (CL) is defined. For this criterion a water level will be used. If the Predicted Maximum Sea Level (PMSL) exceeds the CL the barrier has to be closed. A very low CL would cause the largest reduction of the RWLs but also many barrier closures. Since the frequency of closure of the Waterway barrier is restricted the CL has to be chosen carefully.

The safety against flooding depends on the height of the dike with respect to the water levels in front of it. Even with the barrier in operation, dikes will still be necessary. In the safety system that describes the protection of the land against flooding the barrier is only one but vital part. In dike design the water levels are represented by the DWL. Dikes are constructed in such a way that there is a fixed safety margin as long as the DWL is not exceeded. The barrier reduces the influence of the sea. The water levels behind the barrier are therefore to some extent man-made. This also implies that the reliability of the barrier, its operation and human errors are of influence. A probabilistic approach that integrates functional and structural reliability is therefore indispensable if an overall assessment of the safety system and the related DWLs is to be achieved. This is described in the next sections.

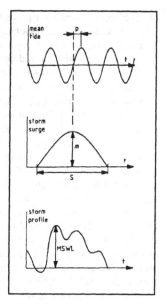

Figure 4   *Storm parameters and profile*

3.2 *Probabilistic calculations for the one-barrier system*

In the Netherlands it is common practice to calculate DWLs using extrapolated hydraulic boundary conditions and a numerical model to simulate the water levels. The alternative, extrapolating measured water levels, cannot be applied when major changes (like adding a barrier) are made to the hydraulic system.

To simulate the effect of the barrier on the RWLs a one-dimensional mathematical model is used. Boundary conditions are water levels at sea and river discharge. Sea water levels result from a superposition of the astronomical tide and a storm surge. Three parameters are used for characterization of the storm surge (figure 4): the maximum storm surge height (m), the storm surge duration (s) and a phase difference (p). A mean tidal curve is used for the astronomical tide. One storm profile (a combination of m, s and p) has a Maximum Sea Water Level (MSWL). The river discharge (q) is treated as a constant during the storm since it varies on a much longer time scale than storm surges. One combination of boundary condition parameters leads to a Maximum River Water Level (MRWL) at every location along the river. The hydraulic model is used as a transfer function from boundary conditions to MRWLs at various locations.

Using the transfer function to solve the integration boundaries, the probability of exceedance of the MRWL can be calculated from the following integration:

$$P(MRWL > mrwl) = \iiint\limits_{MRWL > mrwl} f_{m,s,p,q} \, dm \, ds \, dp \, dq \tag{1}$$

in which f denotes the probability density and d the differential. The frequency distributions of all boundary condition parameters are known by extrapolating data from long periods of registration. Assuming statistical independency, which is realistic, the probability density function of parameter combinations ($f_{m,s,p,q}$) is also known. The integration procedure is illustrated schematically in figure 5 for the simple case of only two boundary condition parameters, the discharge (q_{river}) and the MSWL. The figure shows the probability density functions of the boundary conditions (MSWL, q_{river}), the probability of the combination (p_h*p_q), and lines of equal MRWL at a certain

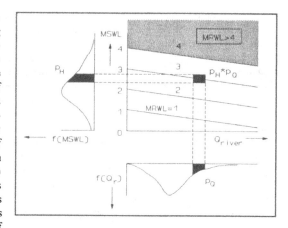

Figure 5 *Schematic integration procedure*

location, which is a result of the transfer function. The hatched area indicates all combinations that lead to MRWLs above the level MRWL=4. Integrating the probability density function over this area results in the frequency of exceedance for this level.

3.3 Integrating risk analysis

In section 3.1 it was explained that the effect of the barrier is influenced strongly by the functional and structural reliability of the barrier. To describe these influences more quantitatively a risk analysis has been carried out. Figure 6 shows the event tree with all possible branches that may lead to a MRWL exceeding the DWL. The same principle applies to the exceedance of any other water level or any other result. The upper part of the figure shows the cause of events if the system functions properly. There are four paths where things can go wrong:

1 PMSL < CL, so the barrier is not closed, but MRWL > DWL
2 PMSL ≥ CL, but the barrier is not closed, and MRWL > DWL
3 PMSL ≥ CL, the barrier is closed but is not strong enough to withstand the loads from wind and water; it collapses, and MRWL > DWL
4 the system functions well but still MRWL > DWL

The first two paths reflect the functional reliability, the third path reflects the structural reliability and the last path reflects fate. The four probabilities related to the paths (indicated as P1 to P4) should be calculated for every boundary condition combination. However, in the calculation procedure a more general and efficient approach is followed. This is discussed now.

For path 1, 2 and 3 the water levels are calculated with an open barrier. In principle the resulting water levels of path 3 will be lower than in case of an open barrier. However, to simplify

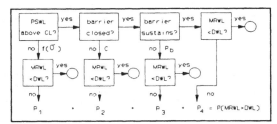

Figure 6 *Basic event tree*

168

matters it was assumed that in case of collapse the hydraulic system would react as if no barrier was present. This means that for each hydraulic boundary condition two hydraulic states are relevant: the barrier is open or it is closed. Reliability can then simply be integrated by performing the integration from formula (1) for both open and closed barrier situations, taking into account the probabilities of the barrier being open or closed. This is done by weighing the probability function f with the probabilities of each state (P_{open} and P_{closed}). In formula for some parameter X (e.g. water levels):

$$P(X>x) = \int\int\int\int f_{m,s,p,q} * P_{open} dmdsdpdq \; + \int\int\int\int f_{m,s,p,q} * P_{closed} dmdsdpdq \qquad (2)$$
$$X_{open}>x \qquad\qquad\qquad\qquad\qquad X_{closed}>x$$

The (conditional) probability of the barrier being open depends on the PMSL, the failure of not closing and the failure due to collapse. The relations are described in section 3.4.

In the calculation procedure the hydraulic model is run for all boundary condition parameter combinations, once with the barrier open and once with the barrier closed. All parameters of interest (water levels, hydraulic head etc.) are stored in databases. These databases are used to perform the integration (2). With this procedure it is very easy to perform a sensitivity analysis on parameters related to the functional and structural reliability since it only involves the weighing factors (P_{open} and P_{closed}) and the databases remain unchanged.

3.4 Determining the probability of the state of a barrier

The three effects leading to an open barrier situation will now be described and translated to a probability.

Open barrier due to inaccuracy of the PMSL

Predicted water levels are often inaccurate. The difference between the PMSL and the MSWL has a stochastic nature. In case of the PMSLs at Hoek van Holland this difference can be modelled with a Gaussian distribution (mean μ, standard deviation σ). Then PMSL only depends on MSWL, μ and σ. If the PMSL is below the CL the barrier will not be closed. Now for every MSWL (i.e. combination of boundary conditions) the contribution to P_{open} can be calculated from:

$$P_{open|PMSL} = Pr\{PMSL(MSWL)<CL\} \qquad (3)$$

This probability is easily derived from the normal distrubution.

Open barrier due to not closing

If the PMSL exceeds the CL then the barrier should be closed. However, the barrier may not be closed because of human or technical errors. The probability of these errors is schematized with a constant value C. The contribution to P_{open} then becomes:

$$P_{open|C} = Pr\{PMSL(MSWL) \geq CL\} * C \qquad (4)$$

Open barrier due to collapse

If the barrier is really closed and the loads on the barrier become too high it collapses. The barrier's strength (r) is a stochastic variable. Normally a lognormal distribution is assumed. Comparing each hydraulic load (l) on the barrier with the strength r results in a probability of collapse. The contribution to P_{open} then becomes:

$$P_{open|Pb} = Pr\{PMSL(MSWL) \geq CL\}*(1-C)*P(l > r) \tag{5}$$

The total probability of collapse (P_b) results from intgrating $P(l > r)$ over all hydraulic boundary conditions. Since P_b is a design parameter which in this project is set to a very low target value (10^{-5} to 10^{-6}/year) it is easily seen that the influence of collapse on DWLs with a frequency of exceedance of 10^{-4}/year or more will be negligible.

For each boundary condition P_{open} is determined as the sum of the contributions described above. The probability of the barrier being closed is then simply $P_{closed} = 1 - P_{open}$.

3.5 Extending the procedure to a two-barrier system

In the two-barrier system the area is protected by two barriers: one barrier in the New Waterway and one in the Hartelchannel. In principle the procedure to calculate the hydraulic effect of two barriers is the same as followed for the one-barrier system. However, a few differences have to be mentioned. First the event tree illustrated in figure 6 now applies to both barriers. It should also be realized that functional reliabilities may be correlated. For example: the PMSL for both barriers, used to decide on a closure, will originate from the same source. Secondly, instead of two, there are now four possible states of the system. Each barrier may be open or closed and four combinations are possible.

As with the one-barrier system, the hydraulic effects are calculated with the hydraulic model, which is now extended with the Hartel barrier. The results of all four states are stored in databases. A typical example of these results is shown in figure 7. Here the MRWL in Rotterdam is shown for all four states as a function of only two boundary condition parameters (MSWL and river discharge). The figure illustrates the different results obtained from the four states of the system for the same boundary condition (e.g. H_1, Q_2). If the probabilities of the four states are marked P_a, P_b, P_c, P_d respectively then the integration procedure is described by:

$$P(X > x) = \sum_{i=a}^{d} \underset{X_i > x}{\int \int \int \int} f_{m,s,p,q} * P_i \, dmdsdpdq \tag{6}$$

The results of this approach are discussed in the next section.

4 Effects of introducing a second storm surge barrier on water levels

4.1 Introduction

From the characteristics of the hydraulic system, some basic principles of the effect on water levels have become clear without having made one computation (see section 2). This section will elaborate on this, using some preliminary results of the safety analysis. These results are emanating from a sensivity analysis carried out on the base of the calculation procedure as described in section 3, but with two instead of four hydraulic boundary condition parameters (MSWL and q). In the characterization of the storm surge two parameters (storm duration and phase difference) were set to representative values (29 and 4.5 hours respectively). The results are used in an indicative sensivity analysis which will be extended carrying out a detailed study. This is in preparation now.

The analysis incorporates the characteristics of both regions and refers to the situation without the extra harbour entrance, with only a barrier in the New Waterway.

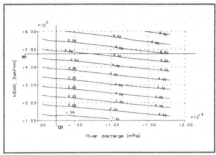

S_W and S_H open.

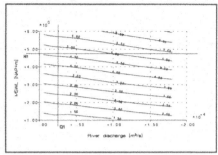

S_W open and S_H closed.

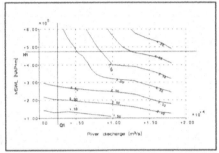

S_W closed and S_H open.

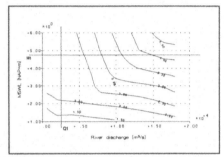

S_W and S_H closed.

Figure 7 *Iso-level lines for Rotterdam*

When figures are shown this concerns comparison of the situation with both Beerdam and Rozenburg lock removed (BD+RL OPEN) to the reference situation (REFERENCE).

4.2 Maximum reduction on design water levels by closing the barrier(s)

A first insight into the sensitivity analysis can be gained by depicting the situation when both barriers (or one in case of the reference) are closed 100% effectively at each high water period. In this way uncertainties in closure are excluded. The results show - with respect to the situation without a barrier - the maximum reduction on DWLs. In figure 8 the iso-level line of the DWL is represented for a characteristic location in each of the two regions. Rotterdam is situated in the tidal region and Hellevoetsluis is located along the southern branch in the transition region (figure 2). For each location the iso-level line is shown for the reference situation and for the situation when two barriers are present and closed each high water period. In section 3.2 is explained that the form and size of the area above the iso-level line is representative of the frequency of exceedance of the water level.

With the aid of equation (1) - of course reduced to two variables - the frequency of exceedance line of water levels can be calculated. The result is illustrated in figure 8 below the iso-level lines, for each location for the two concerned situations. The point corresponding with the iso-level line in the figure above is explicitly indicated. In both figures it can be seen that the maximum reduction on DWLs will be less by creating a second connection with the sea. For both locations the area above the iso-level line is larger and thus, the frequency of exceedance of this water level is larger.

171

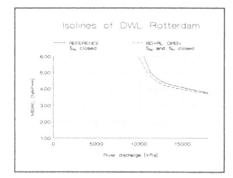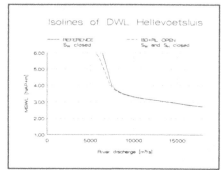

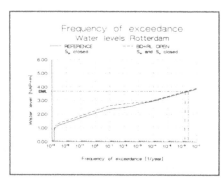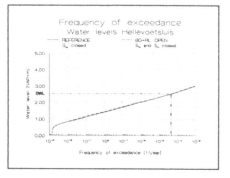

Figure 8   *Isolines and frequencies of exceedance of two locations for one (REFERENCE) and two barriers (BD + RL OPEN) closed.*

The difference in the two locations can be found in the exact magnitude of the frequency of exceedance of the DWL, again: in a situation with (a) closed barrier(s). In the case of Rotterdam this frequency is very small: by closing the barrier(s) the water level at this location is reduced substantially to a frequency that lies far below the required safety. For Rotterdam the frequency of exceedance of the safety standard amounts to $1 \cdot 10^{-4}$/year. As was stated before: the conditions when both barriers are closed do not contribute to situations that determine the height of the DWL.

This is not the case for Hellevoetsluis. The frequency of exceedance of the safety standard amounts to $2.5 \cdot 10^{-4}$/year. As can be read from figure 8 this safety standard is almost achieved under conditions with (a) closed barrier(s). Indeed, as was stated before, DWLs for locations in the transition region are determined by accumulation of river discharge, with (a) closed barrier(s). These characteristics are important to realize: with a second connection to the sea the increase in frequency of exceedance of the DWL with closed barriers has to be minimized. Every significant increase does not comply with safety requirements and cannot be compensated for by closing the barrier(s) more often.

4.3 Requirements for criteria for closure

For every location in the delta area iso-level lines can be made for the four states of the system (see figure 7). Because DWL is fixed for each location it is possible to derive some requirements for closure of the barriers. A reasonable assumption is made that maximum water levels anywhere in the delta area are equal or higher when both barriers are open, compared with the case when only the Hartel barrier is closed. Secondly, the requirement is given that the frequency of closure of both barriers has to be minimized. This implies that the Hartel barrier has to be closed more often (section 2). Having represented the iso-level lines of DWL for each location in the area in a situation when the barriers are still open, the envelope of the lowest lines (i.e. of conditions most occurring) will indicate the conditions when the Hartel barrier has to be closed (figure 9). A storm surge barrier cannot be closed 100 % effectively. By quantifying the uncertainties in the closing procedure however one can derive a criterion for closure (the CL). For this the CL has to be somewhat lower than the DWL envelope. A similar reasoning concerning the hydraulic conditions that require the Waterway barrier to be closed is legitimate. In this derivation however, one has to focus on conditions that only the Hartel barrier is closed. This relationship is also represented in figure 9.

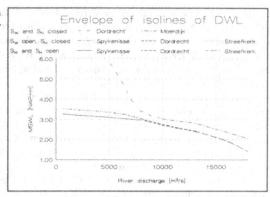

Figure 9   *Envelope of conditions requiring the closure of Hartel barrier and Waterway barrier.*

Locations that are part of the relationships are dominant in determining safety requirements. Figure 9 shows that in case of high discharges these locations are situated in the transition region (Dordrecht and Streefkerk). For lower discharges the tidal region is dominant (Spijkenisse). Notice that in case of high discharges both relationships do not differ much.

The envelope of conditions that lead to DWL somewhere in the delta area when both barriers are closed is illustrated by the upper line in figure 9. Above this line DWL exceedance cannot be avoided by closing barriers more often.

4.4 *Effects of changing the strategy of operating the barriers*

By studying the effects of closing one or two barriers on RWLs many of the sensitivities in the system have become clear. This can also be used in indicating points of significance in the formulation of an operational control of the barrier. The most important operational aspect regarding safety is the moment at which the barrier has to be actually closed. To close the barrier the PMSL must be higher than the CL <u>and</u> a Closing Condition (CC) must be <u>actually</u> reached. This section will deal with the choice of a CC.

In formulating a strategy of operational use, one can optimize for many aspects. Within the scope of this paper, only aspects that affect safety against flooding will be dealt with. Optimization of e.g. technical aspects of the barrier are clearly left aside.

In the transition region the situation with closed barriers is dominant for DWLs (because of the accumulation of river discharge). From figure 8 can be derived that in the case of Hellevoetsluis, safety endangering conditions are most likely to appear at periods of high discharge. At discharges of approximately 7000 m^3/s or less DWL will only occur (with closed barriers) in combination with very severe storm surges (MSWL of at least NAP + 4.5 m). The combined probability of occurrence of these conditions is so limited, that it does not contribute to the frequency of the safety standard. At high discharges however it is essential that only minimal inflow during closure of the barriers is allowed into the delta area. This can be accomplished by closing the barriers at the last turn of the tide (with minimal inflow) before the storm surge rises. In this way minimization of MRWLs and by that of DWLs is achieved. The iso-level lines in figure 8 are indeed based upon computations wherein both the Waterway barrier and - if applicable - the Hartel barrier are being closed at the turn of the tide (local zero inflow) at discharges equal or greater than 6000 m^3/s. For lower discharges a part of the storm surge is allowed into the delta area and the barriers are closed at a (local) fixed water level. Comparing both iso-level lines of Hellevoetsluis it can be seen that here the lines are deviating from each other. The probability added by the lower iso-level line is very limited (compare the change in frequency of exceedance) so it barely affects DWL. Applying the same strategy at extreme high discharges would certainly increase DWL.

For DWLs in the tidal region the CC is of much lesser significance. In this region it is more important that the top of a storm surge that will cause exceedance of DWL with (an) open barrier(s) is cut off to some extent. As safety requirements regarding the closing condition at discharges less than 6000 m^3/s are not as strict, it is possible to optimize for other aspects. For example a CC of local zero inflow may reduce bottom protection near the barrier and a CC at a high water level may reduce hydraulic head over the barrier.

5 Conclusions

- To meet the requirements of both safety against flooding and minimal closure of the barriers the use of an integral safety concept is indispensable. In this concept both hydraulic modelling and the functional and structural reliability of the two-barrier system are to be handled probabilistically and integrated to assess safety conditions.
- Characterization of the hydraulic system of the Rhine and Meuse delta area is very important in understanding the overall effectiveness of the barriers in protecting an area from flooding. In this way aspects that affect the safety system can be assessed.

6 Acknowledgements

It is obvious that the study presented in this paper involves an enormous number of computations. Most of this work was done by mr. Henk de Deugd and by mr. Ton Visser also from Rijkswaterstaat. The authors are duly grateful for this.

Abbreviations and symbols

BD	= Beerdam	P_b	= probability of collapse
C	= probability of not closing	PMSL	= predicted maximum sea level
CC	= closing condition	q	= river discharge
CL	= criterion for closure	RL	= Rozenburg lock
DWL	= design water level	s	= duration of the storm surge
m	= storm surge height	S_H	= Hartel barrier
MRWL	= maximum river water level	S_W	= New Waterway barrier
MSWL	= maximum sea water level	μ	= average of error in PMSL
NAP	= reference water level	σ	= standard deviation of error in
p	= phase difference		PMSL

15 Hydrodynamic modelling of the Mersey Estuary for a tidal power barrage

E. A. Wilson and J. H. Porter

SUMMARY

Details are presented of the development and application of numerical hydrodynamic models for a proposed Mersey Tidal Power Barrage which have been used to support parallel energy yield, shipping, accommodation works and environmental studies. The output from these models has proved invaluable in comparing alternative Barrage layouts and in refining the design of the shipping facilities. This modelling was successfully undertaken in-house using the DIVAST basic model mounted on personal computers.

1   Introduction

The use of modelling techniques to study the hydrodynamics of an estuary is not new and this is particularly true of the Mersey Estuary in North West England. In 1885 Osbourne Reynolds constructed in his laboratory at Manchester University a physical model of the River Mersey which was based on scientific principles (Reference 1). It was used to study the influence of eddies on the formation of water channels and movement of sediment using a mobile bed. Following completion of the Manchester Ship Canal in 1893 another physical model, constructed by Professor Gibson also of Manchester University, was used to examine proposed improvements

175

of the navigable approach to the Mersey and Port of Liverpool by extending the training walls in Liverpool Bay. Such walls were subsequently built resulting in an approach channel that has remained stable to this day. The use of the physical model to predict increased scour of sediment in the shipping channel was borne out.

More recently, since 1983 a series of studies has been carried out on a proposed barrage to extract tidal energy from the estuary of the River Mersey.

Stage III of these studies commenced late in 1990 and has been extended for completion by the end of this year. It is envisaged the next phase of studies, Stage IV, will carry the proposal to a final decision on approval and capital funding. This paper, therefore, reports upon hydraulic modelling which is substantially complete but has not yet been drawn to a final conclusion. These studies are being progressed by the Mersey Barrage Company ("MBC") which, in addition to subscriptions from its own shareholders, receives funding assistance from the Department of Trade and Industry (formerly Department of Energy).

The Mersey is a particularly suitable natural site for such a project having a large tidal range, 8.4 m on Mean Springs, and having a large surface area of 70 km^2 discharging through a Narrows of less than 2 km width.

At the outset of these studies the importance of sedimentation and hence hydrodynamic modelling was recognised since large parts of the Estuary have a sandy or silty bed. The resulting hydraulic modelling for sedimentation studies has been reported elsewhere (Reference 4). However, as the studies progressed, hydrodynamic modelling has also been required for energy yield predictions, shipping studies and environmental studies generally. This paper describes the development and results so far obtained from the latter hydrodynamic modelling.

2   The Mersey Barrage

The Mersey Barrage will harness the energy of the tides by retaining the tidal prism of an incoming flood tide in the basin behind a permeable controllable barrier. A potential head is created by delaying the release of water from the basin. This provides a head during the ebb tide to drive turbines within the Barrage. The net energy yield may be increased by utilising the turbines in reverse briefly after high water, pumping a volume of sea-water into the basin. This volume is later released at a greater head difference so creating more electricity than is consumed. During the flood period, flow into the basin is provided by channel

176

sluices and the idling turbines.   Channel sluices
were chosen in preference to hydraulically more
efficient submerged venturi sluices because of poor
ground conditions.
    The selected Barrage location is designated Line 3F
and shown in Figure 1 between Dingle and New Ferry.

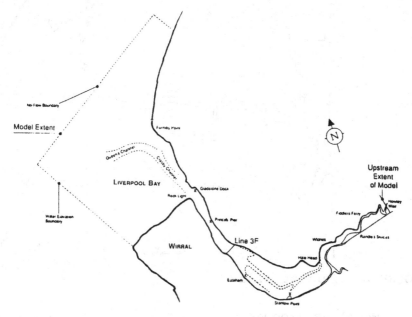

FIGURE 1   Barrage location

    Some leading parameters of the presently preferred
layout of the Barrage which has been developed
through three stages of feasibility studies are given
in Table 1.   A general plan of this layout is shown
in Figure 2 which slightly differs from that
previously proposed with the smaller lock at the
Liverpool shore (Reference 4).

TABLE 1   Mersey Barrage : Leading Parameters

Length between river banks	1,900 m
Turbines	28 No each 8 m diameter
Channel Sluices	46 No each 17 m wide
Installed Capacity	700 MW
Net Energy Yield	1.4 TWh/annum
Peak ebb (generating) flow	14,000 m^3/sec (approx)
Peak flood (sluicing) flow	28,000 m^3/sec (approx)
Locks:-	1 No 270 m x 36 m
	1 No 215 m x 23 m

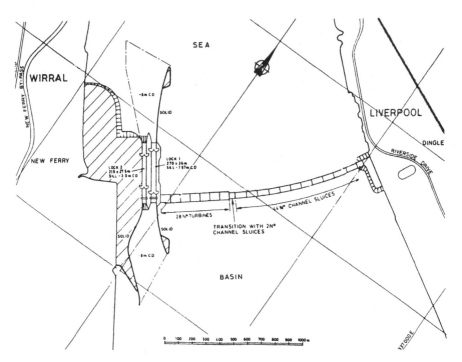

FIGURE 2   Barrage plan

3   Model development

3.1   Selection

MBC recognised the need to be a technically competent
and informed Client because of the many interactions
between the disciplines involved in assessing the
potential impacts. The focus of these interactions is
the change in tidal propagation which in turn effects
energy yield, shipping and the general Estuary
environment.

Therefore, it was decided to establish an in-house
hydraulic modelling capability as an essential
management tool allowing rapid evaluation of
alternatives and providing consistent hydraulic data
as required by the various specialist studies.

To meet the technical needs of flexibility and
accuracy, a two dimensional ("2-D") depth averaged
numerical model was required. To meet the operational
needs of simple operation, ease of development and
study team centralisation, a personal computer based
model was preferred.   DIVAST, a 2-D alternating
direction implicit finite difference model developed
at Bradford University (see for example References 2
and 3), was found to be suitable.

178

For hydraulic modelling to support sedimentation studies only, it was recognised that the two require specialist expertise in their coordination and application. HR Wallingford was retained for this work.

A DIVAST Mersey Barrage model was initially created using a 150 m square mesh size by Rendel Parkman on behalf of MBC in 1989. However, during Stage III the increasing demands of accuracy and versatility made necessary extensive further development which was undertaken directly by MBC.

3.2   Bathymetry

In Stage II bathymetric data was taken from a 1977 survey by the Mersey Docks and Harbour Company. A partial bathymetric survey had been completed in 1984 as part of a study by Cheshire County Council and a satellite image of the Estuary became available in 1989. These showed clear differences from the 1977 data.

Therefore, a full bathymetry survey was commissioned in 1990 by MBC and at the same time field water level and velocity data was collected. The 1990 bathymetry now modelled shows that although extensive movement of the low water channels has taken place the total Estuary volume has within a few percent remained steady at approximately 700×10^6 m^3 over the last 40 years.

3.3   Boundary conditions

The model uses a rectilinear grid, which has been orientated to align with the centre line of the Barrage. To eliminate any significant effect of the Barrage operations upon the assumed boundary conditions the model boundaries in Liverpool Bay were extended seawards. In addition, the mesh size has been reduced to 75 m so that the model now contains a total of 83,000 wetted cells. Water levels are defined along the western and eastern boundaries in Liverpool Bay with no flow across the northern boundary (Figure 1). This assumption is regarded as being justified as flow data from Liverpool Bay has been shown to be mainly tangential to this northern line. (Reference 5.)

A flow rate at the tidal limit, Howley Weir, is nominally set at 50 m^3/s which is the modal flow of the Mersey, outfalls and the primary tributaries in the Upper Estuary combined.

The Barrage itself is simulated by an internal flow boundary at the Barrage location. The Barrage flows are calculated by reference to the head difference across the Barrage in accordance with the predetermined operating regime and using turbine and pump performance characteristics obtained from scale model testing. In addition, the hydraulic performance of the high level channel sluices was determined by flume model tests.

The choice of internal boundary definition has meant that the flows through the Barrage are calculated explicitly within an implicit model. This can lead to instability at low head differences across the Barrage and as a consequence it was found necessary to damp the response of flow rate to change in head difference.

3.4 Local model

In order to provide more accurate flow patterns in the vicinity of the Barrage locks a local model with 37½ m mesh has also been developed. This model extends some 15 km from the mouth of the Narrows to a line between Eastham and Garston, approximately 5 km upstream of the Barrage. The upstream and downstream limits are level boundaries driven by output from the main model with the downstream boundary additionally having the flow orientation, though not magnitude, specified.

3.5 Validation

Validation was carried out against water levels and velocities measured during the 1990 survey at sites from Liverpool Bay to Runcorn Bridge. The model provides an acceptable representation of the water levels and current velocities across a range of tides. Figure 3 and Figure 4 show a comparison of predicted and measured water levels and velocities.

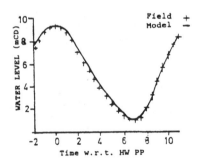

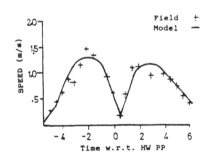

FIGURE 3 Water levels at Liverpool

FIGURE 4 Velocities in approach channel

4   Model application

For its application the model boundaries were calibrated to produce five representative tides at Prince's Pier without a Barrage present; High Spring, Mean Spring, Mean Tide, Mean Neap and Low Neap.

These tides were then used to simulate a large number of alternative Barrage layouts and operating strategies. Construction stages as well as the completed structure have been examined. By this means alternatives were compared in terms of their hydraulic and consequent other impacts and the feedback obtained has allowed a progressive refinement of the preferred layout and operating regime. Some 200 model runs have so far been completed. To a certain extent these have focussed upon using the Mean Spring tide as a compromise between identifying extreme impacts and being representative. For operation and construction of the finally preferred Barrage arrangement, however, all five tides have been examined.

4.1   Energy Yield

To predict the annual energy yield obtainable from a given Barrage layout, MBC has separately developed a zero dimensional ("0-D") model which assumes the basin water surface level is horizontal and simulates the downstream approach channel by simple open channel flow (Reference 4). The Barrage flow algorithms it contains are identical to those in the 2-D DIVAST model, but the 0-D model is able to optimise and select its own operating regime. However, the 0-D model does not correctly predict the response of water levels to changes in Barrage flow rate and for this reason the procedure adopted to accurately predict annual energy yield is to take the optimised operating regime from the 0-D model as input to the 2-D model. Figure 5 shows the Barrage water levels and discharge rates compared to the open river values for a Mean Spring tide. The power flows are also shown. The rise in downstream water level at low water due to the continuing generating ebb discharge is apparent and this has a significant effect upon energy yield. The results for all tides are summarised in Table 2. Using the predicted histogram of tidal range frequency for a typical year (1992) and allowing for various other factors the typical annual energy yield is calculated to be 1.4 TWh with a reduction in annual tidal prism upstream of the Barrage site from 205 m^3 x 10^9 at present to 170 m^3 x 10^9 with an operating Barrage.

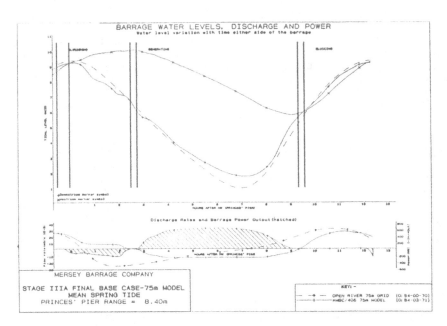

FIGURE 5   Open River and Barrage results compared

TABLE 2   Energy yield predictions per tide

TIDE	PRINCE'S PIER TIDAL RANGE (m)	IMPORTED ENERGY (MWh)	NET OUTPUT (MWh)	TIDAL PRISM ($m^3 \times 10^6$)
Low Neap	3.2	327	368	150
Mean Neap	4.5	289	860	181
Mean	6.5	229	1,675	247
Mean Spring	8.4	258	2,641	279
High Spring	10.0	71	3,404	298

4.2   Shipping

Shipping is of prime importance to Merseyside and upstream of the Barrage location are the ports of Eastham and Garston. Fundamental to assessing the impact Barrage locks may have upon shipping traffic is an understanding of the hydraulic conditions which limit lock capacity.

The local model was used to provide flow data input to a real time shiphandling simulator. In addition, comprehensive tabulated data of predicted levels and cross-currents at the entrances and approaches to locks and jetties was provided.

A typical detailed velocity field at the Barrage
lock approach during sluicing for an early Barrage
layout is shown in Figure 6. Strong and unacceptable
cross-currents at the lock entrances are apparent.
In Figure 7 the same area is shown for the final
Barrage layout. Marked improvements are clear.
These arise firstly from locating turbines rather
than sluices adjacent to the lock so reducing the
local flood velocities and secondly from the
introduction of lead-in jetties to provide areas of
slack water.

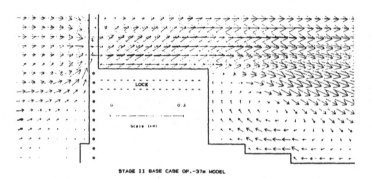

FIGURE 6   Line 3A velocity field at lock

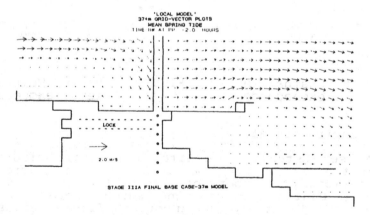

FIGURE 7   Line 3F velocity field at lock

4.3   Accommodation works

The accommodation works necessary because of the
impact of a Barrage upon tidal propagation
principally arise from flood defence and land
drainage requirements, although groundwater effects
must also be addressed. The principal input to these

studies is annualised level exceedence graphs created for selected stations along the Estuary. Examples for stations at Eastham and Howley Weir are shown in Figure 8 and Figure 9 respectively. It may be calculated that immediately upstream of the Barrage the average water level is raised by approximately 2.5 m. This effect attenuates with increasing distance upstream, such that at Howley Weir the average water level is raised by approximately 0.5 m.

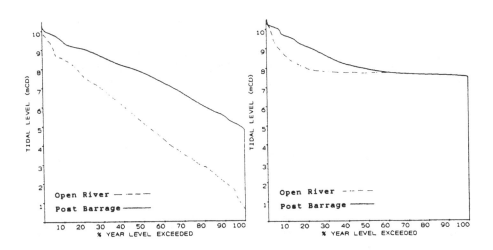

FIGURE 8   Eastham level   FIGURE 9   Howley Weir
           exceedence   level exceedence

4.4   Environmental

The Mersey Estuary is a Site of Special Scientific Interest and in particular is currently internationally important for Shelduck, Teal, Pintail, Dunlin and Redshank which winter on the Mersey. In Figure 10 the important feeding areas in the Middle to Upper Estuary are shown. To assess the impact upon these feeding grounds, exposed area plots for intervals throughout the tide were prepared comparing the with and without Barrage predictions. An example is shown in Figure 11. Comparison to the previous figure shows that much of the intertidal area lost due to a Barrage is of little importance to wintering waterfowl, although allowance must also be made from the reduced time periods for which the remaining intertidal mudflats may be exposed.

184

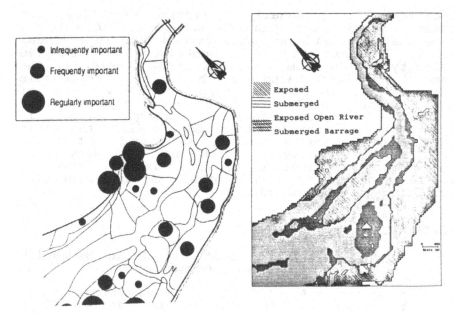

FIGURE 10   Bird feeding   FIGURE 11   Exposed areas
            areas from   from Stanlow to
            Stanlow to   Runcorn at
            Runcorn   HW +9 hours

   Another prime environmental concern is the impact a
Barrage may have upon water quality particularly
since the National Rivers Authority wish to see the
Mersey improve from its present status of Class 4
(badly polluted) to Class 1 (good) or Class 2 (fair)
by the year 2010.   Environmental Resources Limited
are presently setting up on behalf of MBC the DIVAST
model to simulate multiple water quality parameters.

5   Concluding remarks

The personal computer based hydraulic modelling
system adopted by MBC for its studies has performed
most satisfactorily and has endorsed the view that
such a capability must be retained for future
studies.   Indeed it is already envisaged that
developments of this system coupled with development
of the in-house 0-D optimising program may eventually
be used in managing operation of the completed tidal
power barrage.
   Nevertheless, there are significant dangers in over
reliance upon the wealth of detailed output provided
by such a model, simply because it is readily
available and cannot easily be cross-checked.

185

Therefore, it is equally considered essential that outside specialist expertise should continue to be retained to provide authoritative advice in critical areas. In this respect, comparisons of detailed velocity fields from other numerical models for the same layout and bathymetry have shown noticeable discrepancies in eddy formation in the vicinity of the Barrage whilst retaining excellent level and general flow rate agreement. Consequently, it is presently anticipated that a local undistorted scale physical model may be required to provide definitive detailed flow patterns which are necessary when considering shiphandling at lock entrances. However, it is expected that all other Estuary hydraulic modelling will be numerical.

REFERENCES

1   Allen, J, (1947), Scale Models In Hydraulic Engineering, 1st ed, Longmans, Green and Co Ltd, London.

2   Falconer, R A, (1984), A Mathematical Model Study Of The Flushing Characteristics Of A Shallow Tidal Bay, Proceedings of the Institution of Civil Engineers, Part 2, Vol 77, September 1980, pp 311 - 332.

3   Falconer, R A and Chen, Y, An Improved Representation Of Flooding And Drying And Wind Stress Effects In A 2-D Tidal Numerical Model.

4   Mersey Barrage Company, (1992), Tidal Power From The River Mersey: A Feasibility Study Stage III Report, Mersey Barrage Company, Liverpool.

5   Prandle, D and Ryder, D K, (1985), Measurements Of Surface Currents In Liverpool Bay By High-Frequency Radar, Nature, Vol 315, pp 128 - 131.

Part 3
SALINITY INTRUSION MODELLING

16 An estuarine salinity model

*B. Pearce, V. Vivek, H. McIlvaine, E. Simek
and P. Sucsy*

Introduction

An SO$_2$ scrubber is to be installed at the site of B. L. England Generating station in Great Egg Harbor Bay, New Jersey. As part of the design process for this new installation, it was necessary to establish long term salinity information for the site. This information will in turn allow for more precise design of a planned SO$_2$ scrubber at the site. This is important for several reasons, in particular, the production of CaSO$_4$ precipitate (Gypsum) suitable for use as wall-board. This raw material can be sold to a manufacturer, avoiding the cost and environmental impact of landfilling. The salinity of water to be used in the scrubber is important for the design, especially for this site, where the quantities of fresh water from wells are limited by state regulation.

Egg Harbor is a tidal estuary fed by fresh water inflows from the Tuckahoe and the Great Egg Harbor rivers. Although, the volume of tidal water input in the harbor in a tidal cycle is very large compared to the fresh water inflow from the rivers and it is not so important for velocity calculation, the salinities in the harbor are directly related to the fresh water inflow from the Tuckahoe and the egg harbor rivers. The effect of fresh water inflow can be seen in Figure 1. The average salinity recorded at the B.L. England Generating station has substantially decreased after a rainfall event, which is understood to have increased the fresh water inflow. Hence the flow duration curve for these two rivers was determined. The methodology is dealt with in the hydrology section of this paper. The flow field in the model was calculated for various fresh water inflow conditions. The resulting flow field was then used in the advection-diffusion equation to find the salinities.

The Hydrology

As there is no gaging station near the mouth of the rivers to determine the flow from the rivers into the bay, the boundaries of the watershed for the two rivers were drawn along the topographical ridges which appeared to be flow divides. The area of the watershed was determined by the dot grid method. A factor of flow per unit area was determined for the sub basins flowing into each USGS gaging station on the rivers. This factor was determined by taking the historical flow from the gaging station divided by the area of the gaging basin. The flow factor was multiplied by the total basin area to find the total basin discharge. Great Egg Harbor and Tuckahoe river basins were treated separately so as to better incorporate the variability of the whole watershed. Then with the calculated flows for one year a combined flood duration curve for the two rivers was determined, which is given in Figure 2.

The Flow Model

Overview

The currents in Great Egg Harbor were required to ultimately calculate the salinities in the harbor. Great Egg harbor is a shallow, well mixed, tidal estuary. Tide currents thus dominate the flow field in the harbor. The wind stress is also an important parameter in the equation of motion and it certainly changes the flow field significantly, which in turn would affect the salinities. The effect of a strong wind event is clearly reflected on the measured salinities at the B. L. Generating station in Figure 1. But since the aim of the project was to determine more or less long term results, wind stress was not included in the equation of motion.

A two dimensional, vertically-averaged, hydrodynamic model forced by the ocean tide was used to simulate the current in the harbor. A source of mass was included to simulate the influx of fresh water from the two rivers into the bay. The mass input does, however, allow for the appropriate volume of fresh water to enter the model, which is ultimately important for calculating the salinities.

The tide model used to simulate the currents in the Great Egg Harbor was '3DENS', (Sucsy, et. al. (1991) which solves a three dimensional equation of motion with different forcing mechanisms. The numerical technique is largely based on the method of Pearce et. al. (1979) and Pearce and Cooper (1981). '3DENS' was run in the two dimensional mode and thus solved the vertically averaged nonlinear, shallow water equations of motion. These equations are the x and y directed equations of conservation of momentum

$$\frac{\partial U}{\partial t} + U\frac{\partial U}{\partial x} + V\frac{\partial U}{\partial y} - fV = -g\frac{\partial h}{\partial x} - \frac{kU\sqrt{U^2+V^2}}{D} + A_H(\frac{\partial^2 U}{\partial x^2} + \frac{\partial^2 U}{\partial y^2}) \qquad (1)$$

$$\frac{\partial V}{\partial t} + U\frac{\partial V}{\partial x} + V\frac{\partial V}{\partial y} + fU = -g\frac{\partial h}{\partial y} - \frac{kV\sqrt{U^2+V^2}}{D} + A_H(\frac{\partial^2 V}{\partial x^2} + \frac{\partial^2 V}{\partial y^2}) \qquad (2)$$

and an equation of conservation of mass,

$$\frac{\partial h}{\partial t} + \frac{\partial(DU)}{\partial x} + \frac{\partial(DV)}{\partial y} = 0 \qquad\qquad (3)$$

in which U = velocity in x direction; V = velocity in y direction; f = Coriolis parameter; k = coefficient of bottom friction; D = total depth from bottom to free surface; A_H = horizontal eddy viscosity parameter; h = deviation of free surface from mean sea level.

The model was forced by specifying the elevation at the inlet of Great Egg Harbor. The model was forced at the M_2 tidal frequency only. The M_2 tide, of period 12.42 hours, is the dominant tidal component and essentially represents the mean tidal regime. An amplitude of 0.7 meters was used to simulate the large tide range on 11/3/90. For simulation of salinities during high river flow, low river flow, and average river flow conditions, the mean tidal amplitude of 0.58 meters (3.8 foot range) was used.

The geometry and bathymetry of Great Egg Harbor was resolved using square finite-difference cells of dimension 200 x 200 meters. The resulting model grid contained 60 x 38 cells. Figure 3 shows the resulting geometry obtained at this resolution. The direction of North is approximately diagonally down and towards cell (1,1) in the figure. The model boundary where the tide was specified is labeled as 'Inlet' in the figure. The points A, B, C, and D correspond to points where time series of salinities were constructed and will be dealt with in the diffusion model section

This grid was overlaid onto the NOAA Nautical Chart 12316 of scale 1:40000. Mean low water depths were then entered into the model at the center of each model grid. Great Egg Harbor is quite shallow, with the exception of a few prominent narrow channels.

Tuning of Flow Model

As mentioned above, the tide model was forced at the model boundary with a 0.58 meter amplitude wave of period 12.42 hours, representing the mean tidal range of 3.8 feet. A total river influx of 420 cfs was used to simulate the mean river conditions. The total flow was divided in a 3:1 ratio between Great Egg Harbor River and Tuckahoe River. The model was tuned to match the observed mean tide at three locations. The locations used for comparison were cell (31, 34) southeast of Cowpens Island near Ocean City, cell (57, 32) at the extreme end of Peck Bay, and cell (40, 11) west of Drag Island in the inner harbor.

Tuning was done in the usual trial-and-error fashion of varying the bottom friction coefficient and horizontal eddy viscosity coefficient over their normal range of values and comparing the resulting model amplitudes and phases with the observed amplitudes and phases. For Egg Harbor, the model was most sensitive to changes in the value for horizontal eddy viscosity. This fact is fortunate as it allowed to estimate a value for this parameter to within a fairly narrow range, and it could be anticipated before hand that the value of the diffusivity coefficient required for the dispersion model will be of the same order as the value for the eddy viscosity coefficient. The final values used for the flow model were 0.002 for the bottom friction coefficient, and 20.0 m^2s^{-1} for the horizontal eddy viscosity coefficient. Using these values, mean tide ranges within 0.1 foot was obtained at all three locations used for comparison. For phase, the observed time lag in high water and in low water between each station and a fourth station at grid (22,32) in the Rainbow Channel was compared with the corresponding lag between highs and lows from the model. The results are shown in Table 1.

Location	Grid	Lag of High water		Lag of Low water	
		Observed (Minutes)	Model (Minutes)	Observed (Minutes)	Model (Minutes)
Cowpens Island	(31,34)	+12	+12	+24	+16
Peck Bay	(57,32)	+39	+40	+67	+64
Inner Great Egg Harbor	(40,11)	+32	+36	+62	+60

TABLE 1 *Comparison of the lag of high and low water between three stations and a in Rainbow channel at grid (22,32) as obtained from observation and predicted by the flow model.*

The most interesting feature of the tide in Great Egg Harbor is the difference in lag time for inner harbor as compared to Rainbow channel. The high tide in the inner harbor occurs approximately 40 minutes after high tide in Rainbow Channel, but low tide occurs about an hour later. This shallow water effect is correctly predicted by the flow model.

A series of velocity vector plots were obtained from the model. Figures 4 and 5 show these plots for high tide and low tide respectively. Maximum tidal velocities approach 1 m/s. Depth averaged velocities in x and y direction were also obtained as data files for use in the advection-diffusion equation.

The Advection-Diffusion Model

Overview

The velocities and elevations from the flow model were used in 'DIFFUSE', a two-dimensional advection-diffusion model to predict salinities in Great Egg Harbor for different river flows. 'DIFFUSE' is a standard finite difference dispersion model which solves the advection-diffusion equation of the form

$$\frac{\partial(DC)}{\partial t} + \frac{\partial(DUC)}{\partial x} + \frac{\partial(DVC)}{\partial y} = K_D \left[\frac{\partial}{\partial x} (D \frac{\partial C}{\partial x}) + \frac{\partial}{\partial y} (D \frac{\partial C}{\partial y}) \right] \qquad (4)$$

In which, C = the mass of salt per unit volume of water; K_D = vertically averaged diffusivity constant and D = total depth.

For the boundary conditions to the dispersion model, it is only necessary to specify the salinity of water entering or leaving the model through the model boundaries. Water enters the model at river boundaries (continually) and at Great Egg Harbor Inlet on the incoming tide. Water leaves the model only at Great Egg Harbor Inlet on the out going tide. At the river boundaries the salinity is simply set to zero, as the mass flux at river boundaries are all fresh water. At Great Egg Harbor inlet the salinity of the water leaving the inlet is calculated from the solution of equation (4). The salinity of water entering the model at the inlet cannot be determined from the model equations and must be specified. An average value for this parameter of C_{in} was chosen. The value of C_{in} is given by:

$$C_{in} = e \times C_0 + (1-e) \times C_{out} \tag{5}$$

where C_0 = the salinity of ocean water; C_{out} = average salinity of the water leaving the model at the inlet on an outgoing tide; Q_f = the freshwater flowrate into the model; e = fraction of ebb volume not returning. Equating the total mass of salt entering and leaving the model boundary provides the following relation:

$$e \times TP \times [\frac{C_0 - C_{out}}{C_0}] = Q_f \times T \tag{6}$$

where TP = volume of tidal prism; T = tidal period. The tidal prism was estimated from the flow model to be 3.9×10^7 m^3. Reported values for e range from 0.24 - 0.6. Using this range of values for e, a value of 10 m^3/s. for Q_f and 44712 s for T gives a range of 0.019 - 0.047 for the bracketed ratio in (6). This means that the salinity difference between the receiving waters and outgoing waters at the harbor inlet must be very small, less than 1.5 parts per thousand. By assuming a value of 30 ppt for C_0, the salinity of the receiving waters, the above formula predicts values in the range 28.6 - 29.4 ppt for C_{out}. Arbitrarily choosing 29.0 ppt for C_{out} allows us to estimate e for Great Egg Harbor inlet as 0.34. C_{in}, then, is 29.5 ppt rounded to the nearest half. This value for C_{in} introduced realistic results in the model calculations. Note, however, that because of the large volume of the tidal prism compared to the volume of river flow per tidal cycle the value of C_{in} will not be more than one or two ppt different in salinity from the value of the receiving waters regardless of the choices made for C_{out}.

Tuning of Advection-Diffusion model

The salinity data for 11/3/90 was chosen to tune the dispersion model because these data were collected during a time of relatively static hydrodynamic conditions. The winds had been modest for over four days, the tide was in a period of fairly constant amplitude, and there had been no rainfall for ten days allowing for steady river flowrates. The flow model was run with a tide of amplitude 0.7 meters (simulating the high tides for that day) and river flowrates of 171 cfs (4.84 m^{3}s$^{-1)}$ for Great Egg Harbor River and 53.7 cfs (1.52 m^{3}s$^{-1)}$ for Tuckahoe River. The total river flowrate for 11/3/90, then, was 224.6 cfs. The only tuning parameter for the dispersion model was the value of diffusivity coefficient, K_D. The K_D value was so adjusted that the model salinities at the B.L. England Station were close to those observed on 11/03/90. Figure 6 shows the model salinities predicted by the model compared to the four observed values for 11/03/90 for K_D of 20m^2/s. The timing of the measured and calculated salinity peak and trough is off by few hours - not the values - just the peak. This is probably due to "sub-grid-scale" phenomena, that is, circulation and

193

or mixing that happens at a scale smaller than the grid can scale. The other reason for error in phase could be the assumption of isotropic diffusivity, which is expected to be more in the direction of the inlet than in the other direction. The other reason for this error may be assumption of an isotrophic, rather than non-isotrophic K_D.

Conclusions

The model results compare quite well with data at the B.L. England plant. Although data for salinity were not available elsewhere in the harbor for that day, Figure 7 shows predicted model salinities at three other locations. These locations are point A at the model boundary, point B inside Rainbow Channel, and point C in Peck Bay. As anticipated, the salinity at the model boundary varies only slightly. The salinity in Peck Bay is also extremely uniform, a phenomena supported by data. The salinity inside Rainbow Channel mirrors that at the B.L. England station, but is more saline, as expected. The average salinities of August 1971 were available. The flow data for August 1971 were not available but were evidently near the average flow at that time. So the model was run at the average flow of 420 cfs. (11.9 m^3/s.). Figures 8 and 9 show the comparison of salinities at low tide and high tide as obtained by the model with the salinities at ebbing tide and flood tide obtained from the summary of August 1971 data. The figure shows a reasonable comparison considering the fact that the true wind condition and the exact fresh water flow are not known. The aforesaid limitations imposed on the study can also cause minor differences.

REFERENCES

(1) Pearce, B. R., Cooper, C. K. and Doyle, E., (1979), Hurricane Generated Currents, *Proceedings of ASCE Speciality Conference on Civil Engineering in Oceans IV*, San Francisco, CA., U.S.A., pp. 398-415.

(2) Pearce, B.R. and Cooper, C. K., (1981), Numerical Calculation Model for Wind Induced Flow, *Journal of Hydraulics*, 107, HY3, pp. 285-302.

(3) Sucsy, P., Pearce, B. R. and Panchang, V., (1991), *3DENS - A Three - dimensional Tide and Wind Model for Coastal Application*, Technical Report, Dept. of Civil. Engineering, Univ. of Maine, Orono, U.S.A.

(4) New Jersey Dept. of Environmental Protection, (1972), *Studies of The Great Egg Harbor River and Bay*, Misc. Report No. 8M, Bureau of Fisheries, Div. of Fish, Game and Shellfish, New Jersey, U.S.A.

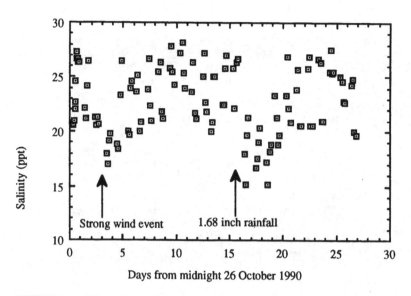

FIGURE 1   *Effect of rainfall and wind on the observed salinities at B. L. England Station*

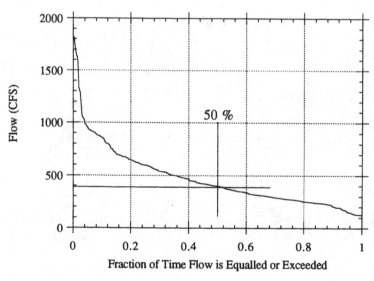

Figure 2 *Combined flow duration curve for Great Egg Harbor and Tuckahoe Rivers.*

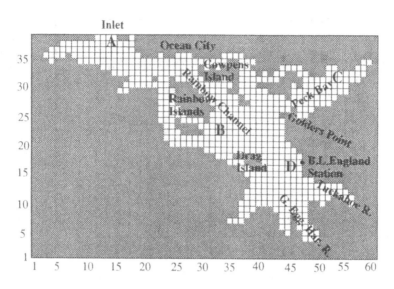

FIGURE 3. *Model grid of Great Egg Harbor, New Jersey*

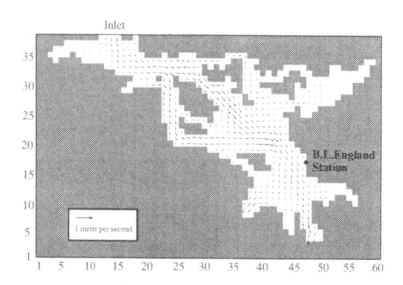

FIGURE 4 *Velocity vector in the Great Egg Harbor during flood tide at inlet*

196

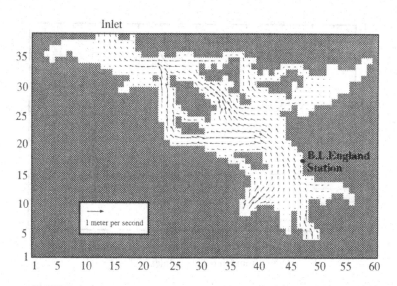

FIGURE 5 *Velocity vectors in the Great Egg Harbor during low tide at inlet*

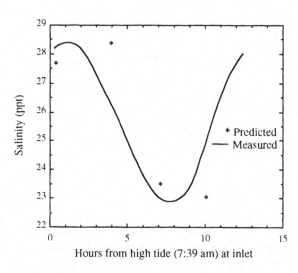

FIGURE 6 *Predicted versus measured salinities at plant for 11/3/90*

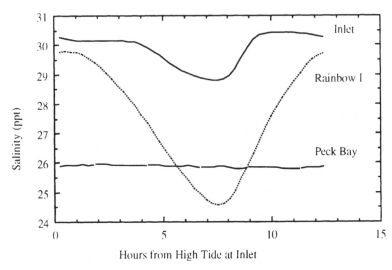

FIGURE 7   *Salinities Predicted by Model for Flow of 11/3/90 at Locations A, B, and C*

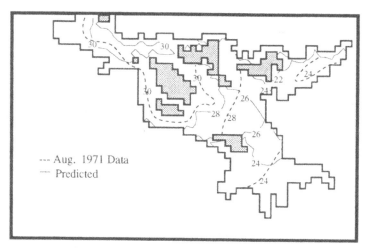

FIGURE 8   *Comparision of predicted salinities at high tide (420 cfs. - mean flow) with a summary of flood tide data taken during August 1971 (New Jersy. Dept. of Environmental Protection. 1972).*

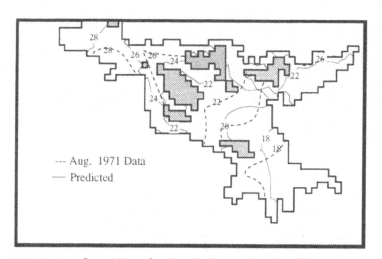

FIGURE 9   *Comparision of predicted salinities at low tide (420cfs. - mean flow) with a summary of ebbing tide data taken during August 1971 (New Jersey, Dept. of Environmental Protection 1972).*

17 3-D finite element simulation of salt-water intrusion in an estuarine barrage

T. Kawachi, M. Ban and K. Hiramatsu

ABSTRACT

A Galerkin finite element model that employs the linear and polynomial basis functions in the horizontal and vertical respectively is used to simulate the three–dimensional transient behaviors of the salt–water injected through navigation locks. An overview of the model is given, and a few demonstrative simulation examples for a real estuarine barrage (Nagara–Gawa Barrage) are presented.

1. Introduction

The increasing water use activities in the upper and middle course of a river reduces the seaward fresh water flow in the estuary, and thus enhances an upstream advance of the saline wedge by which the existing river water use in the lower course may incur serious salt nuisances. On the other hand, the increasing industrial and/or agricultural developments in lower alluvial areas require an increasing amount of fresh water available in estuaries. In order to meet such requirements, i.e., removal of salt nuisances and water resources developments, the estuarine barrage or reservoir is often built that works as a barrier against salt–water intrusion and/or a fresh water storage. The key to making a success of the proposed hydraulic project is to keep the pool salinity at an appropriate level for water use practice. Thus, prediction of pool salinity variations resulting from intrusion and retreat of the salt–water inevitably injected from the navigation locks is of primary importance. This is also the first stage in an effort to predict the ecological environment inside a pool that will dramatically be changed by barrage construction.

This paper is associated with numerical simulation technique for assessing the impact of the salt–water from the locks on the salinity environment of a barrage pool. The physical process occurring in the pool may be characterized as a 3-D density-stratified flow on account of significant salinity variations in the vertical and horizontal, thus entailing to couple water motion and salt transport. For this, the complete Galerkin finite element model with linear and depth-dependent polynomial basis functions in the horizontal and vertical respectively, of which basic framework was previously presented [Kawachi(1991)], is employed. The model is an extension and improvement over the earlier polynomial function models [Heaps(1971), Koutitas(1978), Davies *et. al.*(1979)] which consider water with homogeneous densities in all directions and apply the conventional finite difference method to the horizontal, and therefore is capable of producing continuous distributions of salinity as well as current velocity over the vertical column of water and makes it simpler in all dimensions to treat geometrical irregularities. In the following, an overview of the model is given and demonstrative simulation examples for a real barrage, the Nagara-Gawa estuarine barrage, are presented.

2. Basic Equations

Assuming the hydraulic shallowness and neglecting the convective acceleration and the horizontal momentum exchange, the equations describing a coupled phenomenon of water motion and salinity transport in a barrage pool can be expressed in the following form.

$$\frac{\partial \zeta}{\partial t} + \frac{\partial}{\partial x}\int_0^h u\,dz + \frac{\partial}{\partial y}\int_0^h v\,dz = 0 \tag{1}$$

$$w = \frac{\partial}{\partial x}\int_z^h u\,dz + \frac{\partial}{\partial y}\int_z^h v\,dz \tag{2}$$

$$\frac{\partial u}{\partial t} - \Omega v = -g\frac{\partial \zeta}{\partial x} - \frac{g}{\rho}\frac{\partial}{\partial x}\int_0^z \rho\,dz + \frac{\partial}{\partial z}\left[N_{zx}\frac{\partial u}{\partial z}\right] \tag{3}$$

$$\frac{\partial v}{\partial t} + \Omega u = -g\frac{\partial \zeta}{\partial y} - \frac{g}{\rho}\frac{\partial}{\partial y}\int_0^z \rho\,dz + \frac{\partial}{\partial z}\left[N_{zy}\frac{\partial v}{\partial z}\right] \tag{4}$$

$$\frac{\partial s}{\partial t} + \frac{\partial(us)}{\partial x} + \frac{\partial(vs)}{\partial y} + \frac{\partial(ws)}{\partial z} = \frac{\partial}{\partial x}\left[K_{xy}\frac{\partial s}{\partial x}\right] + \frac{\partial}{\partial y}\left[K_{xy}\frac{\partial s}{\partial y}\right] + \frac{\partial}{\partial z}\left[K_z\frac{\partial s}{\partial z}\right] \tag{5}$$

$$\rho = \bar{\rho} + \alpha s \tag{6}$$

where x,y,z = Cartesian coordinates positive eastward, northward and downward, respectively, t = time, u,v,w = respective current velocity components, s = salinity, h = undisturbed water depth, ζ = water surface elevation above undisturbed water depth, ρ = water density, Ω = Coriolis parameter, N_{zx}, N_{zy} = vertical eddy viscosity coefficients, K_{xy}, K_z = horizontal and vertical turbulent diffusion coefficients, respectively, g = gravitational acceleration, $\bar{\rho}$ = reference density, and α = dimensionless coefficient.

To obtain unambiguous solutions to Eqs. (1) through (6), the following boundary conditions are considered.

(a) Open Boundary (Γ_o) ;

$$\zeta = \zeta^*(t) \tag{a-1}$$

$$s = s^*(z,t) \quad for \ un_x + vn_y < 0 \quad or \quad K_{xy}\frac{\partial s}{\partial x}n_x + K_{xy}\frac{\partial s}{\partial y}n_y = 0 \quad for \ un_x + vn_y > 0 \tag{a-2}$$

(b) Inflow/Outflow Boundary (Γ_q) ;

$$u = u^*(z,t), \quad v = v^*(z,t) \tag{b-1}$$

$$s = s^*(z,t) \quad for \ inflow \quad or \quad K_{xy}\frac{\partial s}{\partial x}n_x + K_{xy}\frac{\partial s}{\partial y}n_y = 0 \quad for \ outflow \tag{b-2}$$

(c) Salt Injection Boundary (Γ_s) ;

$$un_x + vn_y = 0 \tag{c-1}$$

$$K_{xy}\frac{\partial s}{\partial x}n_x + K_{xy}\frac{\partial s}{\partial y}n_y = R^*(z,t) \tag{c-2}$$

(d) Land Boundary (Γ_n) ;

$$un_x + vn_y = 0 \tag{d-1}$$

$$K_{xy}\frac{\partial s}{\partial x}n_x + K_{xy}\frac{\partial s}{\partial y}n_y = 0 \tag{d-2}$$

(e) Free Surface Boundary (Λ_f) ;

$$-\rho N_{zx}\frac{\partial u}{\partial z}\Big|_{z=0} = \tau_{sx}^*(t), \quad -\rho N_{zy}\frac{\partial v}{\partial z}\Big|_{z=0} = \tau_{sy}^*(t) \tag{e-1}$$

$$K_z\frac{\partial s}{\partial z}\Big|_{z=0} = 0 \tag{e-2}$$

(f) Bottom Boundary (Λ_b) ;

$$-\rho N_{zx}\frac{\partial u}{\partial z}\Big|_{z=h} = k_b(\rho u)_{z=h} \ or \ \lambda_b(\rho u|U|)_{z=h}, \quad -\rho N_{zy}\frac{\partial v}{\partial z}\Big|_{z=h} = k_b(\rho v)_{z=h} \ or \ \lambda_b(\rho v|U|)_{z=h} \tag{f-1}$$

$$K_z\frac{\partial s}{\partial z}\Big|_{z=h} = 0 \tag{f-2}$$

where n_x, n_y = components of the outward normal to the boundary surface, ζ^*, s^*, u^*, v^* = respective prescribed boundary values, τ_{sx}^*, τ_{sy}^* = components of the prescribed shearing stress at the free surface, R^* = prescribed salt flux per unit area of boundary surface, $|U|(=\sqrt{u^2+v^2})$ = magnitude of current velocity and k_b, λ_b = bottom resistance coefficients for linear and quadratic inadhesive conditions, respectively.

The diffusivities, N_{zx}, N_{zy}, K_z , can be assumed to exponentially decrease with the increasing local Richardson number [Mamayev(1958)] to have

$$N_{zx} = N_z^* \exp\left\{-\mu_1\frac{g\frac{\partial \rho}{\partial z}}{\rho\left(\frac{\partial u}{\partial z}\right)^2}\right\}, \quad N_{zy} = N_z^* \exp\left\{-\mu_1\frac{g\frac{\partial \rho}{\partial z}}{\rho\left(\frac{\partial v}{\partial z}\right)^2}\right\} \tag{7}$$

203

$$K_z = K_z^* \exp\left\{-\mu_2 \frac{g\frac{\partial \rho}{\partial z}}{\rho\left[\left(\frac{\partial u}{\partial z}\right)^2 + \left(\frac{\partial v}{\partial z}\right)^2\right]}\right\} \tag{8}$$

where μ_1, μ_2 = undetermined empirical constants (μ_1 = 1.5, μ_2 = 3.0 [Leendertse *et. al.*(1975)], μ_1 = 4, μ_2 = 18 [Perrels *et.al.*(1981)]), and the neutral diffusivities of homogeneous fluid, N_z^*, K_z^*, are given resorting to the Reynolds analogy and the mixing length theory as

$$N_z^* = K_z^* = L^2\left[\left(\frac{\partial u}{\partial z}\right)^2 + \left(\frac{\partial v}{\partial z}\right)^2\right]^{\frac{1}{2}} \tag{9}$$

with $L = \kappa(h-z)(z/h)^{1/2}$ (κ : *Von Karman* constant).

The diffusivities, K_{xy}, can also be, in an analogy to the momentum transfer, identified with the sub-grid scale (SGS) eddy viscosity [Smagorinsky(1963), Smagorinsky *et.al.*(1965)] expressed in terms of the local velocity deformation and the mesh spacing, which leads to

$$K_{xy} = (c\Delta)^2\left[\left(\frac{\partial u}{\partial x} - \frac{\partial v}{\partial y}\right)^2 + \left(\frac{\partial v}{\partial x} + \frac{\partial u}{\partial y}\right)^2\right]^{\frac{1}{2}} \tag{10}$$

where c = dimensionless constant (close to 0.1 [Deardorff(1970), Leonard(1974)]), and Δ = mesh spacing.

3. Model Description

3.1 Galerkin finite element modelling

Modelling strategy is briefly reviewed. For more details and basic features of the model, the reader should refer to the previous work [Kawachi(1991)].

For finite element representations of Eqs. (1) through (5), the primitive variables to be determined are first approximated by the following series including the Chebyshev polynomials as depth-dependent basis functions.

$$\begin{aligned}
u &= A_r(x,y,t)\,T_r(H) \\
v &= B_r(x,y,t)\,T_r(H) \\
s &= C_r(x,y,t)\,T_r(H)
\end{aligned} \tag{11}$$

where r = repeated subindex implying summation (r = 0, 1, 2,···, m) and $T_r(H)$ = Chebyshev polynomials given by the recurrence relation ;

$$\begin{aligned}
T_0(H) &= 1 & &: r=0 \\
T_1(H) &= H & &: r=1 \\
T_r(H) &= 2HT_{r-1}(H) - T_{r-2}(H) & &: r\geq 2
\end{aligned} \tag{12}$$

with the normalized depth $H = (2z-h)/h$ $(-1\leq H\leq 1)$.

Introducing Eq. (11) into Eqs. (3) through (5) and applying the Galerkin process to the resulting equations where $T_k(H)$ (k = 0, 1, 2,···, m) is taken as weighting functions achieves vertical integrations of these equations with elimination of the partial derivatives with respect to depth coordinate. The natural boundary conditions related to momentum and mass transfer through free surface and bottom, involved in Eqs. (e-1), (e-2), (f-1) and (f-2), are then taken in by integrating the depth-dependent terms by parts. The remaining equations Eqs. (1) and (2) in a two-dimensional form are also rewritten by a variable conversion from z to H, that is, by direct introduction of Eq. (11).

Secondly, an entire set of the resulting partial differential equations is integrated over the horizontal domain by reuse of the Galerkin process. To this end, the k-th components of the unknown coefficients A_r, B_r and C_r, ζ, and w at a point over the vertical column of water are approximated by the following linear expressions for a triangular element.

$$A_k = G_j(x,y)A_{kj}(t), \quad B_k = G_j(x,y)B_{kj}(t), \quad C_k = G_j(x,y)C_{kj}(t)$$
$$\zeta = G_j(x,y)\zeta_j(t), \quad w = G_j(x,y)w_j \tag{13}$$

where j = repeated subindex implying summation (j = 1, 2, 3) and G = linear basis function The Galerkin integration is performed with the weighting functions $G_i(x,y)$ (i = 1, 2, 3), and the natural boundary conditions associated with horizontal salt flux, involved in Eqs. (a-2), (b-2), (c-2) and (d-2), are then treated so as to be automatically satisfied. Thus we can arrive a a set of ordinary differential equations in time.

3.2 Time-marching schemes and Stability

Next, the piecewise discretized equations are stepwise integrated through time with the aid of the explicit time-marching schemes of a singe step where only two time stages are considered ; the Kawachi scheme [Kawachi et.al.(1990a), Kawachi et.al.(1990b)] for a pair of momentum and continuity equations and the well-known forward-time scheme for the salt transport equation (See Figure 1). For this, the consistency mass matrix at the advanced time stage $t=n+1$, which prevents explicit determination of the unknowns, is diagonalized by the lumped

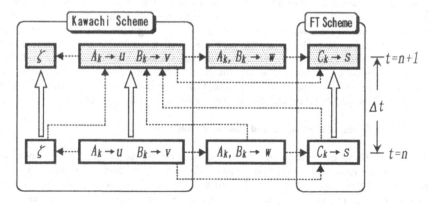

Figure 1 Time-marching procedure

mass approximation. In addition, the mass matrix at the current time stage $t=n$ is replaced with a generalized mass matrix [Kawahara(1982)] including the consistency parameter ω that permits selective damping (i.e. dissipation) of the useless short waves generated by numerical disturbances. With proper assembly of the resulting finite element algebraic equations over an entire domain of interest, all the solutions at the advanced time stage are one after another obtainable according to the algorithm shown in Figure 1.

The respective stability criteria for the adopted schemes that serve as rough guides for determination of the time increment Δt, derived from uncoupled linear one-dimensional equations, are given as what follows. The time increment that simultaneously satisfies both inequalities may ensure stable computations.

$$\Delta t \leq \frac{2l}{\sqrt{gh_{max}}} \left[1 - \frac{\omega_1}{3} \right]^{\frac{1}{2}} \qquad for\ Kawachi\ Scheme \qquad (14)$$

$$\Delta t \leq \frac{K}{u^2_{max}} + \left[\frac{K^2}{u^4_{max}} + \frac{\omega_2 l^2}{3u^2_{max}} \right]^{\frac{1}{2}} \qquad for\ Forward\text{-}time\ Scheme \qquad (15)$$

where l = element length and ω_1, ω_2 = consistency parameters ranging from 0 (full lumping) to 1 (full consistency), differently defined for each scheme.

4. Simulation Examples

A sample of solving the salt–water intrusion problem in an estuarine barrage is shown by dealing with a real case of the Nagara–Gawa barrage.

The river Nagara–Gawa that enters the Bay of Ise communicating with the Pacific, in the mid–southern part of Japan, had, whenever hit by typhoon, caused serious flood damages to the riverine low–lying area. For this, attempts had long been made to find an effective and inexpensive hydraulic engineering strategy for reducing the risk of flooding. Finally the proposal was approved in 1973 that the river near the coast should be deepened by the dredging, and the barrage should be constructed across the mouth of the river to push back the salt–water penetrating much more upstream due to the river deepening and to form and be an important source of fresh water for the existing and increasing water demand. For successful implementation of the project, elaborate prior investigations of the biotic and abiotic aspects of the barrage construction were, until the construction works started in 1988, being made with an interdisciplinary approach. The computational work described here is a part of our contributions with the aim of gaining basic knowledge for design and operation of the navigation lock and of predicting the salinity environment in the barrage pool.

4.1 Finite element discretization and Boundary conditions

The Nagara–Gawa estuarine barrage, 661 metre wide, consists of ten main gates, a navigation lock, two fish roads and a road bridge. The main gates are of double–leaf roller type that permits underflow, overflow or both. The gates of the same type are also built in across the lock to form the 15 metre wide, 80 metre long and 4.8 metre deep chamber. The fresh water storage behind the barrage is around 16.8 million m^3 (6000 metre long × 7 metre deep × 400 metre wide).

For simulations, first, the barrage pool is horizontally discretized into a total of 324 triangular elements jointed at the 204 common nodes (See Figure 2). Needless to say, no depthwise discretizations are needed because of the use of the continuous polynomial functions over the depth. Thus, water depths at the nodes are only required in feeding depth–related data into a computer.

The boundary conditions specified are also depicted in Figure 2, except for free surface and bottom. For the essential boundary conditions involved, i.e., $s^{\cdot}(z,t)$ at the upstream end, $u^{\cdot}(z,t)$ and $v^{\cdot}(z,t)$ at the outlet gates and the intake, and $R^{\cdot}(z,t)$ at the lock gate, their vertically continuous distributions that may change with time must be given in the form of the

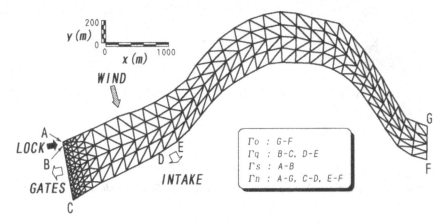

Figure 2 Finite element spatial discretization

Chebyshev polynomials like Eq. (11). This can be relatively easily done by use of the Chebyshev interpolation technique which determines polynomial coefficients from the boundary values specified at discrete points over the vertical column of water.

4.2 Demonstrative computational results

To date, a great deal of computations have been conducted under various conditions stemming from the combinations of ; intake discharge, gate discharge (opening and the number of the moving gates, and outflow regime), speed and direction of the wind acting on the free surface, salt content of the lock chamber, time–varying characteristics of salt amount injected from the lock. To show illustrative examples, however, only a few typical computational results are now referred. The key parameters equally adopted for these are as ; Δt = 3.0 s, Ω = 0.0 rad/s, $\bar{\rho}$ = 1000 kg/m^3, α = 0.75, k_b = 0.0025 m/s, μ_1 = 1.5, μ_2 = 3.0, κ = 0.4, and c = 0.1. The parameter, m, indicative of the order of solution approximation over the depth, is taken as m = 6. The mesh spacing Δ for evaluation of the horizontal mass diffusivity is calculated by $\Delta = (l_1 \cdot l_2 \cdot l_3)^{1/3}$ where l_1, l_2, l_3 = respective side lengths of a triangular element.

At first, two comparative cases are illustrated ; (1) free penetration without any pool water withdrawal (no intake and no gate discharge) and (2) restricted penetration with withdrawal from the outlet gates (but no intake discharge). The former allows for zero river discharge which may cause the crucial rise of pool salinity, while the latter for normal river discharge which requires the gates to be appropriately opened to retain a targeted pool surface level and therefore may significantly restrict salt–water penetration. For both, it is assumed that the lock chamber is filled with seawater (s = 32.5 kg/m^3) to have the salt content $1.872 \times 10^5 kg$, and it is, as a plume without momentum flux, injected into the pool according to the time–varying vertical distributions of salt flux $R^*(z,t)$ that are shown in Figure 3. Here note that salts are in a half–parabolic form plumed through a lower half–depth at the lock gate to allow for densimetric water exchange between the lock chamber and the barrage pool, and that the frequency of such a 30–minutes injection is restricted to only once.

Free penetration This case is computed from the initial conditions ; hydrodynamically cold

207

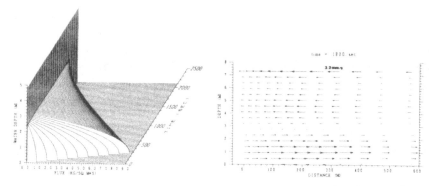

Figure 3 Injected salt flux distributions   *Figure 4 Density-induced counterflow*

situations (zero current velocities over entire domain and horizontal free surface) and spatially uniform salinity $s = 0.066 \, kg/m^3$. All the outlet gates are fully closed during the entire computation, so no fluvial water virtually comes in at the upstream end where the water surface elevation is specified to be invariant. However, if density-induced counterflow occurs there, fluvial water may less dominantly come in as a compensation. Transient penetration process of the injected salt-water, of major interest, is presented in Figure 5(a) which illustrates the salinity profiles along the central pool axis at selected time stages including the period after a stoppage of the salt supply from the lock. Note that the vertical axis of coordinates is normalized with respect to local water depths. To show horizontal salinity distributions, the bird's-eye view Figure 6(a) is inserted that includes those over the pool bed at the typical time stages $t = 600 \, s$ (maximum injection) and $t = 1800 \, s$ (injection subsidence). In addition, density-induced counterflow is illustrated in Figure 4 for the central axis of a limited area just behind the outlet gates.

Restricted penetration For this, the steady-state flow computation is performed in advance to generate the initial hydrodynamical conditions for the objective computation. In this prior computation, the salt-water injection from the lock is not considered to keep the pool salinity at a prescribed constant $s = 0.066 \, kg/m^3$, and the vertical current velocity distributions at the outlet gates to be opened are specified so that the normal river discharge $Q = 25.0 \, m^3/s$ may be withdrawn. Starting from the initial conditions obtained, salinity variations due to the same salt-water injection as in the above case are computed retaining the outflow from the gates. The results presently demonstrated are those for withdrawal from the two underflow sluices with half opening, which are illustrated in Figure 5(b) and Figure 6(b) in comparison with the previous case. Comparison clearly implies that the continuous withdrawal of lower pool water indeed restrict the rise in salinity due to the convective removal of salts, but makes the injected salts perceptibly disperse because of the increase in turbulent diffusivities. Namely, the pool water withdrawal indispensable for salt-water removal may enlarge the body of brackish water and eventually delay its perfect retreat from the pool. It thus appears that taking a well-linked operation of the gates and the lock so as to alleviate such a paradoxical dilemma is a viable policy on a realistic base.

Three-dimensional water motion Finally a disparate case of the restricted penetration is referred to show an overall flow structure. The computation considers blow of the wind 10 m/s with the direction indicated in Figure 2, and water withdrawal from both the intake and the

208

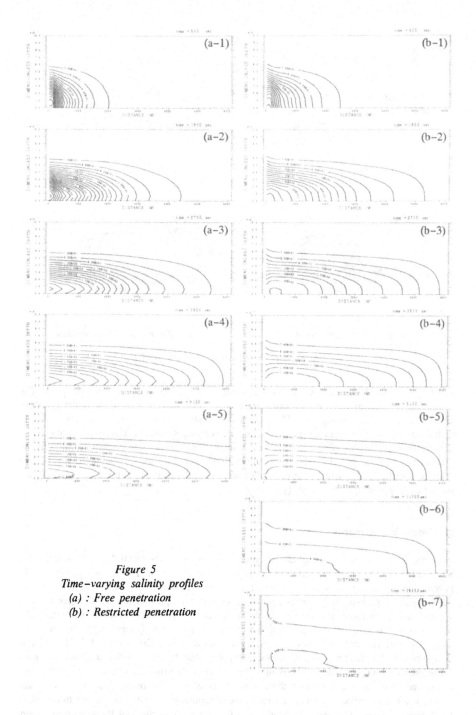

Figure 5
Time-varying salinity profiles
(a) : Free penetration
(b) : Restricted penetration

209

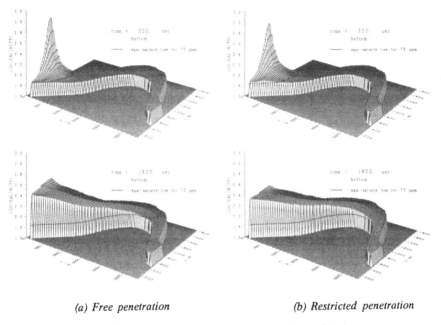

(a) Free penetration   (b) Restricted penetration

Figure 6 Salinity distributions over the pool bed

outlet gates that is in total equal to the normal river discharge entering the pool. The withdrawal is done by overflow for both, and then the middle 7-sluices of the outlet gates are lowered to have the opening 1 m. The resulting flow structure does not significantly change with time because of its less subjection to the time-varying baroclinic effects produced by the salt-water injection, therefore the three-dimensional water motion can be demonstrated by the flow at any time stage which is, for example, illustrated in Figure 7 including the current velocity distributions at selected three levels ; surface, mid-depth and bottom, and marking the velocities larger than 0.05 m/s by (+).

5. Concluding Remarks

In comparison with the stacking models often referred to as the multi-layered model [e.g., Simons(1973)] or multi-leveled model [e.g., Leendertse et.al.(1973), Wang(1977)], the polynomial function models handle the vertical physical variations not via discretization process but via purely mathematical process, and thus permit free specification of a degree of their approximations without any change in programming. In particular, use of the 0-th order polynomials that assumes vertically uniform variations decreases dimensionality by one, such as 3-D model reduced to 2-D depth-averaged model or laterally averaged 2-D model to depth-averaged 1-D model. In this respect, the polynomial function models are also attractive.

The present 3-D model for analysis of a coupled physical process, indifferent to continuous or layered stratification, possesses these advantages of the polynomial function models and further enables versatile and accurate treatments of geometrical configuration due to complete finite element representations. The model is in fact a truly powerful tool for solving in a more

210

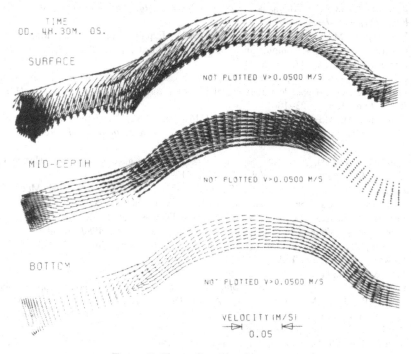

TIME
00. 4H.30M. 0S.

SURFACE
NOT PLOTTED V>0.0500 M/S

MID-DEPTH
NOT PLOTTED V>0.0500 M/S

BOTTOM
NOT PLOTTED V>0.0500 M/S

VELOCITY (M/S)
0.05

Figure 7 Three-dimensional water motion

refined fashion the salinity problems in relatively narrow estuarine water bodies such as estuarine barrages, inlets or embayments, well reproducing irregular land boundaries which may be dominant factors in solution determination. In addition, by increasing the consistency parameter, the model can effectively impede numerical oscillations often caused by purely convective or convective dominated salt transport, which leads to substantial diffusivity increase.

As a follow-up study, applicability and validity of the model is at present investigated through solving the salt-water intrusion problem in the Nakaumi-Shinjiko estuarine basin subject to tidal excitation for which the observed data are available.

REFERENCES

[1] Davies, A.M. and A. Owen, (1979), "Three Dimensional Numerical Sea Model using the Galerkin Method with a Polynomial Basis Function", *Appl. Math. Modelling*, Vol.3, pp 421–428

[2] Deardorff, J.W., (1970), "A Numerical Study of Three-Dimensional Turbulent Channel Flow at Large Reynolds Numbers", *Jour. Fluid Mech.*, Vol.41, part 2, pp 453–480

[3] Heaps, N.S., (1971), "On the Numerical Solution of the Three Dimensional Hydrodynamical Equations for Tides and Storm Surges", *Mémoires Société Royale des Sciences de Liège*, 6e série, tome I, pp 143–180

[4] Kawachi, T., M.H.Shajari and I.Minami, (1990a), "Explicit Finite Element Models for Pipeline Transients, – Principles of Computation –", *Trans. Jap. Soci. Irri. Drain. Recla. Eng.*, **147**, pp 27–34

[5] Kawachi, T., M.H.Shajari and I.Minami, (1990b), "Explicit Finite Element Models for Pipeline Transients, – Computational Stability and Accuracy –", *Trans. Jap. Soci. Irri. Drain. Recla. Eng.*, **147**, pp 35–44

[6] Kawachi, T., (1991), "Three–Dimensional Finite Element Model of Density–Stratified Flow using Depth-Dependent Polynomial Basis Functions", *Trans. Jap. Soci. Irri. Drain. Recla. Eng.*, **156**, pp 93–100

[7] Kawahara, M., H.Hirano, K.Tsubota and K.Inagaki, (1982), "Selective Lumping Finite Element Method for Shallow Water Flow", *Int. Jour. Num. Meth. in Fluids,* Vol.2, pp 89–112

[8] Koutitas, C., (1978), "Numerical Solution of the Complete Equations for Nearly Horizontal Flows", *Adv. Water Resour.*, Vol.1, No.4, pp 213–217

[9] Leendertse, J.J., R.C.Alexander and S.K.Liu, (1973), "A Three–Dimensional Model for Estuaries and Coastal Seas : Volume I, Principles of Computation", *R–1417–OWRR*, Rand

[10] Leendertse, J.J. and S.K.Liu, (1975), "A Three–Dimensional Model for Estuaries and Coastal Seas : Volume II, Aspects of Computation", *R–1764–OWRT,* Rand

[11] Leonard, A., (1974), "Energy Cascade in Large–Eddy Simulation of Turbulent Fluid Flows", *Adv. Geophy.*, **18A**, pp237–248

[12] Mamayev, O.I., (1958), "The Influence of Stratification on Vertical Turbulent Mixing in the Sea", *Bull. Acad. Sci. USSR, Geophys. Ser.,* translated by V.A. Salkind, pp 494–497

[13] Perrels, P.A.J. and M.Karelse, (1981), "A Two–Dimensional, Laterally Averaged Model for Salt Intrusion in Estuaries", *Transport Models for Inland and Coastal Waters*, Academic Press Inc., pp 483–534

[14] Simons, T.J., (1973), "Development of Three-Dimensional Numerical Models of the Great Lakes", Scientific Series 12, Inland Waters Directorate, CANADA Center for Inland Waters

[15] Smagorinsky, J., (1963), "General Circulation Experiments with the Primitive Equations, I. The Basic Experiment", *Monthly Weather Review*, Vol.91, No.3, pp 99–164

[16] Smagorinsky, J., S.Manabe and J.L.Holloway, (1965), "Numerical Results from a Nine-Level General Circulation Model of the Atmosphere", *Monthly Weather Review*, Vol.93, No.12, pp 727–768

[17] Wang, H.P., (1977), "Multi–Leveled Finite Element Hydrodynamic Model of Block Island Sound", *Finite Elements in Water Resources*, Pentech Press, pp 4.68–4.93

18 Model study on combating salt water intrusion for Hangzhou city

Z. Han, Y. Shao and X. Lu

ABSTRACT

Mathematical and Physical models have been established and verified to study the salt water intrusion problem for Hangzhou city situated by the tidal river Qiantangjiang.By using both models, several factors affecting the chlorinity of the river water were studied quantitatively. Several measures to combat the salt water intrusion have been analyzed and practised with favorable results.

Introduction

Hangzhou is a large city with a population over one million, it's located on the northern bank of the tidal river named Qiantangjiang(Fig 1). About 85% of the water supplied for the city's domestic and industrial uses and the total irrigation water of 100,000 hectares of farmlands are taken from the river. During dry seasons, the chlorinity of the river water often exceeds the critical standard (250 mg/l for industrial and domestic uses and 1,200 mg/l for irrigation) thus resulting in a direct economic loss amounting to several million Yuan per day. Therefore, the study and prediction of the salt water intrusion problem receive great attention from the municipal government. In recent ten odd years, the authors were engaged in the relevant study by mathematical and physical models and the practice of combating the salt water intrusion for Hangzhou city. In this paper, a brief introduction is given to this problem.

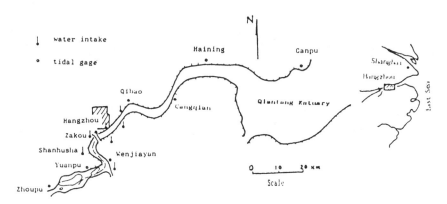

Fig.1 Sketch map of the Qinntang Estuary

1.Brief description of the models

1.1 mathematical model

The Qiantang estuary is a well-known macrotidal estuary. During dry season and spring tide,the chlorinity of the river water is uniformly distributed over the cross sections (the difference of the chlorinity at any point and the cross section mean value being less than 30%), it is, therefore, the one dimensional unsteady flow equations could be used to simulate this problem, the equations are:

$$\frac{\partial Z}{\partial t}+\frac{1}{B}\frac{\partial Q}{\partial X}=0 \tag{1}$$

$$\frac{\partial Q}{\partial t}+2u\frac{\partial Q}{\partial X}+Ag\frac{\partial Z}{\partial X}-u^2\frac{\partial A}{\partial X}-Bg\frac{|Q|Q}{C_z^2 R^2}-\frac{AHg}{2\rho}\frac{\partial \rho}{\partial X} \tag{2}$$

$$\frac{\partial S}{\partial t}+u\frac{\partial S}{\partial X}=\frac{1}{A}\frac{\partial}{\partial X}(AD\frac{\partial S}{\partial X}) \tag{3}$$

$$\rho = 1000 + 1.35 S \tag{4}$$

In which Z,Q,A,S,u,H denote the water stage, discharge, cross section area, chlorinity,cross-sectional mean velocity and water depth respectively. ρ and D denote the density and dispersion coefficient of chlorinity respectively. By dimensional analysis and the field data in the Qiantang estuary, the following relationship is established:

$$\frac{D}{uH} = [1 - \exp(-\frac{u^2}{gH}\frac{Q_f\,T_f}{Q_o T})]\frac{S_i}{S} + 10 \tag{5}$$

The term

$$\frac{u^2}{gH}\frac{Q_f\,T_f}{Q_o T}$$

is called the "estuary number", in which

$$\frac{u^2}{gH}$$

is Froude number,

$$\frac{Q_f\,T_f}{Q_o\,T}$$

is the ratio of flood tidal volume to fresh water volume within a tidal cycle.

$$\frac{S_i}{S_o}$$

is the ratio of the chlorinity at a section to that of fresh water. Because of the high turbulence of the tidal bore and flood flow, the dispersion coefficient D in the Qiantang estuary is some 10 to 100 times as great as that in other estaurines.

Verifications of the mathematical model for different tidal ranges have been carried out (fig 2),the discrepancy of water stage is smaller than 0.1 m, that of the chlorinity is smaller than 30%, the tidal duration exceeding the critical standard is smaller than 20%. The discrepancy of chlorinity is within the variation range in the cross section and hence the model acceptable.

1.2 Physical model

Strictly speaking, a physical model should be undistorted to simulate the flow field, because of very wide-shallow channel and the high turbulence of the flow,it is almost impossible and also unnecessary to build an undistorted model. The model with λ_L=1:1500 and λ_H=1:100 already constructed is used for study(photo 1). The Lower boundary is located at Jinshan, 42 km below Kanpu. The similarity criteria required are:

215

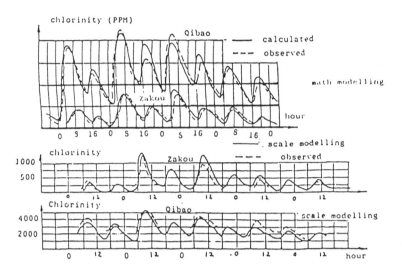

Fig.2 Verifiaction of chlorinity

$$\begin{aligned} \lambda_u &= \sqrt{\lambda_H} \\ \lambda_D &= \sqrt{\lambda_H} \quad \lambda_H \quad \lambda_L \\ \lambda_s &= 1 \end{aligned} \tag{6}$$

here λ_u denote Froude low, λ_D dispersion and λ_s chlorinity. Heimut Kobus [2] pointed out that when the Froude law is satisfied, other similarity criteria would be automatically satisfied if

$$\lambda_H = \lambda_L^a \quad (a = 0.75 \sim 0.66) \tag{7}$$

In our case, the calculated a equals 0.64 which is approximate to the required lower limit 0.66. The verification result is agreeable as shown in fig 2. More detailed verification data and its application in prediction may refer to the reference literature[3].

2.Quantitative analysis of various factors affecting the chlorinity

The tide varies in a semi-lunar cycle and semi-diurnal cycle and the chlorinity varies accordingly. Besides the river fresh water discharge, the discharge directed for use and the chlorinity in lower stretch affect the chlorinity at Zakou - the intake site. They will be discussed separately below.

2.1 Tidal range (ΔH)

The average tidal range of the larger one in each day within half month is used for an indication parameter showing the strength of tide for statistical analysis. Near Hangzhou city, the value varies between 1.25 to 2.7. However, in calculation and model test, the hourly variation of tide and chlorinity is considered.

2.2 River discharge (Q_r)

After the regulation of the upstream large reservoir,the river discharge during dry season varies between 200 and 600 m³/s.

2.3 Diversion discharge (Q_d)

The total diversion discharge is about 120 m³/s at present and will be increased to 240 m³/s in the future.

2.4 Chlorinity in downstream section

The chlorinity at Cangqian is taken as a parameter in analysis which will render direct sensitive effect to the chlorinity at Zakou (intake site) and long term observation data is available there.

2.5 Quantitative Analysis

The quantitative relation between the chlorinity,total duration exceeding permissible limit and the successive exceeding duration (T_s) at the intake site and the average tidal range (ΔH) at Zakou, net discharge Q_n (which equals Q_r-Q_d) and the chlorinity at Cangqian are analyzed and shown in the correlation curves in fig

3. The correlations between various parameters are considered agreeable.

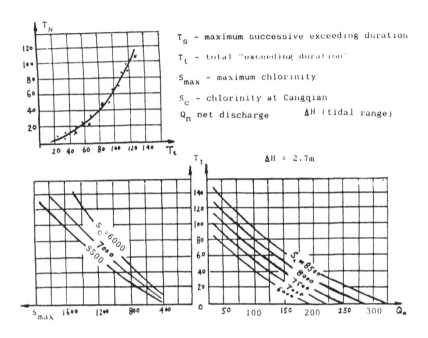

T_s - maximum successive exceeding duration

T_t - total "exceeding duration"

S_{max} - maximum chlorinity

S_c - chlorinity at Cangqian

Q_n net discharge ΔH (tidal range)

$\Delta H = 2.7m$

Fig.3 Correlation curves of Q_n, T_t, T_s and S_{max}

3.The function of upstream reservoir for combating salt water intrusion

Upon analysis Various relations for different tidal ranges, the relation between the tidal range (ΔH). maximum successive exceeding duration (T_s) and the net discharge (Q_n) could be obtained and shown in fig 4.

The tidal range at Qibao is changed as the downstream river configuration changes.After the rainy season, the scouring effect of the downstream river is already known and the tidal range in the following dry season may be predicated. The required net discharge Qn for a definite Ts may be determined by fig 4. When the desired diversion discharge Q_d is given, the required river discharge released from the reservoir Q_r is then Q_n+Q_d. As could be seen in fig 4, when $T_s=0$, the Q_n is about 100 to 400 m³/s corresponding to the tidal range of 1.2-2.7m. Before the construction of the large reservoir, the average monthly discharge during dry season is only 50-100 m³/s, while after the

218

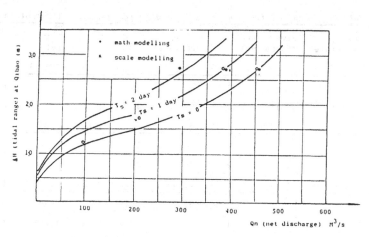

Fig. 4 Correlation curve of ΔH - Q_n - T_s at Shanhusha

construction of the reservoir, the average discharge in the same period is increased to 350 m³/s and may be raised to 600 m³/s in a shorter period. This show that the function of the large reservoir for combating salt water intrusion is rather significant.

4. The function of nearby small regulation reservoir

If a small regulating reservoir is built nearby the intake site, river water of low chlorinity generally during neap tide may be pumped in the reservoir ready for use during spring tide. the size of the regulating reservoir may be designed according to the successive exceeding duration Ts and its corresponding amount of water consumption. Since the dilution effect of the released fresh water from the upstream large reservoir is different for different releasing schemes, the stepped type releasing, i.e large discharge during spring tide and smaller during neap tide, would be more effective than uniform releasing. The following table 1 lists the total amount of water required to release from the upstream reservoir in three months for different releasing scheme, different diversion discharge ,different size of the nearby small reservoir and different guarantee percentages to ensure the chlorinity of intake water be kept below the permissible limit. As show in the table, for p = 95%, Q_d = 120 m³/s T_s = 0, the total volume required is 930 x 10⁶ m³, while for T_s= 2 days, the required volume is only 77 x 10⁶ m³, a reduction of 92% has been obtained. For other cases, similar results could be seen. For Ts = 2 day, the capacity of nearby regulating reservoir is only 1.8 x 10⁶ m³ which is only 0.2 % that of the total volume 930 x 10⁶ m³ required to released from the upstream large reservoir. This is

Table 1 Amount of fresh water required to release for different cases

Guarantee percentage P%	Diversion discharge Qd	Step release			Uniform release		
		Ts=0	Ts=0.5	Ts=2	Ts=0	Ts=0.5	Ts=2
	120	0	0	0	0	0	0
85 %	180	0	0	0	1.55	0	0
	220	0.70	0	0	4.6	3.8	0
	120	0	0	0	0.77	0	0
90 %	180	2.33	0	0	5.43	0	0
	220	5.44	2.30	0	8.54	5.43	0
	120	9.3	3.80	0.77	12.4	6.90	3.90
95 %	180	14	8.50	2.33	17.1	11.60	5.40
	220	17	11.60	5.44	20.1	14.70	5.50

because of the fact that the nearby reservoir could be refilled about six times in three months and,moreover, the fresh water released in the river is used to dilute the quite large amount of saline water in the whole stretch of the river and is certainly much more larger than what was pumped for use. It is, therefore, the construction of a nearby regulating reservoir is very economical.

Based on the practice in recent ten odd years , the mathematical model and physical model calibrated and verified using in situ data from the Qiantang Estuary may serve as a reliable measure to predict the chlorinity of the river water. By using these models, optimal regulation is possible to ensure the quality of the intake water and more than ten million Yuan per annum may be saved. Besides, these models may also be used to study other problem such as water pollution in rivers,estuaries and coastal regions.

REFERENCE

[1] Han zencui,Cheng hangpin, The study of Salinity prediction in Qiantang Estuary. Journal of Hydraulic Engineering . 1981.6 (In Chinese)
[2] Heimut Kobus , Hydraulic simulation.1978.
[3] Han zengcui, Shao yaqin et al. Prediction Techniques For Salt Water Intrusion in Qiantang Estuary by Physical and Mathematical Model.ZECER.April 1989.
[4] Study on Salt Water Intrusion for new Water Works of Hangzhou city.ZECER.April 1989.

Part 4
WATER QUALITY MODELLING

19 Water quality modelling in a hypertidal, muddy estuary

H. G. H. Larsen, K. W. Olesen, A. J. Parfitt and W. R. Parker

1. Introduction

The River Usk rises in the mountains of South Wales and flows in an incised, alluviated valley to the Severn Estuary at Newport. It is a spate river with flood events rising above a mean daily flow of 39 cumec. A mean annual flood is of the order of 450 cumec and a 1 in a 100 year flood is 1050 cumec.

The tidal influences from the Severn Estuary penetrate beyond Newbridge-on-Usk (Figure 1 over 25 km upstream from low water springs at No. 1 buoy (Figure 1). The mean spring tidal range at the docks (Figure 1) is about 11.9m and the mean neap range is about 5.38m. Low water neaps is at Spittles Point (Figure 1) some 6 km upstream of No.1 buoy. The intertidal volume varies between about $26 \times 10^6 m^3$ on springs and $19 \times 10^6 m^3$ on neaps. Salt water is detectable about 25 km upstream on springs and about 22 km on neaps. The estuary is generally well to partially mixed, although marked thermohaline stratification is observed, especially in the lower estuary (Figure 1) on neap tides.

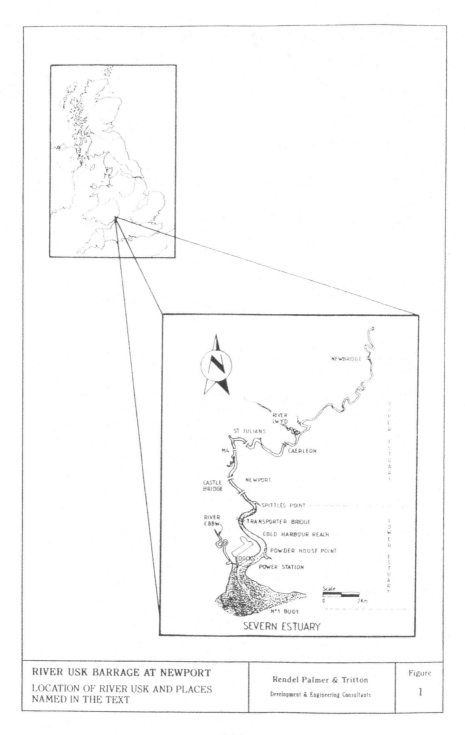

RIVER USK BARRAGE AT NEWPORT	Rendel Palmer & Tritton	Figure
LOCATION OF RIVER USK AND PLACES NAMED IN THE TEXT	Development & Engineering Consultants	1

A prominent feature of the estuary is its high suspended load during spring tides and its much lower suspended load on neap tides, arising from the spring-neap variation in tidal energy. On spring tides, water entering the estuary at low water has a high solids load (8-10 g/l). Concentrations drop exponentially from peak values at the onset of the flood tide to background values of 300-500 mg/l by 2 hours after low water as the seaward margin of the turbidity maximum is advected upstream. During the spring ebb tide, sediment is entrained late in the tide and only reaches the lower estuary about 2 hours before low water. By comparison, high suspended solids levels are only observed during high water in the upper estuary at Caerleon.

Time series of suspended solids levels for locations throughout the estuary (Figure 2) indicate that the Usk has its own turbidity maximum separate from that of the Severn. This turbid water body is advected up and down the estuary during spring tides. During neap tides, much of the mud population of the turbidity maximum settles to the bed below low water neaps to form pools of liquid mud. Suspended solids levels are reduced to 300-500 mg/l throughout the tide. During the neap to spring tidal cycle, these pools of liquid mud are re-eroded. This erosion occurs late in the ebb tide and early in the flood tide as low water levels fall below -4.00m OD. Low water springs is -5.61m OD and low water neaps is - 2.91m OD. During spring tides a high suspended load passes Pier Head during the late ebb and returns early in the flood (Figure 3). Both during the formation of fluid mud pools but more especially during their erosion, the vertical distribution of suspended solids shows marked stratification independently of any salinity stratification and especially in the lower estuary. Although a hysteresis between increasing tidal energy and suspended solids load is well known, the re-erosion of the fluid mud in the Usk occurs much earlier in the neap to spring cycle than has hitherto been apparent. Early sediment modelling indicated that the erosion of the fluid mud was out of phase with tidal amplitude and this was verified by field observations. Accurate representation of this effect and its impact on water quality was a major objective of the modelling.

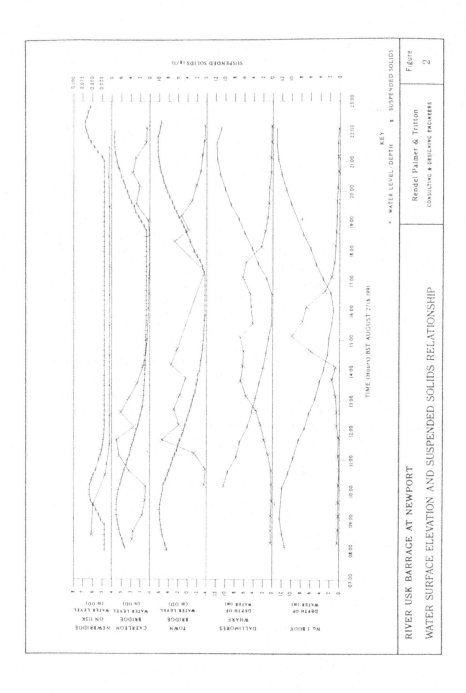

RIVER USK BARRAGE AT NEWPORT

WATER SURFACE ELEVATION AND SUSPENDED SOLIDS RELATIONSHIP

Rendel Palmer & Tritton
CONSULTING & DESIGNING ENGINEERS

Figure 2

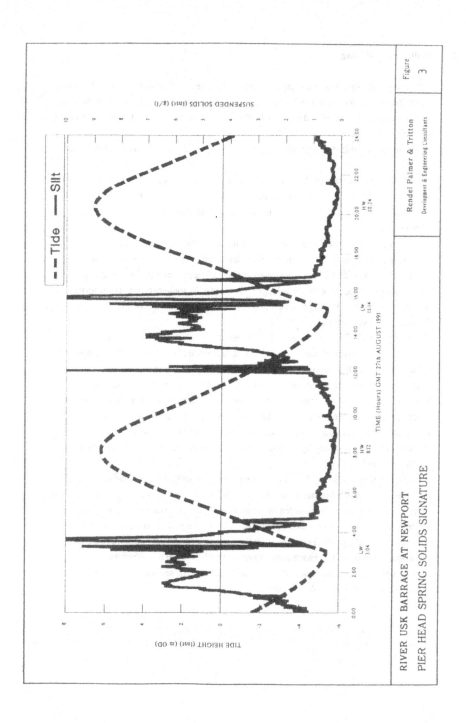

SUSPENDED SOLIDS (Inst) (g/l)

— Tide — Silt

TIME (Hours) GMT 27th AUGUST 1991

TIDE HEIGHT (Inst) (m OD)

RIVER USK BARRAGE AT NEWPORT
PIER HEAD SPRING SOLIDS SIGNATURE

Rendel Palmer & Tritton
Development & Engineering Consultants

Figure
3

227

2.   Sediment modelling

The key requirements of the sediment transport model applied were to describe the cyclic behaviour of the sediment (including fluid mud) during both a tidal cycle and during a spring-neap-spring cycle, and to describe the interaction between sediment and oxygen content in the water. Below a brief description of the sediment transport model is given; more detailed descriptions are given in Olesen & Kjelds (1991) and Olesen et al (1992).

The long period to be simulated in combination with the large tidal range excluded the use of full three and two dimensional models. The modelling was, therefore, based on a one-dimensional model, with some key processes described in either two or three dimensions.

Due to the large vertical and lateral gradients in sediment concentrations, the advection and dispersion of sediment were described in three dimensions. The three-dimensional flow pattern necessary for the advection-dispersion modelling was achieved by assuming the depth averaged flow within a cross-section to be proportional to the square root of the depth (frictional controlled flow) and the vertical distribution of velocity to be logarithmic.

Laterally, the flow velocity, hence also the bed shear stress, varies significantly. A two-dimensional description of erosion and deposition, based on the lateral flow velocity variation described above, was therefore included in the model.

The distribution of sediment at the river bed is highly non-uniform in the Usk Estuary. Soft mud can be found in the deeper parts of a cross-section while there is consolidated mud at the side slopes. Due to consolidation, there will also be a significant vertical variation in sediment properties, such as density and resistance to erosion. A full three-dimensional description of the river bed was, therefore, used (multi-layer bed model).

Deposition rate and settling velocity variation were modelled using standard formulations from the literature (see Krone (1962) and Van Rijn (1989)). Two different erosion formulations were included in the model; viz, an instantaneous erosion model, where all available mud erodes instantaneously when the bed shear stress exceeds a critical value, and a gradual erosion model, where the erosion rate is expressed as a non-linear function of the excess shear stress. The first model described an instability type of re-entrainment of weak fluid mud and the second, gradual erosion of fluid mud and under- consolidated mud.

228

3. Water quality modelling

The model used for water quality simulations is primarily aimed at describing the dissolved oxygen conditions in the river. The following processes affecting the oxygen concentration are taken into account:

- Reaeration
- Decay of BOD
- Nitrification
- Biological respiration
- Photosynthesis
- Benthic sediment oxygen demand (BSOD)
- Oxygenation of suspended sediment (SSOD)

In this paper, only the latter two processes will be discussed in detail. All, except the last, of these processes are further described by Bach et al (1989) and by Olesen et al (1989).

Models describing BSOD and its dependence on different factors such as temperature and dissolved oxygen have been used widely during the last twenty years (Porcella et al, 1986). Most commonly, a zero-order or a first-order formulation of BSOD (a linear function of dissolved oxygen concentration) is applied. In the present model, a zero-order formulation of BSOD, corrected for temperature by a rate constant and for oxygen concentration by a modified Michaelis-Menten expression, is applied:

$$\frac{dC[O_2]}{dt} = \frac{-BSOD\,(20°C)}{H} \cdot \theta^{(T-20)} \cdot \frac{C[O_2]^2}{C[O_2]^2 + K_{O_2}} \tag{1}$$

where :-

- BSOD(20°C)	is the benthic sediment oxygen demand at $20°C$ $[g\,O_2/m^2/d]$
- H	is the water depth $[m]$
- θ	is the temperature coefficient for oxygenation of suspended sediment. Dimensionless.
- T	is the water temperature $[\,^0C]$
- C[O_2]	is the concentration of dissolved oxygen $[g\,O_2/m^3]$
- K_{O_2}	is the square of the half-saturation concentration of dissolved oxygen for benthic sediment oxygen demand $[(g\,O_2/m^3)^2]$.

229

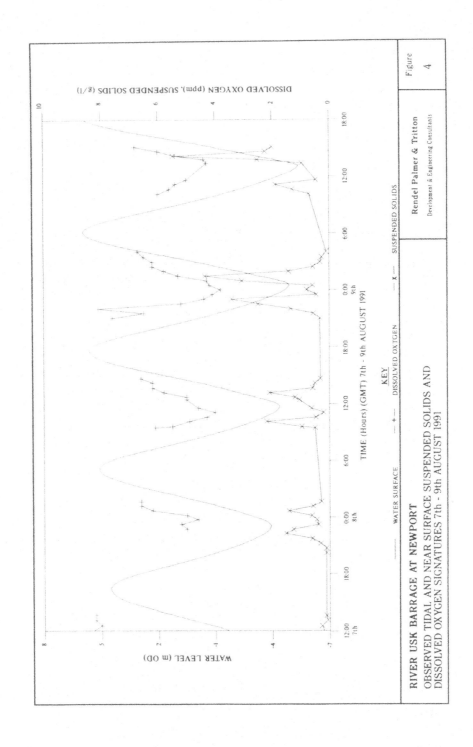

DISSOLVED OXYGEN (ppm), SUSPENDED SOLIDS (g/l)

TIME (Hours) (GMT) 7th - 9th AUGUST 1991

KEY

WATER SURFACE ——— DISSOLVED OXYGEN ——+—— SUSPENDED SOLIDS ——×——

WATER LEVEL (m OD)

RIVER USK BARRAGE AT NEWPORT

OBSERVED TIDAL AND NEAR SURFACE SUSPENDED SOLIDS AND
DISSOLVED OXYGEN SIGNATURES 7th - 9th AUGUST 1991

Rendel Palmer & Tritton
Development & Engineering Consultants

Figure
4

An indication of the significance of resuspension of sediment on the oxygen balance in River Usk is given in Figure 4. The periods of maximum concentrations of suspended solids coincide with periods of minimum dissolved oxygen concentrations. This situation is found late in the ebb tide and early in the flood tide, when the cross-sectional area of the river is relatively small, causing high velocities and thereby erosion of the river bed.

Figure 5   *Oxygen consumption rate for suspended sediment from River Usk, December 1989.*

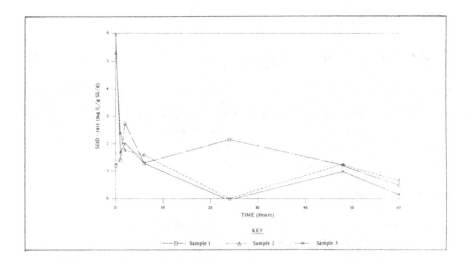

From Figure 5 it is concluded that the SSOD is strongly dependent on the time elapsed after resuspension (one of the samples does not fit this pattern, maybe due to improper sealing of the sample prior to the experiments). Initially, the SSOD is approximately 5mg O_2/g SS/d, but within a few hours the rate of oxygen demand has reached a fairly constant level of approximately 1-2mg O_2/g SS/d. The excess oxygen demand exerted during the first two hours has been calculated as approximately 0.1mg O_2/g SS.

Similar findings, i.e., high initial rate of oxygen demand and subsequent decline in SSOD following resuspension of sediment, have been reported by others, (Barcelone, 1983 and Murphy and Hicks, 1986). This could be explained by initial release of reduced, easily oxidised chemical substances from the resuspended sediment (e.g. sulphide, iron and manganese compounds) causing the high initial oxygen demand rate. (e.g., sulphides and ammonia) causing the high initial oxygen demand rate. The subsequent constant level of SSOD is supposed to be caused by slower chemical oxygenation processes and biological activity in the sediment. This is a well known approach for BSOD modelling (Snodgrass, 1986).

Based on these observations, a model for suspended sediment oxygen demand (SSOD) was established:

$$\frac{dC[O_2]}{dt} = -(A \cdot C[SS] + B \cdot \frac{ER}{H}) \cdot \theta^{(T-20)} \cdot \frac{C[O_2]^2}{C[O_2]^2 + K_{O_2}} \qquad (2)$$

where

- A   *is the steady oxygenation rate of suspended sediment at $20^{\circ}C$ [g O_2/g SS/d]*
- B   *is the initial oxygen demand of suspended sediment at $20^{\circ}C$ [g O_2/g SS]*
- C[SS]   *is the concentration of suspended sediment [g SS/m^3]*
- ER   *is the erosion rate of suspended sediment [g SS/m^2/s]*

Analysis of model sensitivity

In order to assess the influence of various parameters on the simulated oxygen concentrations, a sensitivity analysis including the following key parameters was carried out:

- $K_{BOD,d}$ *decay rate of dissolved BOD [d^{-1}]*
- $K_{BOD,s}$ *decay rate of suspended BOD [d^{-1}]*
- K_{nitr} *Nitrification rate constant [d^{-1}]*
- BSOD   *benthic sediment oxygen demand [g O_2/m^2/d]*
- A   *steady state suspended sediment oxygen demand [g O_2/g SS/d]*
- B   *initial suspended sediment oxygen demand [g O_2/g SS].*

A simulation applying an initial set of parameters was used as a reference to which subsequent simulations were compared.

In the sensitivity analysis, each parameter was sequentially reduced by 50% compared to the reference simulation. The result of the analysis is tabulated in Table 1. The maximum increase of dissolved oxygen concentrations, compared to the reference simulation, during the simulation period is given.

Table 1   *Sensitivity of simulated dissolved oxygen to different parameters. Maximum increase in DO (g/m^3) at different chainages (distances from river mouth) upon reduction of parameter values by 50%.*

Parameter	0.850 km	6.095 km	11.930 km	17.740 km
$K_{BOD,d}$	0·24	0·42	0·47	0.51
$K_{BOD,s}$	0.03	0.05	0.06	0.06
K_{nutr}	0.22	0.42	0.46	0.49
BSOD	0.09	0.14	0.20	0.25
A	1.90	1.57	1.31	0.88
B	0.01	0.01	0.01	0.01

From Table 1, it appears that the most important parameter to the oxygen balance is the steady state suspended sediment oxygen demand rate (A). The initial suspended sediment oxygen demand (B) is the least important of the parameters tested. Hence, determination of suspended solids should be given high priority in field campaigns.

The simulated erosion rates and simulated concentrations of suspended sediment, produced by the sediment-transport model, are used as input to the water quality model.

Calibration and validation of the water quality model

The dissolved oxygen model has been calibrated to describe field data from August 1991 and validated on data from August 1978. The 1991 data was selected for calibration because monitoring of suspended sediment and ammonia was included, this not being the case for the 1978 data.

233

Calibration   In August 1991, monitoring of water quality variables was carried out at 5 stations on two one-day surveys in a spring-neap-spring cycle in order to provide calibration data for the model. Samples were taken every 30 minutes during periods of approximately 12 hours.

The dissolved oxygen results for a neap tide event (20 August 1991) and a spring tide event (27 August 1991) simulated by the calibrated model are shown in Figures 6 and 7 respectively, with the measured data indicated. Results for two stations, located at chainages 4.27 and 14.57 km from the seaward boundary, are shown.

Figure 6   *Calibration of water quality model. Simulated and observed dissolved*
                  oxygen. Neap tide, August 1991.
                  a: Chainage 4.27 km from seaward boundary.
                  b: Chainage 14.57 km from seaward boundary.

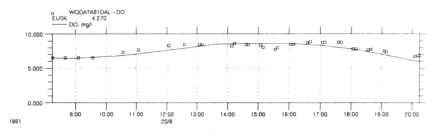

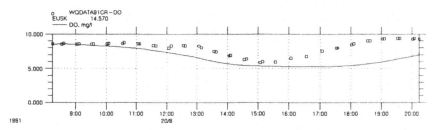

234

Plots of simulated and observed concentrations of suspended sediment for the calibration periods are shown in Figures 8 and 9.

Figure 8   *Simulated and observed concentrations of suspended sediment. Neap tide, August 1991.*
        Calibrated sediment-transport model.- a: 4.27 km & b: 14.57 km from seaward boundary.

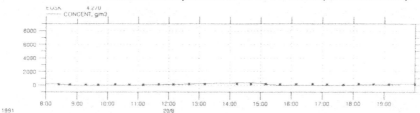

Figure 9   *Simulated and observed concentrations of suspended sediment. Spring tide, August 1991.*
        Calibrated sediment-transport model.- a: 4.27 km & b: 14.57 km from seaward boundary.

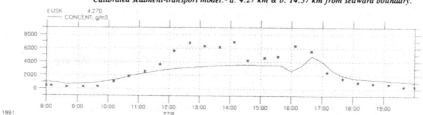

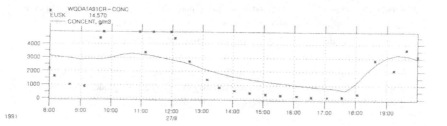

235

Figure 7   *Calibration of water quality model. Simulated and observed dissolved*
                  oxygen. Spring tide, August 1991.
                  a: Chainage 4.27 km from seaward boundary.
                  b: Chainage 14.57 km from seaward boundary.

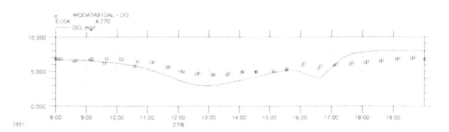

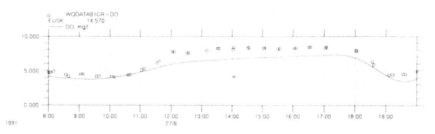

From the figures, it appears that the best correlation with field data is achieved for
the spring tide period. Generally, the deviation between simulated and observed values
for both spring and neap tide is in the range 0-1.5 g/m^3 (corresponding to
approximately 0-20%), and the maximum deviation is approximately 3 g/m^3. The
underestimation of dissolved oxygen at chainage 14.57 km in the neap simulation (Figure
6b) may be caused by an inaccuracy in the applied fresh water boundary conditions, and
has nothing to do with the suspended sediment - dissolved oxygen relationship. This is
revealed by inspection of simulated and observed salinity (not shown here).

Though not incorporated in this model, the SSOD may vary over a spring-neap-spring cycle
due to the fact that the thickness of sediment layers being resuspended depends on the
tidal energy. In deposited sediment, the deeper layers normally have higher potential
oxygen demands than layers closer to the sediment surface (e.g. Lauria & Goodman,
1986). This suggests that the specific oxygen demand (oxygen demand per unit suspended
sediment) caused by resuspension of sediment may be higher at spring tides than at neap
tides. This has not been further investigated in this project.

236

The neap and spring tide situations represent two fundamentally different regimes of suspended solids. While in the neap tide velocities never allow suspended sediment to rise above 500 g/m³, suspended sediment peaks at more than 10,000 g/m³ in the spring tide. Though the detailed dynamics are not fully resolved by the model, it generally estimates the concentration of suspended sediment at the right level. Observed vertical inhomogenous distribution of suspended sediment may contribute to a disagreement between simulated and observed concentrations.

Validation   Results from monitoring of water quality variables at 8 stations on two surveys in August 1978 were used as an independent data set to test the calibration. As mentioned above, monitoring of suspended sediment and ammonia was not included in these surveys.

The accordance between simulated and observed concentrations are considered rather good (results not shown). The deviation from the measured values is generally less than 1 g/m³ (approximately 15%) and the maximum deviation is approximately 2 g/m³. These results are actually better than for the calibration period.

Figure 10 presents simulated and observed values of dissolved oxygen during the entire spring-neap-spring cycle, using the calibrated model. The results for chainages 2.78 km and 14.57 km are shown, for the August 1978 event.

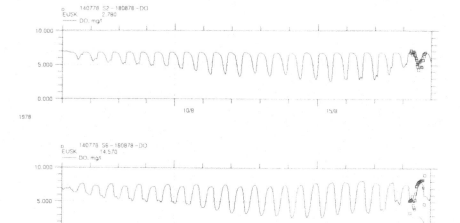

Figure 10   *Simulated and observed dissolved oxygen during a spring-neap-spring*
                   cycle. August 1978.
                   a: Chainage 2.78 km from seaward boundary.
                   b: Chainage 14.57 km from seaward boundary.

Based on these results, it is evaluated that the long-term stability of the model is acceptable. The model adequately describes the variation and dynamics of dissolved oxygen over a period that totally excludes any significance of initial conditions on output results. Moreover, in this long-term simulation, field-based boundary conditions have only been available for the two survey days. In the rest of the cycle, artificial boundary conditions have been applied, interpolating between the two data set. This, of course, tends to decrease accuracy and accordance with measured data.

238

3. Conclusions

Simultaneous measurements of dissolved oxygen and suspended sediment indicated that suspended sediment plays an important role in the oxygen balance of the River Usk.

This was supported by laboratory experiments on sediment samples from the river. In closed bottle tests, the oxygen demand of suspended sediment was found to decline from approximately 5g O$_2$/g SS/d to approximately 2g O$_2$/g SS/d during the initial phase (less than two hours) or resuspension of sediment and 0-2g O$_2$/g SS/d during the following period.

These findings were used to model the influence of suspended sediment on the oxygen balance of the river. Through sensitivity analyses, it was verified that among the factors included in the analyses, oxygen demand associated with resuspension of sediment most significantly affected the oxygen balance.

The model was calibrated and verified against measured data. It is concluded that the model describes the observed relation between suspended sediment and dissolved oxygen reasonably well. Generally, the model predicted dissolved oxygen within an error of 0-20% compared to observed values. The model was shown to be long-term stable, evaluated on a spring-neap-spring cycle scale.

A key aspect of the system has been shown to be the re-suspension of fluid mud shortly after neap tides. In the model runs from 1978 and 1991, this appears as a maximum oxygen sag arising shortly after neap tides rather than at spring tides. Oxygen levels drop as the fluid mud formed during neap tides is eroded and then recover toward spring tides as the tidal exchange of water with the Severn Estuary increases. These model findings were confirmed by field observations (Fig.11).

Figure 11   Simulated and observed sediment concentrations - spring-neap-spring cycle August 1991.

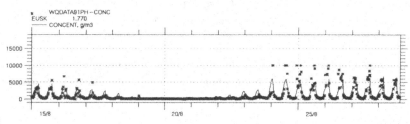

239

References

[1]   Bach, H.K., H. Brink, K.W. Olesen and K. Havno, (1989): "Application of PC-based Models in River Water Quality Modelling", Proc. 2nd Int. Conference on Hydraulic & Environmental Modelling of Coastal, Estuarine and River Waters, Bradford, 1989.

[2]   Barcelone, M.J., (1983): "Sediment Oxygen Demand Fractionation, Kinetics and Reduced Chemical Substances". Wat. Res., 17, No. 9, pp. 1801-1093.

[3]   Krone, R.B.,(1962): "Flume studies on the transport of sediment in estuarial shoaling processes: Final report"; Hyd. Eng. Lab. and Sanit. Eng. Res. Lab., Univ. of California, Berkely (USA); June 1962.

[4]   Lauria, J.M. & Goodman, A.S., (1986): "Measurement of Sediment Interstitial Water C.O.D. Gradient for estimating the Sediment Oxygen Demand", in K.J. Hatcher (editor): Sediment Oxygen Demand: Processes, Modelling and Measurement, University of Georgia, Athens, U.S.A., pp. 367-388.

[5]   Murphy, P.J. & Hicks, D.B., (1986): "In-situ Method for measuring Sediment Oxygen Demand", in K.J. Hatcher (editor): Sediment Oxygen Demand: Processes, Modelling and Measurement, University of Georgia, Athens, U.S.A., pp. 307-322.

[6]   Olesen, K.W., Havno, K. & Malmgren-Hansen, A., (1989): "A Water Quality Modelling Package for fourth-generation Modelling". Proc. IAHR Congress, Ottawa, 1989.

[7]   Olesen, K.W. and Kjelds, J.T.,(1991): "Modelling of alluvial cohesive sediment transport processes"; Int. Symp. on the Transport of Suspended Sediment and its Mathematical Modelling; Florence (Italy). September 2-5, 1991.

[8]   Olesen, K.W., Parker, W.R., Parfitt, A.J. and Enggrob, H.,(1992): "Field laboratory and model investigations of the feasibility of a tidal barrage in a muddy hypertidal estuary"; Tidal Power Conference and Exhibition, Inst. of Civil Engineering, London (UK), 19-20 March, 1992.

[9]   Porcella, D.B., Mills, W.B. & Bowie, G.L., (1986): "A Review of Modelling Formulations for Sediment Oxygen Demand", in K.J. Hatcher (editor): Sediment Oxygen Demand: Processes, Modelling and Measurement, University of Georgia, Athens, U.S.A., pp. 122-138.

[10]   Snodgrass, W.J., (1986): "Comparison of Kinetic Formulations of three Sediment Oxygen Demand Models", in K.J. Hatcher (editor): Sediment Oxygen Demand: Processes, Modelling and Measurement, University of Georgia, Athens, U.S.A., pp. 139-170.

[11]   Van Rijn, L.C.,(1989): "Handbook on sediment transport by current and waves"; Delft Hydraulics, Delft (Netherlands); Report H461, June, 1989.

20 Theoretical and practical aspects of a hydrodynamic and water quality study of the tidal River Axe, Weston-S-Mare

E. Aristodemou, D. V. Smith and C. D. Whitlow

ABSTRACT

One dimensional hydrodynamic and water quality simulations have been carried out in order to determine the concentration levels of B.O.D., ammonia and dissolved oxygen in the tidal part of the River Axe, Weston-super-Mare. The work formed part of series of investigations carried out by Wessex Water Plc after a new outfall was proposed discharging fully treated effluent in the River Axe. The hydraulic simulations were carried out using the ONDA software which solve the open channel flow equations by applying the Preissmann box finite difference scheme. The water quality simulations were based on the STYX software which solves the one-dimensional advection-diffusion equation using an explicit implementation of the SMART algorithm of Gaskell and Lau, [5].

1. Background

Weston Super Mare is a coastal resort on the Severn Estuary in South West England. There is a resident population of 80,000 people which on peak weeks in the summer rises to 120,000. The current discharge regime for the town involves fine screening and disinfection. Flows are discharged into the tidal River Axe through an outfall at Black Rock. Wessex Water proposes to significantly improve the treatment facilities for Weston by providing a new sewage works on the Bleadon Levels. It is proposed that the discharge is made on the ebb tide to significantly improve the dispersion of the discharge into Weston Bay. Mathematical modelling has formed part of a series of investigations in support of these proposals [10]. This work has been in support of the planning applications with its associated environmental assessment and the applications to NRA for a revised consent to discharge. These applications were being considered at the time of writing this paper. The paper concentrates on the theoretical basis of the modelling procedure and its applications to the River Axe channel. The hydraulic model ONDA and water quality model STYX, both developed by Sir William Halcrow and Partners Ltd were used. The tidal regime considered throughout the study was a spring tide and all the necessary information was based on surveys carried out in the summer and autumn of 1991 [9].

2.   Numerical Modelling of Water Quality

2.1   *Introduction*

In recent years, increasing awareness of environmental issues has focused the attention of scientists and engineers on the problem of predicting the dispersion of water contaminants in riverine and estuarine systems. Knowledge of the spatial and temporal distribution of concentrations of the salient species can provide invaluable assistance to those involved in planning environmental improvements to existing situations by locating suitable sites for discharges and appropriate temporal regimes according to the ambient hydraulic or water quality conditions. The mathematical modelling of these processes has taken two directions in recent years, each of which has associated advantages and disadvantages. The approach which has proved most popular in terms of its use in proprietary software and modelling text books is the use of the Advection Diffusion Equation (ADE).

$$\frac{\partial \phi}{\partial t} + U.\nabla\phi = \nabla.(D\nabla\phi) + S \tag{1}$$

Where ϕ is a scalar transport variable
       U   is a known vector velocity field
       D   is a symmetric second rank dispersion tensor
       S   is a source or sink term

The ADE has the great advantage of being physically based, as it is derived from the Navier Stokes equations which describe motion in any Newtonian incompressible fluid. The relationships between each component process involved are clearly stated and the significance of each relative to the others can be readily understood. The process of simplifying the ADE from three dimensions to one dimension reduces equation (1) to

$$\frac{\partial(AC)}{\partial t} + \frac{\partial}{\partial x}(QC) = \frac{\partial}{\partial x}(DA\frac{\partial C}{\partial x}) + S \tag{2}$$

Which is a valid representation of the longitudinal mass balance according to Bowden [2], Chatwin and Allen [3], or Fischer [4].

The equation thus describes the one dimensional dispersion of a contaminant in a flow whose known longitudinal velocity is U(t). This velocity is an averaged quantity both over the width and depth of the flow so cannot represent boundary shear layer effects adjacent to the channel banks and bed, re-circulation zones or effects due to stratification so long as D remains constant. Furthermore since a single concentration is associated with each longitudinal co-ordinate, the contaminant is assumed to be well mixed or at least averaged for the one-dimensional assumption to hold. In general, these assumptions tend to lead to more accurate results in hydraulic modelling than in water quality modelling. The problem in water quality modelling for a particular scenario lies in the calibration which consists of making a suitable choice for the Dispersion Coefficient D. In general this may be at least spatially dependent and may exhibit significant variability, sometimes by an order of magnitude.

The main alternative to the ADE is the Aggregated Dead Zone (ADZ) method pioneered by Beer and Young [1]. Unlike the ADE, the diffusion is not assumed to be Fickian and the dispersion process is assumed to be dominated by the aggregated effect of storage or dead zone processes. A dead zone represents not just areas of dynamic storage but all regions where mixing is taking place such as turbulent eddies or regions of reversed flow associated with bends. The advantages of the ADZ model are that it is often capable of more accurate simulation of observed dispersion events at all timescales than the ADE due to the flexibility of the description of the diffusion process. The disadvantages are that it requires a lot of good quality data which is particularly true for a statistically rather than a physically based model. This may cause difficulties when modelling large or complex hydraulic networks even for steady flows. Furthermore as far as the authors are aware, the ADZ approach has never been used for problems where the flow velocity was non-uniform. The problem of modelling the tidal part of the River Axe could only therefore be addressed by the ADE method.

2.2   *Numerical Modelling of the ADE in One Dimension*

Early attempts at numerical modelling of equation (2) were always plagued by large amounts of (uncontrolled) numerical diffusion associated with a first order finite difference representation of the variables. Despite this, solutions produced using these methods were at least bounded and stable which is more than can be said for any second order finite difference scheme where results can oscillate wildly (Liu and Falconer [7]). The first accurate and reasonably reliable scheme was due to Leonard [6] and christened as QUICK (Quadratic Upstream Interpolation for Convective Kinematics). Researchers from the fields of Mechanical and Aeronautical Engineering such as those at Rolls Royce, Imperial College and Leeds University realised that even a third order scheme such as QUICK had deficiencies for strongly advective problems. Oscillations could be observed which would sometimes give rise to negative turbulent kinetic energy in turbulence models of engine flows and unphysical negative concentrations in water quality models. These problems are due to the unboundedness of computed solutions in the neighbourhood of sharp changes in gradient of dependent variables caused by the discretisation of the advective term in the advection diffusion equation. Despite these difficulties the QUICK scheme and even first order schemes are still employed in present day programs for water quality modelling.

243

A method which avoids these problems is the SMART algorithm, first proposed by Gaskell and Lau [5]. This approach uses a nonlinear switch mechanism to determine the appropriate discretisation of the advective term. The method, based on a control volume formulation is stable, conservative, transportive and essentially third order. It does not exert a high computational cost but is mathematically more complicated than other methods and consequently maybe more demanding to program. Further, it can be shown analytically to satisfy the "convective boundedness criterion" defined originally by Leonard [6]. It has been used successfully in extremely complex two and three dimensional flows (eg Gaskell and Lau [5] and Wright et al [11]). SMART has also been used in conjunction with Reynolds Stress Turbulence Modelling and Multigrids.

The SMART algorithm was first written into the Halcrow Water Quality Model, STYX in 1988 as part of a study to assess salinity in the River Usk. STYX uses an explicit implementation of SMART to solve unsteady problems and uses results generated by Halcrows' ONDA model for the hydraulic modelling of rivers and estuaries. In order to rigorously test the algorithm, STYX was applied to three stringent benchmarks problems whose solutions are shown in Figures 1, 2 and 3:

These three benchmark problems show that SMART and therefore STYX is well suited to solving water quality problems where the dispersion process is dominated by advection. Wallis et al (1989) [8] estimated that approximately two thirds of the dispersion of contaminants in rivers is generated by advection. This effect is even more dominant in the estuaries of the Western Coast of Britain and particularly around the Bristol Channel which has one of the largest tidal ranges in the world.

3.   Hydraulic Simulations

3.1   *Computational Details*

Figure 4 shows the tidal part of the Axe Channel under investigation, starting from the Brean Cross Sluice as the upstream end down to Ferry point at the downstream end. The physical distance between the two ends is 2.65km. The values of the discretisation parameters Δx and Δt, which refer to the space and time increments respectively, used throughout the study were (i) $\Delta x = 25m$ and (ii) $\Delta t = 180$ secs. The physical distance between the first upstream node A1 representing the Brean Cross Sluice and the last downstream node E133 representing the present outfall location is approximately 3.3km. In addition to the discretisation parameters cross-sectional measurements were taken at four locations along the channel. The cross-sectional information for the remaining nodal points is determined by an interpolation routine within the software.

Boundary Conditions In order to carry out the simulations for the present outfall case the model required:- (i) Discharge (m³/sec) versus Time (sec) at the upstream end (node A1) and (ii) Stage (metres above Ordance Datum) versus Time (sec) at the downstream end (node E133). As the intention was to study the worst case possible, a minimal discharge at the upstream end was considered with a value of Q=0.05 m³/sec. Real measurements of tidal heights during a spring tide were taken at the node A107 relative

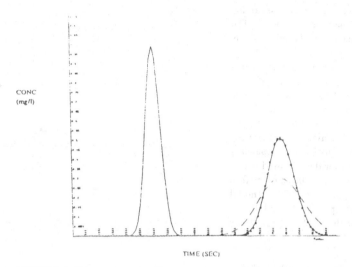

CONC
(mg/l)

TIME (SEC)

FIGURE 1 *Dispersion of pollutant with an initial Guassian distribution. Results shown are for (i) — exact (ii) □ - SMART (iii) V - QUICK (iv) —.—. UPWIND. Δt = 5 sec, Δx = 100m, D = 10.0m²/s, V = 0.5 m/s*

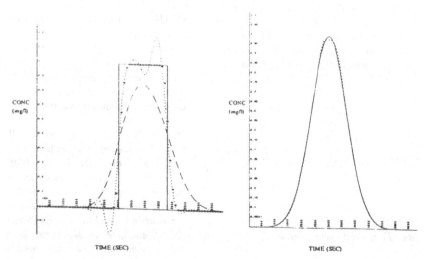

CONC
(mg/l)

TIME (SEC)

CONC
(mg/l)

TIME (SEC)

FIGURE 2 *Dispersion of pollutant with an initial step profile. Results shown are for (i) — exact (ii) □□ - SMART (iii) ..... QUICK (iv) —.— UPWIND Δt = 10.0 sec, Δx = 100.0m, D = 0.0m²/s, V = 0.5m/s.*

FIGURE 3 *Dispersion of pollutant with an initial Guassian distribution under the conditions of a symmetric flow reversal. Results shown are for (i) — exact, and (ii) ---- - SMART Δt = 10.05, Δx = 100.0m, D = 0.0m²/s*

245

to the bed level as measurements at the E133 node could not be obtained. The assumption was made that the tidal heights relative to O.D. at the node E133 were approximately the same as the ones measured at node A107 with the effect of slope between the two nodes taken into account. The tidal heights measured over the spring tidal cycle are shown in Figure 5 where four identical cycles were repeated. The hydraulic boundary conditions required for the proposed outfall case were as above but with an extra flow hydrograph input at C1 (nodal representation of proposed outfall) of 0.23m³/sec during the discharge period and zero otherwise.

Calibration Coefficients/Sensitivity Tests In the dynamic equation describing the hydraulic state of an open channel the term relating to bed friction is usually important and the value of the empirical friction coefficient, known as Manning's coefficient is determined by calibration. This implies that hydraulic data, and in particular tidal heights, at one or two locations within the

FIGURE 4 *Tidal part of river under invesigation*

channel, showing the variation over a tidal cycle has to be collected and compared to simulated values.

During this simulation work half hourly hydraulic information was available at the node A107. Both tidal heights and current meter measurements were collected at this point. The tidal heights were used as the boundary conditions at E133 and the current meter measurements used for comparison with the model results. Further tidal height data at

FIGURE 5 *Tidal heights during a spring tide. Four tidal cycles shown*

the sluice (node A1) was also collected. In order to show the effect of the bed friction coefficient a range of time values was considered. Figures 6 and 7 show the computational tidal height and velocities together with the measured data at the nodes A1 and A107 respectively. From Figure 6 it is seen that the Manning's coefficient has a significant effect only at the beginning of the low water period.

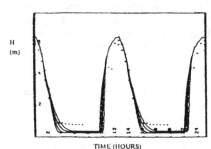

TIME (HOURS)

FIGURE 6 *Simulated Tidal heights using different Manning's n in the range 0.02 to 0.075; crosses represent real data*

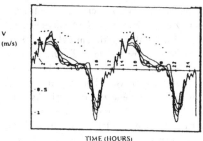

TIME (HOURS)

FIGURE 7 *Simulated Velocities using different Manning's n in the range 0.02 to 0.075; crosses represent real data*

During the remaining part of the tidal cycle the effect on the tidal heights is minimal. From Figure 7 the effect of the Manning's value is more prominent during the low water period with the velocity values increasing with a decrease in the Manning's coefficient. For the remaining part of the tidal cycle, however, the effect of changing the value is again minimal. The lower simulated tidal heights and velocities during the low water period when compared to the measured data is attributed to the lack of knowledge of the discharge through the sluice. A minimal discharge of 0.05 m³/sec was assumed in the models. The discrepancy during the low water period highlights the importance of having continuous measured data at the boundaries.

4 Water Quality Simulations

4.1 *Present Outfall Case*

Boundary Conditions As with the hydraulic simulations the water quality model requires the definition of boundary conditions at both the upstream and downstream ends. Therefore the time variation of the three variables of interest over a tidal cycle had to be established. Real measurements comprised of four data points during the spring tidal cycle for the upstream end (node A1) and three data points for the downstream end (node E133). For the upstream end information was available at the times of (a) high water (b) 2.00 hours after high water (c) 5.5 hours after high water and (d) 9.5 hours after high water. For the downstream end, information was available at the same times with the exception of the time at low water ie 5.5 hours after high water. Interpolations had to be carried out between the available data points in order to form the continuous time variation of the four variables at the two end points of the model. Ideally data from continuous monitoring would be available for the specification of boundary conditions. As no information was available at the downstream node E133 during the low water period, it was assumed that the concentrations of B.O.D. and

247

ammonium would have the maximum values as that of the discharged untreated effluent which are B.O.D. of 400 mg/litre and ammonium of 30 mg/litre. The dissolved oxygen value was taken to be as low as 2.0 mg/litre. Tests are presently being carried out to determine the dissolved oxygen levels of crude sewage and these will be incorporated in the study, if necessary, at a later stage.

Calibration Coefficients   Three empirical coefficients are used, associated with the processes describing loss of oxygen due to oxidation of B.O.D. loss of oxygen due to oxidation of ammonium and gain of oxygen due to surface re-aeration.

Sensitivity tests were carried out using different values within recommended ranges but no significant variations occurred in the results, indicating that the most important processes affecting the concentration levels are the advection and diffusion processes. An example is shown in Figure 8 where a comparison of the results using two different values of the re-aeration coefficient is shown. The simulated output is dissolved oxygen at the node A107 and the re-aeration coefficients used were 1.0 cm/hour and 7.0 cm/hour. The effect of the diffusion coefficient D on B.O.D. and NH4 was clearly seen when water quality simulations were carried out using three different values, these being 1.0 m²/sec, 10.0 m²/sec and 50.0 m²/sec. Results are shown in Figure 9 and from the comparisons it was obvious that increasing the value of the diffusion coefficient, the peak at around 9.5 hours after high water increased. For the remainder of the tidal cycle the effect of changing the diffusion coefficient was minimal. As the peaks observed using the diffusion coefficient of 50.0 m²/sec were much higher than the measured peaks, the water quality simulation for the proposed outfall were carried out using diffusion coefficients of 1.0 m²/sec and 10.0 m²/sec only. The effect of the diffusion coefficient on dissolved oxygen is shown in Figure 10. As can be seen the increase of the diffusion coefficient affects the values during the low water period only.

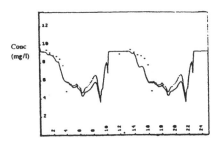

TIME (HOURS)

FIGURE 8   *Simulated D.O. using re-aeration coefficients of (i) —— 1.0cm/h and (ii) ..... 7.0cm/h (iii) + - real data*

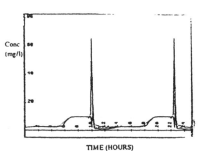

TIME (HOURS)

FIGURE 9   *Simulated B.O.D. at node A12 for diffusion coefficients of (i) —— 1.0m²/sec, (ii) ----- 10.0m²/sec and (iii) ..... 50m²/sec (iv) + - real data*

4.3 Proposed outfall Case

Boundary Conditions   The proposed outfall case requires, in addition to the definition of the upstream and downstream boundary conditions, the definition of the input

248

variables at the discharge location. These were B.O.D. of 25.0 mg/litre, ammonium (NH4$^+$) of 30.0 mg/litre and dissolved oxygen of 4.5 mg/litre. For the upstream and downstream boundary conditions certain assumptions had to be made as it is not possible to measure the water quality variables at the boundary nodes. These were ideal water quality conditions at both ends. These assumptions were based on the fact that the effluent was to be discharged for a certain period of time after high water only and thus it would not affect the upstream node A1 and by the time it reaches the downstream node it would be heavily diluted.

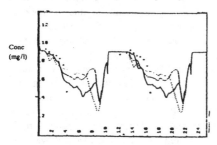

TIME (HOURS)

FIGURE 10 *Simulated D.O. at node A107 for diffusion coefficients of (i) —— 1.0m²/sec, (ii) ----- 10.0m²/sec and (iii) ..... 50m²/sec (iv) + - real data*

Calibration Coefficients The values of the various oxidation rates and re-aeration rates used in the present outfall case were also used in this case. These were oxidation rate of B.O.D. of 1.0 per day, oxidation rate of ammonium of 0.16 per day, oxidation rate of nitrite of 0.43 per day, and oxygen re-aeration coefficient of 1.0 cm/hour. Simulations were carried out for the different diffusion coefficient of 1.0m²/sec, and 10.0 m²/sec.

4.4 Simulation Results

The simulations used for the comparisons between the simulated present and proposed cases together with the measured data were based on a diffusion coefficient of 1.0 m²/sec, oxidation rate of B.O.D. of 1.0 per day and oxidation rate of ammonium of 0.16 per day and oxygen re-aeration coefficient of 1.0 cm/hour. Examples of the comparisons are shown in Figures 11 to 13.

Simulated B.O.D. From the simulations it was clear that for the nodes upstream of the proposed discharge the effect is to reduce the B.O.D. levels by approximately a factor of 5. The measured data under the present conditions and the simulated present situation give B.O.D. values of around 10.0 mg/l while the proposed scheme gives B.O.D. levels of around 2.0 mg/l. For the nodes downstream of the discharge location the effect is slightly different. For these nodes the discharge period of the proposed discharge has a prominent effect. For the discharge period of 0.5-5.50 hours after high water, peaks occur at a different time in the tidal cycle. For some nodes the peaks are comparable to the simulated present situation (approximately 10.0 mg/l) but lower than the measured ones. For the remaining nodes the peaks are still approximately 10.0 mg/l but they are now lower than the simulated present and measured values. This indicates that although for this discharge period there are peaks, these peaks are not higher than the present measured values. An example of the results is shown in Figure 11. For the discharge period of 0.5-4.0 hours after high water the B.O.D.

concentration levels are still as low as around 2.0 mg/l even for the downstream nodes. Thus a reduction in B.O.D. levels occurs under the proposed scheme particularly when the discharge period is 0.5-4.0 hours after high water.

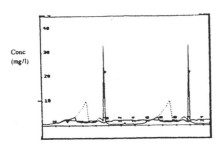

TIME (HOURS)

FIGURE 11 *Simulated B.O.D. at node A87 for (i) —— 0.5-4.00 discharge, (ii) ---- 0.5-5.50 discharge and (iii) ..... present (iv) + - real data*

Simulated Ammonium (NH4) From the simulation results it was clear that for the nodes upstream of the proposed discharge location, the concentration levels remained similar to the present situation with values not greater than 2.0 mg/l. However for the downstream nodes (Figure 12) the effect is again different. For the discharge period of 0.5-5.50 hours after high water ammonium levels rise above the simulated present situation and above the measured values. However for the discharge period of 0.5-4.00 hours after high water, ammonium peaks do not rise above 3.0 mg/l and for the greater part of the tidal cycle concentration levels are the same as for the simulated present

situation. From Figure 12 it is also seen that there is good agreement between the simulated present situation and the measured values. Thus, again, when the appropriate discharge period is chosen, the proposed scheme retains the ammonium levels within the ranges of the present situation.

Simulated D.O. From the study it was found for the nodes upstream of the proposed discharge location, the discharge period had no effect on the D.O. levels. However as it is seen in Figure 13, for the downstream nodes, the effect of the proposed scheme is to increase the D.O. levels during the low water period. The increase occurs for both proposed discharge periods, with the D.O. levels higher for the discharge period of 0.5-4.0 hours after high water.

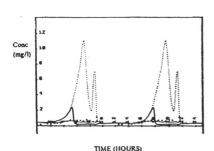

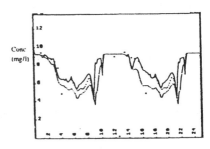

TIME (HOURS)

FIGURE 12 *Simulated NH4 at node A107 for (i) —— 0.5-4.00 discharge, (ii) ----- 0.5-5.50 discharge and (iii) ..... present (iv) + - real data*

TIME (HOURS)

FIGURE 13 *Simulated D.O. at node A107 for (i) —— 0.5-4.00 discharge, (ii) ----- 0.5-5.50 discharge and (iii) ..... present (iv) + - real data*

250

5 Conclusions

5.1 *Present Outfall Case*

The results from the calibration procedure of the models, using the present outfall case, showed that: (a) an increase in the diffusion coefficient affects significantly the values of B.O.D. and ammonium (NH4) only at the time when flooding begins (9.5-10.0 hours after high water). For the remaining of the tidal cycle the concentration values are not greatly affected. Very high values of diffusion coefficients such as 50.0 m²/sec give peaks much higher than the observed ones. The values of diffusion coefficients of 1.0 m²/sec and 10.0 m²/sec seem more appropriate, (b) the values of the simulated dissolved oxygen seem to increase with an increase during the low water period with the increase in the diffusion coefficient, (c) varying the empirical coefficients does not have any significant changes in the concentration values, indicating that the most important processes are the advection and diffusion processes. In addition, the fact that the variation of diffusion coefficient affects the majority of the variables only at one time during the tidal cycle, implies that the most important process is advection.

5.2 *Proposed Outfall Case*

For the proposed outfall case the conclusions were: (a) The length of time over which a discharge is made on the ebb tide is critical to an improvement in water quality, (b) B.O.D. levels are greatly reduced when compared to the simulated present discharge regime. The reduction is observed at all model nodes and for both ebb discharge period, although for the longer of the two discharge periods the simulated levels are greater, (c) ammonium levels are reduced at the upstream nodes but at the downstream nodes levels remain within the same range as the present discharge regime for the shorter tidal release. When the longer tidal release is considered, 0.5 to 5.0 hours after high water, simulated concentrations at the downstream nodes are higher than both the observed and the simulated present discharge concentrations, (d) Dissolved oxygen levels increase with the proposed scheme. For the upstream nodes, the profiles for the two discharge periods are very similar. For the downstream nodes, values are higher for the shorter tidal release, (e) Of the two cases studied the discharge period of 0.5 to 4.00 hours after high water is preferred. Field trials may define that this period can be extended. The present modelling indicated that the period would not be as long as 5.5 hours after high water. It is recommended that a release time of 0.5 to 4.5 hours be investigated.

The modelling indicates that for a proposed controlled discharge on the ebb tide, water quality of the tidal part of the Axe Estuary is significantly improved. The levels of BOD are reduced at all nodes, and the levels of ammonium and nitrite are reduced at the nodes upstream of the proposed discharge location. Dissolved oxygen levels also increase particularly during low water.

ACKNOWLEDGEMENTS.

The author would like to thank C.F. Skellett, Managing Director of Wessex Water Plc, for permission to publish this paper. The opinions expressed herein do not necessarily reflect the views of Wessex Water Plc.

REFERENCES.

[1]   Beer,T. and Young,P.C. (1983), "Longititudinal Dispersion in Natural
        Streams". *J Env.Eng, ASCE, 109, 1049-1067.*

[2]   Bowden,K.F. (1982), "Theoretical and Practical Approaches to Studies of
        Turbulent Mixing Processes in Estuaries". *Paper presented at NERC Workshop
        on Estuarine Processes, University of East Anglia.*

[3]   Chatwin,P.C. and Allen,C.M. (1985), "Mathematical Models of Dispersion in
        Rivers and Estuaries". *Ann. Rev. Fluid Mech. 17, 119-149.*

[4]   Fischer,H.B. (1976), "Mixing and Dispersion in Estuaries". *Ann. Rev. Fluid
        Mech. 8, 107-133.*

[5]   Gaskell,P.H. and Lau,A.K.C. (1988), "Curvature Compensated Convective
        Transport; SMART, A New Boundedness Preserving Transport Algarithm".
        Int. J. Num. Meth. Fluids, 8, 617-641.

[6]   Leonard,B.P. (1979), "A Stable and Accurate Convective Modelling Procedure
        Based on Quadratic Upstream Interpolation". *Comp. Meth. Appl. Mech. Eng,
        19, 59-88.*

[7]   Liu,S.Q. and Falconer,R.A. (1989), "Application of the QUICK Scheme for
        Two Dimensional Water Quality Modelling". *Pro. Int. Conf. Hydraulic &
        Environmental Modelliing of Coastal Estuarine & River Waters, Bradford
        Gower Press.*

[8]   Wallis,S.G., Young,P.C. and Beven,K.J. (1989), "Experimental Investigation
        of the Aggregated Dead Zone Model for Longtitudinal Solute Transport in
        Stream Channels". *Proc. Inst. Civ. Engrs, Part 2, 87, 1-22.*

[9]   Wessex Internal Report. (1991), "D7013 Weston Super Mare, Axe Channel
        Discharge Studies 1991, Water Quality Investigations".

[10]   Wessex Internal Report. (1991),"1-Dimensional Study of the Axe Channel :
        Hydrodynamic and Water Quality Simulations". *Report No.
        WWT/1051/91/MOD.*

[11]   Wright,N.G, Gaskell,P.H.and Sleigh,P.A. (1991), "Multigrid Solutions of
        Practical Engineering Flow with Local Grid Refinement," *International
        Conference on Industrial and Applied Mathematics (ICIAM) 1991, Washington
        D.C.*

21 A one-dimensional microbiological pollution model of the Upper Clyde Estuary

J. M. Crowther, J. M. Bennett, S. G. Wallis,
J. C. Curran, D. P. Milne, J. S. Findlay
and B. J. B. Wood

ABSTRACT

We summarise the work done so far to produce a computer model of the dispersion of *Escherichia coli* bacteria in an estuary. A biological model has been combined with hydrodynamic, sediment transport, and salinity models and applied to the Clyde estuary. The model was calibrated for friction factor and dispersion coefficient and gives good agreement with measured levels and depth–averaged velocities. We discuss two runs of the model and compare the results with field data for bacterial and sediment concentrations. Considering the stratified nature of the estuary, the model results are very encouraging.

1   The Clyde Estuary

The Clyde estuary (Figure 1) has been an important factor in the economic development of the city of Glasgow but, with the decline of both shipping and ship–building, the estuary is being used increasingly for recreation and tourism.

The mean tidal range of the estuary is about 1.9m for neaps and 3.0 for springs, as measured at Greenock, and there is an amplification of the tidal range by about 15% between Greenock and Broomielaw in the centre of Glasgow.

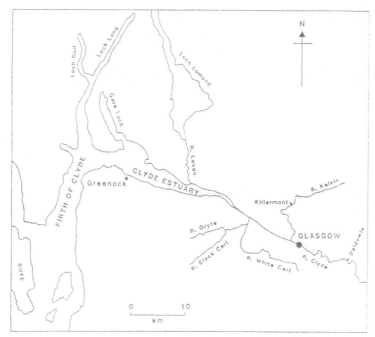

FIGURE 1   *Map of the Clyde Estuary and Firth of Clyde*

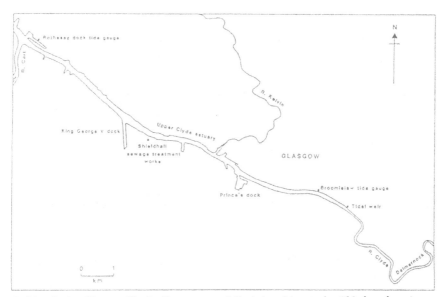

FIGURE 2   *Upper Clyde Estuary modelled in this study (11 km long).*

The River Clyde provides the main fresh water input to the estuary and the flow has varied from as low as 5.7 m^{3}s^{-1} to as much as 831 m^{3}s^{-1}.

The mathematical model developed during this project has been applied to an 11 km section of the Clyde estuary from Rothesday Dock to the Tidal Weir in the centre of the City (Figure 2).

This part of the estuary is a narrow, dredged channel varying in depth at high water from 12 m at Rothesday Dock to about 3 m at the tidal weir. As will be discussed later, this estuary is typically highly stratified with relatively fresh water overlying salt water.

2 The Mathematical Model

The model is based on the following equations, which have been averaged over depth and cross-section. Conservation of mass:

$$(W + W_s) \frac{\partial h}{\partial t} + \frac{\partial Q}{\partial x} = \frac{\rho_L}{\rho} L , \tag{1}$$

conservation of momentum:

$$\frac{\partial Q}{\partial t} + \frac{\partial}{\partial x} \left[\frac{Q^2}{A} \right] + gAS_f + gA \frac{\partial h}{\partial x} + \frac{gAR}{2\rho} \frac{\partial \rho}{\partial x} = 0, \tag{2}$$

the equation of state:

$$\rho = \rho_o (1 + \lambda S) , \qquad \lambda = 0.00075 , \tag{3}$$

conservation of salt:

$$\frac{\partial}{\partial t} (AS) + \frac{\partial}{\partial x} (QS) = \frac{\partial}{\partial x} \left[AD \frac{\partial S}{\partial x} \right] + LS_L , \tag{4}$$

conservation of sediment:

$$\frac{\partial}{\partial t} (AC_s) + \frac{\partial}{\partial x} (QC_s) = \frac{\partial}{\partial x} \left[AD \frac{\partial C_s}{\partial x} \right] + LC_{sL} + \ldots$$

$$\ldots + E - D_p C_s , \tag{5}$$

and conservation of bacteria:

$$\frac{\partial}{\partial t} (AC_b) + \frac{\partial}{\partial x} (QC_b) = \frac{\partial}{\partial x} \left[AD \frac{\partial C_b}{\partial x} \right] + LC_{bL} - KAC_b \tag{6}$$

In these equations x measures distance upstream (in metres) and t measures time (in seconds). The unknowns are :

$h(x, t)$	the water level in metres relative to a datum,
$Q(x, t)$	the discharge $(m^3 s^{-1})$,
$S(x, t)$	the salinity, measured in parts per thousand,
$C_s(x, t)$	the concentration of suspended solids $(kg\ m^{-3})$
$C_b(x, t)$	the concentration of bacteria in millions m^{-3}.

$W(x, h)$, $R(x, h)$ and $A(x, h)$ are known functions of x and h and are defined by the coss-sectional profile of the river. They are, respectively, the width of the estuary at the surface (in metres), the hydraulic radius (m) and the cross-sectional area (m^2).

$W_s(x)$ is the average width of the static storage areas, and allows for flows into and out of docks and basins, which are assumed not to effect the momentum balance within the estuary.

The source term in Equation 1, $L(x, t)$, represents the lateral inflow per unit length (measured in $m^2 s^{-1}$), and caters for small tributaries. We assume these add no momentum to the river. On the part of the Clyde for which the model has been tested there are two sources, the river Kelvin and Shieldhall Sewage works, flow data being supplied from the Killermont gauging station (operated by the Clyde River Purification Board) and the Shieldhall Inlet Works (operated by Strathclyde Regional Council).

$\rho_L = \rho_0(1 + \lambda S_L)$, is the density of the inputs, while the terms S_L, C_{sL} and C_{bL} appearing in Equations 4 – 6 are the salinity, suspended solids and bacterial concentrations of the inputs.

In Equation 2 the momentum coefficient has been taken to be 1 (i.e. the average momentum flux has been assumed to be equal to the flux of the average momentum). Similar assumptions are implicit in Equations 4 – 6.

S_f, the friction slope, includes effects of bed roughness, bends, viscosity and Reynolds stresses. It is given by the empirical law

$$gAS_f = \frac{fQ|Q|}{8AR}, \tag{7}$$

where f is the Darcy-Weisbach friction coefficient. So far we have treated f as a constant, equal to 0.03, although the program has been written to allow for f to be a function of x, if desired.

The last two terms in Equation 2 arise from the pressure gradient. The coefficient AR in the last term is used to approximate the integral over the river width of the depth squared.

The equation of state (3) links the hydrodynamic equations to the salinity equation through the density ρ. When the spatial variation in the salinity is small, the last term in (2) is small and could be ignored.

We have chosen to treat the suspended solids and bacterial concentrations in a similar manner to the salt concentration; see Equations 4, 5 and 6. In these equations we have modelled all the dispersive effects by a Fickian diffusion term with a longitudinal dispersion coefficient $D(x, t)$ $(m^2 s^{-1})$. The main contribution to this term arises from different advection rates owing to the velocity shear perpendicular to the direction of flow. In our one-dimensional model we assume this is proportional to the average velocity, although this is not always true, especially when the flow is highly stratified. In fact the model currently runs with D a function of x only, increasing linearly from 100 $m^2 s^{-1}$ at the tidal weir to 900 $m^2 s^{-1}$ at Rothesay Dock where the velocity is greatest.

In the suspended solids model, Equation 5, the terms $- D_p C_s$ and E represent the amount of sediment lost through deposition and gained through erosion. The deposition rate is given by the Krone formula:

$$
D_p = \begin{cases} (1 - \dfrac{u_*^2}{v_d^2}) \, W_b \, v_s & \ldots \quad u_* < v_d \\[2em] 0 & \ldots \quad v_d < u_* \end{cases} \tag{8a}
$$

while the erosion is determined from

$$
E = \begin{cases} 0 & \ldots \quad u_* < v_e \\[2em] (\dfrac{u_*^2}{v_e^2} - 1) \, W_b \, \lambda & \ldots \quad v_e < u_* \end{cases} \tag{8b}
$$

and the shear velocity u_* can be calculated from the average speed:

$$
\left| \frac{Q}{A} \right| = \frac{u_*}{k} \, \log_e \left[30.2 \, \frac{H}{k_s} \right], \tag{8c}
$$

where k_s is the Nikuradse sand roughness, H the depth of water and the von Karman constant k for flow with sediment is 0.174 [4].

Deposition occurs when the shear velocity u_* is less than a critical value v_d, while erosion occurs at shear speeds in excess of v_e. W_b is the bed width in metres, v_s the settling velocity and k the erosion rate. The model is similar to one used by Le Hir, Bassoulet and L'Yavanc [5], but simpler in that we have chosen v_d, v_e, v_s and k and k_s to be constants:

v_d = 0.01 m s^{-1}, v_e = 0.012 m s^{-1}, v_s = 0.01 m s^{-1},

λ = 2.10^{-5} kg m^{-2} s^{-1}, k_s = 0.002 m.

The number of bacteria in the estuary depends on the relative rates at which they are reproducing and dying. The estuary is a hostile environment for *E. coli* and very few will actually reproduce. Bacterial concentrations are measured by growing samples from the estuary on agar plates and counting the number of colonies that are formed. Thus C_b in fact measures the concentration of bacteria that retain their ability to reproduce, should favourable conditions arise, and the inactivation rate K (s^{-1}) in Equation (6) measures the rate at which they lose this ability. This inactivation rate depends on a multitude of variables: for example, the amount and/or the quality of sunlight and hence the turbidity of the water, nutrient and oxygen levels within the water, and the salinity and temperature of the water. Thus the actual value of K can vary widely. Laboratory experiments carried out in the absence of light [1] have shown that at 20°C and with a salinity of 32 ppt, 90% of the bacteria become inactive in just over 7 hours, while at 10°C and 10 ppt salinity it takes about a month, a hundred times longer. Findlay, Curran, Milne, Crowther and Wallis [2] have shown that the inactivation rate is also reduced for very high concentrations of bacteria, a result important close to outfalls. Further experiments show that the amount of suspended solids, apart from affecting the turbidity, is important in two other ways, firstly through adsorption and sedimentation [1], [7] and secondly through a self-protection mechanism of the bacteria [6]. Up to 30% of the bacteria can be adsorbed on suspended solids, some of which then sediment out. During daylight hours, however, the most important effect is the sunlight. Fujioka, Hashimoto, Siwak and Young [3] indicate that 90% of *E. coli* will die in about an hour in field conditions that include sunlight.

In our model we specify values of temperature and sunlight which vary with time, but not in space. The die-off rate is then determined from experimental results as a function of the local salinity and suspended solids concentration.

The boundary conditions for the model are as follows. The height, salinity, suspended solids and bacteria concentrations are specified at a downstream site ($x = x_0$):

$h = h_0(t)$, $S = S_0(t)$, $C_s = C_{s0}(t)$, $C_b = C_{b0}(t)$,

while upstream ($x = x_1$) we have:

$Q = Q_1(t)$, $S = S_1(t)$, $C_s = C_{s1}(t)$, $C_b = C_{b1}(t)$.

When using the model on the Clyde, the water levels h_0 have been taken from an automatic tide gauge at Rothesay Dock operated by the Clyde Port Authority, while the discharge Q_1 is obtained from the Daldowie gauging station, run by the Clyde River Purification Board. S_1, the salinity at the

258

upstream boundary has been assumed to be zero, while values of S_0, C_{so}, C_{bo}, C_{s1} and C_{b1} need to be measured.

The hydrodynamic model is started up from rest at a high tide at least one full tidal cycle before the start of the salt, suspended solids and bacterial model,

$$h = h_0(t_0) \quad , \quad Q = Q_1(t_0) \quad \text{for all } x \text{ at } t = t_0.$$

This ensures that errors arising from inaccurate guesses of the initial conditions have had time to disappear. Initial conditions for the transport models also need to be measured. While the salinity and suspended solids distributions can be predicted from a few measurements, this is not true of the bacterial concentrations, which can be highly localized.

3 Numerical Methods for Solving the Model Equations

The hydrodynamic equations (1) and (2) were solved by using a Preissmann finite difference scheme. This is implicit in time and therefore allows large time steps to be used. An irregular grid with 29 points was used to represent the estuary. The average step-size was a little less than 400 m. Figure 3 shows the widths at each point of the grid.

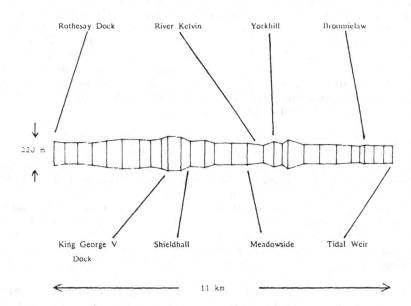

FIGURE 3 *Schematic diagram of the Upper Clyde Estuary showing the maximum widths at each point of the grid and the places mentioned in the text. (Widths drawn on a different scale to lengths, do not include storage widths W_s.)*

259

For the transport equations (4), (5) and (6), a linear Galerkin finite element method was used to treat the x derivatives [8], [9]. This is easier to implement on a non-uniform grid than a finite difference scheme since the equations are second order in x. When the grid is uniform the discretization is almost the same in both cases. The time derivatives in the transport equations were treated by finite differences in a similar manner to the hydrodynamic equations. Large timesteps can be used without effecting the stability of the solution, although they will of course effect the accuracy. To determine the best time step the model was run with various time steps between 30 seconds and 15 minutes. Little significant difference was observed between the results, and 15 minutes has generally been used for speed.

Boundary conditions are interpolated to allow for any time step. Output consists of the values of the height, velocity, salinity, sediment and bacterial concentrations at specified intervals, set by the user.

4   Verification of the Model

The hydrodynamic model gives good agreement (Wallis et al. (1989)) with recorded heights at Broomielaw and with depth averaged measurements of velocities.

Figures 4,5 show model runs compared with observations for two days in 1989: the 5th of April and the 16th of August. The continuous line represents output from the model, the + signs show field data. On both days a survey was carried out for comparison with the model.   Data were collected at the boundaries (Rothesay Dock and the tidal weir), at the inflows (the River Kelvin and Shieldhall Sewage Treatment Works), and at two other points roughly a third and two thirds of the way between the end points. Measurements were taken of sunlight, salinity, suspended solids concentration and bacterial concentration.

The state of the tide was similar for both days, Figures 4a,5a, with the transport models running from an hour or two before high tide to about four hours afterwards.   Figures 4b,5b show the velocities which are also similar, though larger in April. The salinities are shown in Figures 4c,5c, and suspended solids in Figures 4d,5d.   Bacterial concentrations were larger in August (Figures 4e,5e), when there was found to be a high concentration at the Rothesay Dock boundary. coming either from the Daldowie Sewage Treatment Works further downstream, or from the River Cart, just outside our model. As the tide turned, bacterial concentrations throughout the river were reduced.

There are a few outliers on the graphs, which may arise from inaccuracies in measurements or initial guesses. The graphs shown were run with an inactivation term made up of two parts, the smaller one depending on the salinity, suspended solids and temperature, varying from zero to 2.0×10^{-5} s^{-1} and a larger contribution depending on the sunlight and turbidity, which went up to 5.0×10^{-5} s^{-1}.

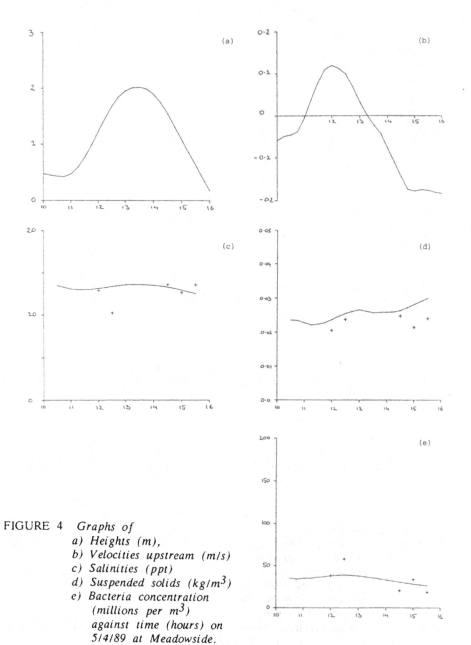

FIGURE 4   *Graphs of*
     a) Heights (m),
     b) Velocities upstream (m/s)
     c) Salinities (ppt)
     d) Suspended solids (kg/m^3)
     e) Bacteria concentration
       (millions per m^3)
       against time (hours) on
       5/4/89 at Meadowside.

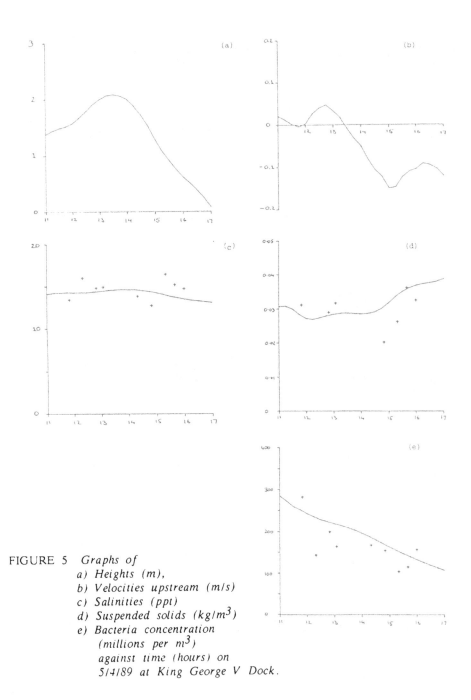

FIGURE 5 Graphs of
a) Heights (m),
b) Velocities upstream (m/s)
c) Salinities (ppt)
d) Suspended solids (kg/m^3)
e) Bacteria concentration
(millions per m^3)
against time (hours) on
5/4/89 at King George V Dock.

262

During both surveys the estuary was observed to be highly stratified, with fresh water from the river flowing over salty sea water. The bacterial concentrations also appeared to be slightly stratified, being greater in the fresh water, either because of buoyancy or because of higher inactivation rates in the sea water. This stratification effects the accuracy of both the model and the field data. Overall, however, the agreements with observation are encouraging for all the models.

5 Acknowledgements

The authors wish to thank the Department of the Environment for its financial support for the project, and also the Project Officer, Mr C E Wright, and his colleaague Dr Otter for their advice and assistance. We also wish to thank Mr D Hammerton, Director of the CRPB, for his support and for permission to publish this paper, the crew of the CRPB survey vessel Endrick 2, and Mr D Thomson of the Clyde Port Authority for helping us with tidal data and for giving us the benefit of his detailed knowledge of the Clyde estuary. Finally we wish to thank Strathclyde Regional Council for assistance with flow data and with sampling at Shieldhall Sewage Treatment Plant

6 References

[1] Findlay J.S. (1990), Ph.D. Thesis, University of Strathclyde.

[2] Findlay J.S., Curran J.C., Milne D.P., Crowther J.M. and Wallis S.G., (1990). The Self-Protection of *E. coli* in seawater. *Journal of the Institution of Water and Environmental Management*, Vol. **4**, No. 5, pp 451-456.

[3] Fujioka R.S., Hashimoto H.H., Siwak E.B. and Young R.H.F. (1981), Effect of sunlight on indicator bacteria in seawater. *Applied and Environmental Microbiology*, Vol. **81**, No. 3, pp. 690-696.

[4] Graf W.H. (1971). *Hydraulics of sediment transport*. McGraw-Hill.

[5] Le Hir P., Bassoulet P. and L'Yavanc J. (1989), Modelling mud transport in a macrotidal estary. *Advances in water modelling and measurement*, Ed. M.H. Palmer, BHRA, Cranfield.

[6] Milne D.P., Curran J.C., Findlay J.S., Crowther J.M. and Wallis S.G. (1989), The effect of estuary type suspended solids on survival of *E. coli* in saline waters. *Water Science and Technology*, Vol. **22**, No. 3, pp 61-65.

[7] Milne D.P., Curran J.C. and Wilson L. (1986), Effects of sedimentation on removal of faecal coliform bacteria from effluents in estuarine water. *Water Research*, Vol. **20**, No. 12, pp 1493-1496.

[8] Wallis S.G., Crowther J.M., Curran J.C., Milne D.P. and Findlay J.S. (1989). Consideration of a one dimensional transport model of the upper Clyde estuary. *Advances in water modelling and measurement*, Ed. M. H. Palmer, BHRA, Cranfield.

[9] Williams D.J.A. and Nassehi V. (1980), Mathematical tidal model of the Tay Estuary. *Proceedings of the Royal Society of Edinburgh*, Vol. **78B**, pp s171-s182.

22 Models for coastal and estuarine problems: A review of ICI's approach on Teesside

R. E. Lewis

ABSTRACT

For more than twenty years, ICI has employed mathematical modelling techniques in seeking to assess the effect on the environment of liquid effluents from its Teesside sites. An historical review is given of the variety of modelling techniques employed and the ways in which they have been used in guiding environmental improvement measures and in setting up monitoring procedures. Consideration is also given to the types of model which will be required for perceived future problems.

1 Introduction

The estuary of the River Tees in north-east England began to develop into a major centre of industrial activity in the 19th century (Figure 1). It was the juxtaposition of the coal fields of County Durham and the iron ore mines of north Yorkshire which led to the setting up of blast furnaces on land reclaimed from the estuary mudflats. Slag from the smelters was used to form a tide training wall along the main channel of the river and this channel was deepened by dredging to provide access for larger ships, thus increasing the export trade in coal and iron products. By 1900 over half the area of intertidal mudflats had been reclaimed. With the coming of steel ships, locally produced steel was used by the shipbuilders and there was an appreciable growth in the populations of the two main towns of Stockton and Middlesbrough. In the absence of an obvious alternative, effluent from the industries and sewage from the towns were discharged to the estuary.

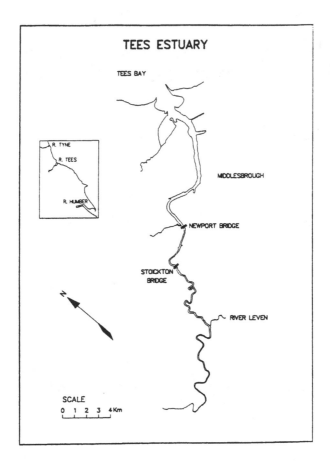

FIGURE 1   *Location of the estuary of the River Tees and Tees Bay.*

In the geological past the location of the Tees estuary formed part of the Zechstein Sea and as this sea dried up appreciable deposits of sodium, potassium and other salts were formed. These salts formed the basis for the chemical industry which grew up on the banks of the estuary after 1920. In 1926 the anhydrite and coal resouces began to be used by ICI to manufacture nitrogenous fertilisers and other ammonia based products. The expansion of the port and the supply of oil, initially by ship and latterly by pipeline from the North Sea Ecofisk field, resulted in the setting up of oil refining plants and petrochemicals works. This led to further reclamation and dredging of the lower estuary, with the concomitant loss of intertidal flats. Today Teesside is one of ICI's principal bases for the manufacture of plastics, nylon and intermediates for other oil based products.

The growth of industry and housing on Teesside over the years from 1930 to 1970 increased the pollutant discharge to the estuary by a factor of about twenty, taking the biochemical oxygen demand (BOD) as a measure of this load. Consequently a central reach of the estuary became totally devoid of oxygen, forming a lifeless zone and prohibiting the passage of migratory fish. Over the period 1950 to 1970 biological studies showed the extent to which the worsening contamination had affected aquatic life and in 1968 the first attempt was made to create a predictive model of the estuary's water quality. Since that time mathematical modelling techniques have been used to assess environmental impact, plan clean-up measures and select monitoring sites in response to the ever changing nature of industrial activity on Teesside. This paper describes the historical sequence of this modelling work, indicating how it has contributed to the achievement of a balance between the needs of industry and the environment. The paper concludes with a brief discussion of the perceived future concerns and suggests which modelling techniques are likely to be most helpful.

2 Historical review of model types and their application

2.1 Estuary models

The first model of the Tees estuary, which was set up by the Water Pollution Research Board [4], assumed a steady state situation with two uniformly mixed layers. Although of limited application for describing details of the water quality, this work concluded that the model could provide "useful guidance for broad engineering decisions". It was appreciated at this early stage that the steady state assumption could result in misleading predictions. This is particularly true for those effluents which are discharged to the seaward end of the estuary, as such wastes have a greater chance of being carried through the estuary mouth into the open sea on an ebb tide. It was evident that a tidally varying model would be able to allow for such losses and work on a one dimensional model of this type, in which concentrations over a cross-section were assumed to be uniform, commenced at ICI in 1968 [7].

The Tees has a marked vertical circulation and an inevitable consequence of assuming cross-sectional uniformity is that this circulation forms part of the longitudinal dispersion coefficient. Thus, the coefficient in a 1-D model varies significantly with changes in the strength of the vertical circulation, as may be brought about by variations in river flow or tidal range, and is therefore far from being a constant parameter. Attempts to relate the dispersion coefficient empirically to flow parameters, such as velocity and depth, have never proved entirely successful. This makes it difficult to set up a calibrated estuary 1-D model which is satisfactory for all conditions.

Most of the inputs to the Tees are discharged with large volumes of fresh water and consequently buoyant waste fields develop at the surface and mix downwards, whilst being transported by the tidal movement. This results in an appreciable variation in concentration with depth. With a 1-D model each concentration is expressed as a depth mean and the model cannot indicate how the concentration varies over depth; this situation is unsatisfactory as, in specifying target criteria for water quality, the controlling authority is interested in the passage of migratory fish or the health of the benthic fauna [2]. Attempts have been made to use a functional relation to express the concentration at a specified depth as a function of the depth mean [16] but this approach cannot allow for the changes in vertical distribution caused by different rivers flows or tidal range, for example. A move into two dimensions obviates this problem and also has the advantage that the dispersion coefficients cease to be dominated by the strength of the vertical circulation. In 1969 it was decided that the ultimate aim should be the creation of a time varying model of the estuary which would be two-dimensional in that it would be able to predict concentration distributions along the estuary axis and over depth. It must be remembered that this aim was regarded as very ambitious at the time because running such a complex model on the the best computers then developed was estimated to be "prohibitively expensive".

During the 1970s considerable effort was put into developing the 2-D model. However progress was slow, partly because of the complexity of modelling a stratified system and partly because the running time on a large computer was long enough to make it expensive, although not as prohibitive as feared. The 2-D model also had the limitation that the velocity distribution was derived from an interpolation of observations made along the estuary axis. For these reasons a working 2-D model was not produced until 1980, when the increase in computing power enabled the velocity field to be calculated by solving the equations of motion.

Once set up and calibrated, the 1-D and 2-D models have proved invaluable for many applications. For example, the 1-D model was useful for demonstrating tidal changes in concentration; tidal displacement of the zone of oxygen depletion in the Tees was clearly shown on a cine film, made by photographing a sequence of frames from the model output.

Another application of these models, even at their developmental stage, proved to be in defining the spatial and temporal requirement for field surveys. It became clear at an early stage that the data on the physical and chemical structure of the estuary were wholely inadequate for modelling purposes and joint studies were undertaken by WPRL and ICI in 1968 and 1969. These included investigations of water movement and dispersion using radio-active Bromine 82 as a tracer [5,6]. An important finding from the tracer studies was that, although a substantial proportion (possibly as much as 80 %) of a labelled ICI effluent remained in the surface waters and escaped to sea on the ebb, some effluent was mixed into the lower layers and transported upstream. The proportion retained in the estuary in this way would remove dissolved oxygen through the process of biodegradation. Further joint surveys of the water quality of the estuary have been undertaken every five years since 1970 and the data obtained have proved useful for quantifying the major processes [9,10].

One of the prime uses of the 1-D and 2-D models of the Tees estuary has been to determine which inputs to an estuary have most effect on water quality. In deciding where expenditure should be directed, such information is vital to ensure that improvement measures represent the best return on investment. By removing all input loads except one in turn, the models could be used to indicate the sources which contribute to the concentration of contaminant at a specific point in the estuary, including the effect of inputs at the seaward end. (Figure 2)

(a)

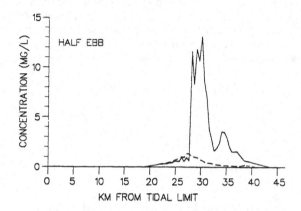

(b)

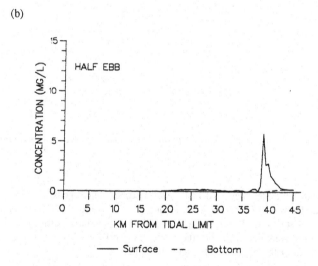

FIGURE 2   *Predicted concentrations of a conserved substance discharged at a rate of 10 te/day at (a) 28 km and (b) 39 km from tidal limit.*

The application of the 2-D model to problems in the eighties will be considered later in the section on the Tees barrage scheme.

2.2 Outflow plume model

As contaminants from the estuary are carried into Tees Bay on each ebb tide, ICI Group Environmental Laboratory commenced studies of benthic fauna distributions in the bay in 1970. In 1973 and 1977 observations were made of the outflow plume formed at the mouth of the estuary on each ebb and these data were used to calibrate a mathematical model of the outflow plume [11,14]. The model was designed to estimate the area of seabed in the bay where concentrations resulting from the downward mixing of the outflow waters would reach a maximum (Figure 3). This is analogous to the maximum ground level concentration which occurs downwind from a chimney stack.

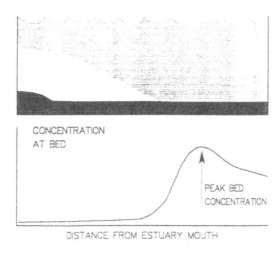

FIGURE 3 *Peak concentration at seabed due to downward mixing of contaminants in outflow.*

The results from the model suggested that areas of seabed where peak concentrations would occur lay between 1.5 km and 4.0 km from the estuary mouth, corresponding to tides of neap and spring range respectively. Although the station positions for the benthic monitoring were set up before the modelling work was instituted, there was sufficient coverage of the bay to test the model predictions.

The results did indeed provide evidence of adverse biological effects close to the predicted areas of maximum concentration. The match with the results of fauna studies not only showed that the monitoring stations were reasonably well selected but it also demonstrated that the methods used to assess biological impact, in this instance the general reduction in biomass and slower shell growth rate for a small bivalve, were sufficiently sensitive to detect changes in water quality [17].

2.3 Sea disposal models

Since 1977 the liquid effluent from ICI's manufacture of methyl methacrylate has been disposed to the North Sea. This waste is released into the wake of a moving ship as it traverses an approved disposal area some 16 km off the mouth of the River Tees. Since the operation started there has been regular monitoring of benthic fauna, fish and plankton in the disposal area and in a 'control' area to the north. In addition to this work, field and modelling studies have been undertaken to estimate the rate of dilution of effluent in the wake and the subsequent dilution of the waste field. The wake dilution was estimated from theoretical considerations of the energy lost to turbulence by the movement of the vessel through the sea [12]. The discharging tanker follows a zig-zag course to spread the effluent over the disposal area and, as the sea disposal operation takes approximately 8 hours, a broad waste field is formed by this procedure. A random walk model was set up and run to predict the changing concentration distribution of this waste field over a 5 day period; fully mixed and stratified conditions were modelled [13].

The predictions of the wake dilution , which were supported by field measurements [3], were related to the volume discharge rate, discharge arrangement and speed of vessel used. By taking account of the results of toxicity studies on marine life in the Laboratory [8], the discharge conditions which would reduce effluent concentrations most rapidly were defined. Although the residual currents may not be strong enough to completely remove effluent from the disposal area before the introduction of another batch of waste, the predicted concentrations suggested that chronic effects on benthic fauna, fish and plankton would be unlikely to occur. This finding is supported by the monitoring studies which have not found any evidence of adverse changes which could be ascribed to the sea disposal operation. Nevertheless, sea disposal of effluent is not regarded as acceptable by the international community and the discharge of the methyl methacrylate waste to sea from a tanker will cease in 1993, once recycling facilities become operational on the ICI Billingham site.

2.4 Tees barrage

As part of its overall plan for improving the social and economic conditions on Teesside, the Tees Development Corporation has proposed that a barrage be constructed across the estuary approximately half way between the Stockton and Newport Bridges (Figure 1). The barrage would act as a weir, allowing river water to spill over into the tidal estuary and preventing the upstream penetration of saline water. It is proposed to use the waters on the landward side of the barrage for recreational purposes such as rowing and canoeing. Power generation is not proposed but a hydraulic system to open flaps in the face of the weir would be included as part of a flood prevention scheme.

Mathematical models of the estuary to assess the effect of the barrage on flooding and contaminant levels under existing discharge loads were set up by Hydraulics Research Limited and the predictions were used to support the acceptance of the scheme by Parliament. When the barrage scheme was proposed, a new version of ICI's 2-D model was created to simulate flow and mixing conditions on the seaward side of the weir.

It is possible that as remedial measures improve dissolved oxygen concentrations in the estuary, nitrification could make an increasing contribution to oxygen removal. For this reason ICI sponsored a reseach student at the University of Dundee to undertake a study of nitrification in the estuary. The results of this study are being used in the 2-D model to ascertain the extent to which oxygen levels are affected by nitrification at present low oxygen conditions and to estimate how much additional treatment might be required to allow for the effect of increased nitrification. Some recent studies of nitrification in the Tees estuary have also been undertaken by Hydraulics Research [15].

Major engineering changes do bring the problem that characteristic flows and mixing conditions may also be altered considerably. When that happens, there are no data for validating the model and reliance has to be placed on the relionships used to describe the system. With respect to the friction and mixing coefficients, the state of knowledge is still weak and confidence can only be gained by demonstrating that the model responds correctly to such parameters as freshwater flow or tidal range. An expected consequence of the construction of the Tees barrage would be a reduction in the strength of the tidal currents. Therefore, the 2-D model of the full estuary was run to show that it still matched the observations when the range was reduced from springs to neaps. This was taken as supportive of the model's ability to simulate a weakening in tidal current strength.

It is essential to collect data to show how well the models could cope with predicting the effect of the barrage on water quality. For this reason, a joint survey funded by the interested parties was undertaken in 1990 to define the present situation with the intention of repeating the survey after construction of the barrage.

The ICI 2-D model was used to provide ICI management and NRA staff with guidance on the possible consequences of the barrage scheme on water quality, taking into account the treatment and recycling measures which ICI had already instituted. The results from the modelling work suggested that the construction of the barrage would greatly reduce the strength of the tidal flow and increase the density difference between the surface and bottom waters. These changes would alter the distribution of dissolved substances, resulting in a deterioration in water quality in some areas and an improvement in others.

2.5   Outfall plumes and dispersing patches

Even after treatment of the domestic and industrial wastes, liquid effluent will still be discharged to the estuary, forming plumes which will change shape and direction with the tidal current. With improvements in the general water quality of the estuary, the populations of resident and migratory fish are likely to significantly increase. Thus, there is a concern that the concentrations of chemicals within these plumes should not be acutely toxic to aquatic life. Although simple Gaussian plume models have been available for several years, they lack the sophistication needed to give a satisfactory prediction of 'mixing zones' and their variation, as required by the NRA for consent purposes. For this reason, models of the plumes produced by the principal discharges are being set up using random walk techniques; in this method elements of the waste field are represented by particles, the locations of which are tracked by the computer. The random walk models can be used to describe the effect on the concentration distribution in a plume due to tidal current reversal, obstacles in the flow such as bridge piers and wharf piles, inhibition of vertical mixing by stratification and degradation of the discharged substance.

In addition to simulating plumes, models are being developed to predict the movement and dilution of patches of material resulting by accidental releases to the drainage system. A random walk approach is again employed for this type of model. Spill models are designed to be run quickly and easily so that with the minimum of input data, an environmental expert on site would be able to assess the scale of the incident. The delay time in the drainage system may be sufficient for the scale of an incident to be estimated so that action, such as temporary storage of the effluent, could be implemented before the contaminant is carried into the estuary. This means that the model has to be located at a central control point on the site and there has to be a data base on the toxicity of the principal chemicals being used in the manufacturing processes. A spill control model currently being used at ICI's Grangemouth site provides an indication of the magnitude of the incident by classifying it as extremely serious, serious, significant or insignificant; this initial response information is then followed by a graphical display of the development of the patch of contamination in the estuary.

3 Discussion

With improvement in the water quality of the estuary, the discharge plumes are likely to become of increasing concern. More detailed models will be required to define 'mixing zones' and their variation during the tidal cycle. Thus, there is a perception that the emphasis will move away from models of the complete estuary to local area models. However, an increasing interest in the more subtle contaminants and their fate within the estuary will still require models of the whole system. Research is likely to centre on the effect of processes, such as sediment transport, biodegradation, nitrification and chemical speciation.

For many practical purposes the 2-D models are adequate, as the canalised shape of the estuary and its relative narrowness mean that contaminants have almost uniform concentration across the estuary, except in the vicinity of the discharge plumes. However, lateral variations may be more significant than previously thought and there may be a move towards 3-D models of the full estuary. It is possible that friction and mixing coefficients will be replaced by a turbulent energy model, as applied to computational fluid dynamics (CFD), and it may be easier to use three dimensional CFD models for this application. However, the problems with using this approach in a natural stratified flow are substantial and, for practical purposes, 'coefficients' are likely to remain in general use.

By analysing data sets in detail, it has been possible to extract more information on the processes which control the system. Analysis of salt fluxes in the Tees estuary showed that an energy balance existed in which the effect of vertical circulation was complementary to the oscillatory effect of the tide [10]. Such data can be used to estimate the magnitude of mixing coefficients along the estuary [9], a step towards selecting mixing rates for the model without the need for numerous trial runs. This approach has much to contribute to understanding estuarine processes and could be even more supportive of model development.

All of the models have provided the biologists with an indication on the suitability of sampling positions for benthic fauna. As the understanding of the processes improves and models become more refined, they will be even more useful in defining where detailed monitoring should be maintained in order to provide the biologists with the best data on the impact of contaminants.

The approach to problems today would differ in some ways from that adopted in 1970, principally because the understanding of the marine environment has advanced and computers, the basic tool for modelling work, have developed beyond all expectation. The 2-D model and the latest plume models, which have a random walk approach, are currently run on a Meiko transputer system. Random walk techniques have considerable potential for the future, especially because their particle tracking basis makes them suitable for suspended solids modelling.

4 Conclusions

The mathematical modelling techniques which have been applied to pollution problems on Teesside over the past 20 years have proved of particular value in:

(a) determining the relative contribution of various discharges to concentrations of toxic substances or dissolved oxygen levels in the estuary;

(b) determining the spread and dilution of effluent plumes from individual sources;

(c) estimating the consequences of engineering changes on water quality in the estuary;

(d) assisting the biologists in the selection of monitoring stations.

As the water quality of the Tees estuary improves, the plumes formed by the various discharges will be seen as areas of particular concern. More emphasis is also likely to put on transport of contaminants by suspended solids and the factors controlling processes such as biodegradation and nitrification. Particle models, assisted by increased computer power, are likely to be of considerable use in all these areas of application.

REFERENCES

[1]   Alexander,W.B, Southgate,B.A. and Bassindale,R.,(1935). Survey of the River Tees, Part II The estuary - chemical and biological. Water Pollution Research, Technical Paper No.5, HMSO.

274

[2] Brady,J.A., Stead,R.G. and Ord,W.O.,(1983). Pollution control policies for the Tees estuary. *Water Polllution Control*,1983, 367-380.

[3] Byrne,C.D, Law,R.J., Hudson,P.M., Thain,J.E. and Fileman,T.W., (1988). Measurements of the dispersion of liquid industrial waste discharged into the wake of a dumping vessel. *Water Research*, **22**, 1577-1584.

[4] Downing,A.L, (1967). Survey of the Tees estuary. *Water Pollution Research 1967*, HMSO, 28-33.

[5] Downing,A.L., (1968). Survey of the Tees estuary. *Water Pollution Research 1968*, HMSO, 27-32.

[6] Downing,A.L., (1969). Survey of the Tees estuary. *Water Pollution Research 1969*, HMSO, 23-28.

[7] Hobbs,G.D., (1970). The mathematical modelling of a stratified estuary. *Advances in Water Pollution Research*, Proc.5th Int.Conf., San Francisco and Hawaii, **2**, III-8.

[8] Hutchinson,T.H. and Williams,T.D., (1989). The use of sheepshead minnow (Cyprinodon variegatus) and a benthic copepod (Tisbe battagliai) in short-term tests for estimating the chronic toxicity of industrial effluents. *Hydrobiologia*, **188/189**, 567-572.

[9] Lewis,R.E., (1981). Estuary mixing. *Chemical Engineer*, **371/2**, 381-383.

[10] Lewis,R.E. and Lewis,J.O., (1983). The principal factors contributing to the flux of salt in a narrow, partially stratified estuary. *Estuarine, Coastal and Shelf Science*, **16**, 599-626.

[11] Lewis,R.E., (1984). Circulation and mixing in estuary outflows. *Continental Shelf Research*, **3**, 201-214.

[12] Lewis,R.E., (1985). The dilution of waste in the wake of a ship. *Water Research*, **19**, 941-945.

[13] Lewis,R.E. and Riddle,A.M., (1989). Sea disposal: modelling studies of waste field dilution. *Marine Pollution Bulletin*, **20**, 3, 124-129.

[14] Lewis,R.E., (1990). The nature of the outflows from the north-east estuaries. *Hydrobiologia*, **195**, 1-11.

[15] Nottage,A.S., Birkbeck,T.H. and Mackie,C., (1991). The distribution, extent and significance of nitrification in the Tees estuary. First International Conference on Water Pollution: Modelling, Measurement and Prediction, Elsevier Applied Science, London, 455-461.

[16] Ratasuk,S., (1972). A simplified method of predicting dissolved oxygen distribution in partially stratified estuaries. *Water Research*, **6**, 1525-1532.

[17] Shillabeer,N. and Tapp,J.F., (1990). Long term studies of the benthic biology of Tees bay and Tees estuary. *Hydrobiologia*, **195**, 63-78.

23 Estuary management: Modelling the impact of variable input loads

K. J. Clark, B. M. Mollowney and B. Harbott

ABSTRACT

Estuary models are needed by regulatory authorities, such as the National Rivers Authority, both as a general quality management tool for the estuary itself and as an aid to establishing consent for discharges to these bodies of water. An estuary may receive polluting loads from many sources; the quality of each discharge will probably fluctuate from day to day which makes more difficult the task of establishing a realistic consent condition. The paper addresses this problem and describes a one-dimensional quality model, QUESTS, which enables the variable nature of the input loads to be accommodated.

A one-dimensional, deterministic, time-dependent, hydrodynamic and water quality estuary model provided the core of the new approach. Around this core was built a statistical shell which provides daily input values for each determinand and predicts estuary water quality using a statistical approach. The inputs are generated from statistical distributions of values for each determinand in each input and from specified correlations between determinands and sites. Predicted water quality is presented as either an instantaneous value or in terms of a mean value and percentiles.

This approach has been applied to a number of estuaries in the United Kingdom, for example, the Colne in Essex. Results from the calibration and application of the model to the Colne are presented.

1. Introduction

Computer models have now been used for some time by both the water industry and regulators as effective tools for managing estuary water quality. These models have often assumed either constant polluting loads or have used a user-specified time series of loads.

Polluting loads to estuaries come from many sources. They may be from diffuse sources, such as agricultural runoff, or from point sources, such as sewage treatment works, industry and fluvial inputs. The quality of these polluting loads will probably fluctuate from day to day. In addition, there will probably be several discharges into any given estuary.

When deriving consents conditions for discharges, the variability in the discharge quality and the combined effect of all discharges need to be considered. Modelling tools used to aid this process therefore need to take such considerations into account. This will become increasingly important as Statutory Water Quality Objectives (SWQOs) and the new classification scheme for estuaries are introduced under the Water Resources Act 1991. The proposals for SWQOs put forward by the National Rivers Authority (NRA) [1] include percentile-based standards of water quality. Consents to discharge should also be defined on a percentile compliance basis [2].

Two models designed to be used for setting consents on river systems, SIMCAT and TOMCAT, consider the variability of inputs. SIMCAT uses statistics from the routine monitoring of discharge water quality. It typically performs independent runs of a water quality model based on the mass-balance equation. Input loads are drawn randomly from the distributions described by the input statistics. Predictions of the statistical measures of river quality are made from the results of these runs [3].

This multi-shot approach has been applied to several estuaries, including the Orwell and the Nene, using a moving-segment steady-state WRc water quality model. When fixed-segment time-dependent models were developed for NRA Anglian Region by WRc, it was suggested that a similar statistical approach should be applied to the new model. This involved incorporating the variability of loads and producing statistical results which could be used directly in the management of the estuary.

The QUESTS model was developed for this purpose. It consists of a deterministic, time-dependent core surrounded by a statistical shell.

2. Statistical shell

Earlier estuary model shells were built around steady-state moving-segment models. These models have short run times and predict a single concentration value for each moving segment. This feature makes them suitable core models for a SIMCAT style multi-shot approach, in which many independent runs are carried out. For each shot, the input loads are randomly drawn from given distributions. The results from each run are accumulated into a set of results, from which means, standard deviations and percentiles are calculated.

The core models in QUESTS are time-dependent. A predicted time series of concentrations, rather than a single result, is produced for each fixed segment per each run. This means that the multi-shot approach adopted in earlier models is unsuitable.

WRc adopts an approach which generates a time-series of daily input loads and boundary conditions from statistics. This time-series is then used to run the core models. Finally, output statistics of predicted estuary water quality are calculated from the predicted time series. In essence, the approach simulates the time varying inputs, the response of the estuary system and a sampling program for the estuary. Figure 1 shows the outline structure of QUESTS.

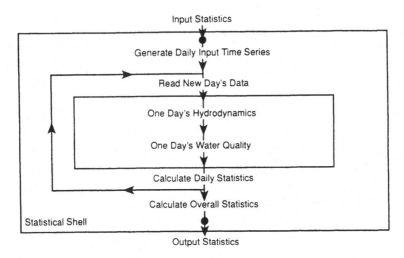

Input Statistics

Generate Daily Input Time Series

Read New Day's Data

One Day's Hydrodynamics

One Day's Water Quality

Calculate Daily Statistics

Calculate Overall Statistics

Statistical Shell

Output Statistics

FIGURE 1. *Outline structure of QUESTS*

QUESTS can also be run directly from a
user-supplied time series of loads. This approach
might be applied to examine a particular scenario,
or to study a given pollution incident, for
example.

2.1 Generation of inputs

Using a method based on earlier work developed by
NRA Anglian Region [4], daily input flows and
concentrations are generated for each determinand.
Values can be generated for all freshwater
boundaries and discharges, including those of
sewage treatments works (STWs), industrial sources
and rivers.

Experience has led to the use of a normal or
log-Normal distribution to describe the
distributions of input concentrations. Dissolved
oxygen, salinity and temperature are assumed to be
normally distributed, whilst all other determinands
are assumed to be log-Normally distributed. Thus
the distributions can be fully described using the
mean and standard deviation. The values of these
two parameters may be based on historical records
or chosen to reflect any scenario being considered.

Each input value is generated by selecting a
value at random from a normal distribution with a
mean of zero and standard deviation of one, then

transforming that value to the normal or log-Normal distribution with the required mean and standard deviation.

The randomly generated values for each determinand can then be cross-correlated with each other if required. Four types of correlation are possible: between the same determinand at different sites, between different determinands at either the same or different sites, and between the same determinand at the same site on successive days.

Correlations between two variables are carried out as follows. The first determinand value is generated as above, and the second calculated assuming bivariate normal distribution with the given correlation. Log-Normal determinands are transformed appropriately before and after this correlation process.

Correlations between more than two determinands are estimated. The approximation to multiple correlation becomes coarser as more variables are included.

The method used to generate a value x, say, which is correlated to all the determinands in a set of other determinands $\{y_1, .., y_n\}$ with correlations $\{r_1, ... r_n\}$ respectively is as follows. Independent values for $\{y_1, ..., y_n\}$ are generated from their distributions. Temporary variables $\{x_1, ..., x_n\}$ are generated assuming independent bivariate normal distributions of x and each of $\{y_1, ..., y_n\}$. The final value for x is then the weighted mean of $\{x_1, ..., x_n\}$ using $\{r_1, ..., r_n\}$ as weights. Thus any correlations within the set $\{y_1, ..., y_n\}$ are neglected.

2.2 Calculation of output statistics

The core models predict variations in estuarine water quality due to tidal movement, daily load variation, the daily cycle and processes within the estuary. An additional aim of QUESTS is to predict the mean, standard deviation and required percentile of the distribution of concentration of each determinand at each segment of the model over the period of the run.

The means and standard deviations can be calculated by the Weibull method (given below) as the run progresses. The nth percentile can be calculated by ranking all the predicted values and finding the value below which n percent of predicted values lie.

In QUESTS the percentiles are estimated in a two step process. In the first step, the means of model predictions for each day are stored. Then the nth percentile of these means is used as an estimate of the overall percentile. Percentiles are also estimated by the reverse process: the nth percentile of the values form each day are stored, and the mean of these is used as the estimate of the overall percentile. During the run, the overall mean μ and sum of squares S are updated at each timestep as follows:

$$\mu = \mu + (x-\mu)/j \tag{1}$$

$$S = S + (x-\mu)^2 (1-1/j) \tag{2}$$

where j is the index number of the timestep, that is, 1 for the first timestep, 2 for the second, .. and x is the current predicted value.

At the end of the run, μ holds the overall mean and the overall standard deviation σ is σ = S/(N-1) where N is the total number of predicted values used.

Daily means and standard deviations are similarly calculated over each day.

3. Core models

The time-dependent core of QUESTS consists of two separate sub-models. One is a hydrodynamic model; the other is a water quality model. The hydrodynamic model simulates the tidal water movement, calculating water levels and flows. These are passed onto the water quality model which then simulates the concentrations of a number of determinands.

3.1 The hydrodynamic model

The hydrodynamic model is based on the one-dimensional equations of conservation of mass and momentum. The equations are formulated in terms of flow Q and water level η and are:

Conservation of mass

$$w \frac{\partial \eta}{\partial t} + \frac{\partial Q}{\partial x} = q \qquad (3)$$

where w = surface width, x = distance along the estuary, t = time and q = lateral inflows per unit length.

Conservation of momentum

$$\frac{\partial Q}{\partial t} + gA \frac{\partial \eta}{\partial x} + \frac{\partial (uQ)}{\partial x} + \frac{gAR.\partial \rho}{2\rho_0 \partial x} + f \frac{|Q|Q}{RA} = 0 \qquad (4)$$

where A=cross-sectional area, g=gravitational acceleration, u=velocity, R=hydraulic radius, ρ=density and f is a friction coefficient.

These equations are solved explicitly on a grid which is staggered in space and time. Upstream differences are used for the advective terms; corrections for small depths are made as necessary [5]. The seaward water level boundary condition for the modelled period is constructed from tidal harmonic components, hence reproducing the spring-neap cycle.

3.2 The water quality model

The water quality model is based on the one-dimensional equation of conservation of mass. This equation is solved for each determinand to advect and disperse the substances within the estuary system.

Conservation of mass

$$\frac{\partial (AC)}{\partial t} + \frac{\partial (QC)}{\partial x} - \frac{\partial}{\partial x} \left\{ AD.\frac{\partial C}{\partial x} \right\} + KAC - L = 0 \qquad (5)$$

where C = concentration of substance, D = longitudinal dispersion coefficient, L = loads per km and K is a decay rate.

Process equations which model the interactions of each determinand are invoked at each timestep [6].

In the present version of QUESTS the determinands modelled are dissolved oxygen, BOD, ammonia, organic nitrogen, oxidised nitrogen, chlorophyll, phosphate, temperature, salinity, coliforms, suspended solids and sediments. The processes modelled include the erosion and deposition of sediments and associated particulate BOD and phytoplankton growth.

The advection dispersion equations are solved explicitly using a Lax-Wendroff type scheme with flux corrections to the advective terms. The process equations are solved explicitly.

4. Application of QUESTS to the Colne

The Colne estuary is 17km long and receives flows from several tributaries. The city of Colchester is situated around the tidal limit, a weir at East Mills. There is some industry along the estuary and cargo vessels travel up to Colchester. There are bathing waters and shell fisheries at the seaward end of the estuary. The Colne supports wildlife reserves and widespread recreational use. The estuary receives effluent from a number of sewage treatment works.

NRA Anglian Region requested WRc to develop a QUESTS model including a sediment sub-model and a facility to model storm events [7]. This development work followed an earlier QUESTS model of the Crouch and Roach estuaries designed in 1989.

A number of surveys were carried out in order to obtain calibration data. These included a bathymetric survey and intensive thirteen hour water level and water quality surveys on spring and neap tides.

Calibration was carried out by running the two core models without the statistical shell. Model parameters were varied until the time and spatial variation of the modelled determinands matched the observations, though standard values were used on the whole. Model predictions were compared with the observed data using time plots and half-tide corrected longitudinal plots. Figure 2 shows the longitudinal plot for dissolved oxygen concentrations during a spring tide. In the figure the positions of observed and predicted data points are adjusted to the position along the estuary that the water would occupy at half tide. This allows all the observed data and predictions to be plotted on one graph. The final calibration parameter set was then used in the full QUESTS model.

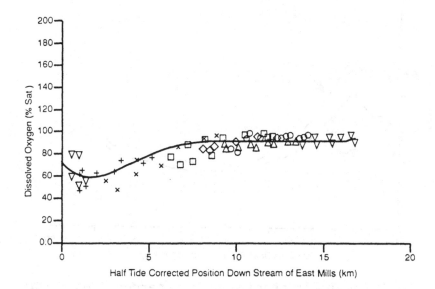

FIGURE 2 *Calibration plot for spring tide dissolved oxygen*

Annual statistics from the 1990 NRA routine discharge sampling program were used as estimates of the input to the full QUESTS model. Each input was assumed to be independent of any other input. A two month run was carried out. Figure 3 shows the predicted mean and 95th percentile of BOD along the Colne estuary.

5. Conclusions

The approach presented here combines deterministic water quality modelling with a statistical element to take account of the natural variability of discharges to estuaries. The QUESTS model can be applied to estuaries for which a one-dimensional approach is suitable. It can be configured for a given estuary by the use of data files.

QUESTS is the subject of continued development, both of its technical content and of its presentation.

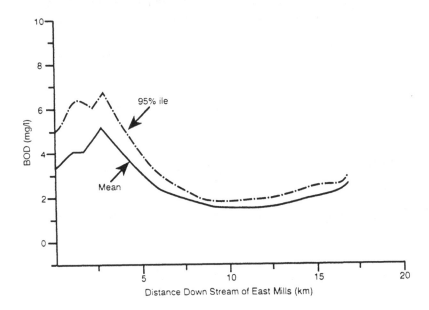

FIGURE 3 *Predicted means and 95th percentile of BOD over two month run*

QUESTS provides a useful tool for consent setting and estuary management.

6. Acknowledgements

This paper is published with the permission of the NRA and the directors of WRc. The authors wish to acknowledge the support of Anglian Water (latterly Anglian Water Services) in early aspects of this study. The authors also wish to acknowledge the help of Dr A Warn and Ms C Ullmer in the preparation of this paper. The views expressed are those of the authors and not necessarily those of the NRA or WRc.

References

[1] National Rivers Authority, (1991), *Proposals for Statutory Water Quality Objectives*, NRA Water Quality Series Number 5

[2] National Rivers Authority, (1990), *Discharge consent and compliance policy: a blueprint for the future*, NRA Water Quality Series Number 1

[3] Warn, A.E., (1987), SIMCAT - A catchment simulation model for planning investment for river quality. *Systems Analysis in water Quality Management*. Ed. M B Beck. IAWPRC.

[4] Warn, A.E., (1982), Calculating consent conditions to achieve river quality objectives. *Effluent and Water Treatment Journal* **22**, (4) pp 152-1554.

[5] Cunge, J.A., Holly, F.M., Verwey, A., (1980), *Practical Aspects of Computational River Hydraulics*, Pitman.

[6] Barrett,M.J.,Mollowney,B.M.,Casapieri,P., (1978), The Thames Model: an Assessment. *Progress in Water and Technology*, **10**, 516, 409-416.

[7] Morgan,N.H. and Slade,S., (1991), *Colne Estuary Water Quality Model*, WRc Report CO2797-M

24 Proposed integrated waste treatment centre, Hull water quality model, River Humber

M. D. McKemey

ABSTRACT

This paper outlines the application of standard water quality models for the prediction of the effects on the River Humber of the emissions from a proposed integrated waste treatment plant to be located south east of Hull.

1   Introduction

The study required the estimation of the effects of the input of metals into the estuary of the River Humber from an aqueous discharge at the site and from outfall from the incinerator stack of a proposed integrated waste treatment plant at Salt End, Hull.   The main effects were anticipated as being on water quality and on sorption of metals into the sediments.   The location and surroundings of the proposed plant are shown in Figure 1.

The proposed integrated treatment plant will have an impact on the Humber through particulate material from the stack being deposited over a wide area onto the river surface and aqueous input from the outfall entering the river in the immediate vicinity of the plant.

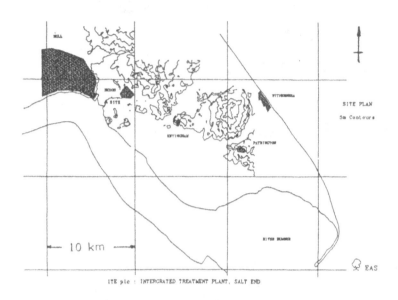

FIGURE 1. Location and surroundings of proposed plant.

Material arriving from these sources will add to that already in the system and will either be retained within the sediments of the river bed or washed into the North Sea by currents and mixing induced by tidal action.

The fate of chemicals deposited from stack emissions and input from the aqueous outfall into the river Humber was investigated using a numerical model. Input and environmental data for use in the simulation was derived from the treatment plant design and from previous studies on water and sediment quality in the Humber and on other published information. The long term averaged input into the estuary resulting from the stack emissions was derived from a model analysis using US EPA approved air dispersion models ISCLT and SCREEN. These models provided inputs based on the anticipated treatment plant output, the local wind conditions and on three possible stack heights. No specific field work was carried out for the creation or calibration of the water quality model.

Following a consideration of the nature of the estuary, the level of available information and the purpose of the study, it was decided to model the system using the U S Environmental Protection Agency's (EPA) WASP4 model suite.

2   Model Overview

The Water Quality Analysis Simulation Program Version 4 (WASP4) model suite was developed and is maintained by the EPA's Centre for Exposure Assessment Modelling (CEAM).   The CEAM acts as a centre of environmental modelling and a clearing house for the work of the seven associated EPA research laboratories and for user feedback for the supported models.

The CEAM also commissions the commercial development and testing of models. WASP4 was developed under an EPA funded research programme in 1981 and has since been continually verified and updated leading to the production of Version 4 in 1988. The total model suite is divided into a several elements which themselves link with other US EPA models to provide a wide ranging and flexible modeling system for the aquatic environment. This model represents one of the main simulation tools of the US EPA for studying impacts on the aquatic environment and has been used for many studies throughout the USA and worldwide.

WASP4 offers an integrated suite of models covering unsteady hydrodynamics, eutrophication, biological cycles and the transport and fate of toxic chemicals. This system was considered suitable because it is sufficiently advanced satisfactorily to model the largely tide driven hydrodynamics of the estuary and provides a third dimension for the settling, scour, deposition and compaction of sediments which is the important mechanism for the binding of metals into the river bed. Eutrophication was not considered in this study.

The main purpose of the model was to simulate the long term rates of accumulation of a range of pollutants in the sediments and to estimate the effects of the treatment plant's inputs on water quality.

The two key points considered when setting up the scheme were that hydrodynamics are driven by short term tidal effects but the water quality and accumulation periods are long term.   Hence the hydrodynamics are calculated from a short time step and largely 1d finite difference scheme with 2d resolution in the vicinity of the treatment plant and the bend in the river. The long term water quality and deposition requirements will tolerate the averaging effects of the large cell sizes that are used to maintain model efficiency though the resulting high numeric dispersion will have a certain distorting effect on any short term water quality results.

Input from the aqueous outfall is assumed to be averaged over the receiving cell since we have no detailed information on the design or exact position of the outfall. The results in the region of the outfall therefore do not take into account any near field effects which will need to be modeled separately when more detailed information is available.

3   Hydrodynamic Model

The hydrodynamics of the Humber estuary were modelled using the DYNHYD4 program from within the WASP4 suite.

DYNHYD4 solves one dimensional equations describing the propagation of a long wave through a shallow water system whilst conserving both momentum and volume. The equation of motion, based on the conservation of momentum, predicts the water velocities and flows. The equation of continuity, based on the conservation of volume, predicts water levels and volumes. This approach assumes that flow is predominately one dimensional, that Coriolis and other accelerations normal to the direction of flow are negligible, that channels are adequately represented by a constant top width with a variable hydraulic depth, that the wave length is significantly greater than the depth and that the bottom slopes are moderate. The equations of momentum and continuity are solved by a finite difference scheme.

The network for the hydrodynamic scheme was based on data from the Associated British Port's River Humber Chart, 1987, and tide levels from the Admiralty Tide Tables. The hydrodynamic network has 16 nodes set out as a largely 1d scheme but with 2d resolution in the vicinity of the outfall site and around the bend in the river where flows cannot be assumed uniform across the river. A 60 second time step was found to be stable and was used throughout the simulation. It is accepted that this scheme involves averaging over large areas of water and therefore will not simulate small scale effects. This however is of limited significance when simulating far field effects, where a high degree of mixing has already occurred, and effects generated over a long period of time.

The network extends from the mouth of the estuary to the upstream tidal limits with flows being driven by river inputs and by tide level data at the estuary mouth. The model was run for mean spring and mean neap conditions and verified primarily against tidal diamond C shown on the Associated British Port's River Humber chart, 1987. This verification point was selected since it is near the centre of the model network and represents average flows suitable for long term simulations. An example plot of the verification achieved is shown in Figure 2.

This simulation generates output files giving water velocities, flows and volumes for each water quality model time step (in this case, 10 minutes) for direct input into the water quality model.

4   Water Quality and Sedimentation Model

The effects on water and sediment quality of inputs from the proposed aqueous outfall and from stack deposition were simulated using the TOX14 model driven by hydrodynamic data generated from the DYNHYD4 simulation.

292

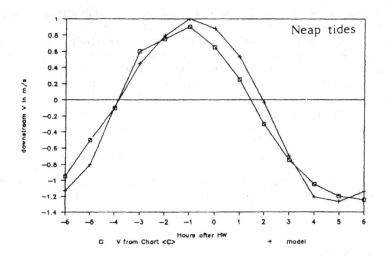

FIGURE 2. Example plot showing model verification achieved.

TOXI4 is a dynamic compartment model for the transport and fate of organic chemicals and metals in all types of aquatic systems. It combines the hydrodynamics generated from the DYNHYD4 simulation with sediment balance and chemical transformation systems discussed below.

Several physical-chemical processes can affect the transport and fate of toxic chemicals in the aquatic environment. The most important of these processes are:

*   Dilution and Transport
*   Dispersion
*   Volatilization and Deposition
*   Adsorption and Desorption on Suspended Sediments
*   Precipitation and Dissolution
*   Reduction and Oxidation
*   Photolysis and Hydrolysis
*   Bioconcentration and Biodegradation

Up to three chemicals and three particulate materials may be modeled at the same time with their general and interactive kinetics being defined by the user.

Heavy metals in the aquatic environment can form soluble complexes with organic and inorganic ligands, sorb onto organic and inorganic particles and precipitate and dissolve. The transport of sorbed material by deposition, scour and settling is also an important mechanism.

Data for inclusion in this model was obtained from publicly available information and no site surveys or specific tests were carried out. It is appreciated that the behaviour of metals in these situations is inherently complex and that site calibration is preferred. The interaction between the dissolved metals and the suspended solids in the water column was seen as the prime mechanism controlling the rate of deposition of metals into the estuary sediments. The model therefore used data obtained from the Humber Estuary Sediment Flux Report 1981 to simulate sediment concentration fields both horizontally and vertically including simplified bed exchange. The generalized nature of the available site data did not allow for a higher level simulation. For the same reason, the main parameters, such as the partition coefficient, sediment range and organic carbon content, were held constant throughout the model field though the sensitivity of the results to normal variations to the values used was investigated.

5   Aspects Modelled

The following aspects were modeled with a summary of the results given below. All long term steady state results are based on mean tide conditions and sediment loads and are calculated at the locations shown on Figure 3.

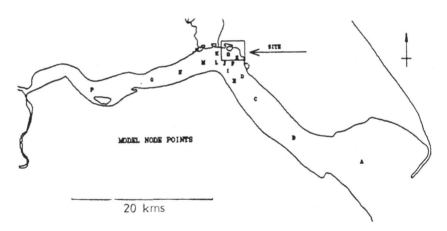

FIGURE 3. Locations used for mean tide conditions and sediment loads.

All inputs are modelled separately and interaction between the various input chemicals and with background chemicals is not considered: it is assumed that superposition will apply since all chemicals are at low concentrations.

All situations are studied for both the rate of accumulation in the bed sediments and for the maximum concentrations in the receiving water.

The inputs modelled are:

a.   an aqueous outfall in the location of the Salt End jetty.

b.   deposition from stack emissions calculated at three stack heights of 60m, 80m and 100m for 5 year average wind conditions.

6   Results

An example of the results of the model study compared with the existing concentration profile is shown in Figure 4.

ARSENIC IN SEDIMENT (typical example)

SEDIMENT QUALITY
Predicted Maximum Concentrations of ARSENIC

:Location	:Deposition per year: Stack mg/kg	Outfall mg/kg	Existing: mg/kg	Years until :Standard
A	0.000000	0.000042	50.00	
B	0.000000	0.000055	60.00	
C	0.000001	0.000085	60.00	
D	0.000001	0.000087	60.00	
E	0.000001	0.000129	60.00	
F	0.000001	0.000081	60.00	
G	0.000001	0.000085	60.00	
H	0.000001	0.000087	60.00	
I	0.000001	0.000091	60.00	
J	0.000001	0.000089	70.00	
K	0.000001	0.000082	70.00	
L	0.000001	0.000089	60.00	
M	0.000001	0.000074	50.00	
N	0.000000	0.000055	60.00	
O	0.000000	0.000034	70.00	
P	0.000000	0.000010	80.00	

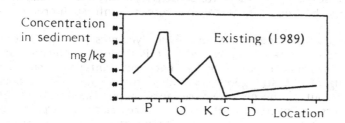

Concentration in sediment mg/kg

Existing (1989)

P   O   K   C   D   Location

FIGURE 4. Example of the results of the model study compared with the existing concentration profile

These results give the model predictions for water quality and sediment accumulation for eight significant metals and compare the results with data on existing concentrations and appropriate quality standards.

The mass balance over the model indicates that 32% of the dissolved input from Salt End remains in the sediments whilst the rest is flushed out past Spurn Head on the ebb tide. It is assumed that material flushed past this point does not return to the system.

7   Uses and Limitations of this Modelling Technique

The limitations and accuracy of numerical simulations are always a matter of concern both to the modeler and to those seeking to use the results. The limitations of the model are usually clearly apparent, at least to those familiar with that kind of analysis. Accuracy is altogether a more difficult concept to address since many factors are simplified or not considered in even the most complex models. There is often a tendency to seek apparent accuracy in elements of the model that are understood and, in doing so, inadvertently introducing further inaccuracies through a failure to look at the environment as a whole.

An example of this might be in a consideration of the model's hydrodynamics. This model undoubtedly simplifies the hydrodynamics and sediment transport mechanisms and more detailed analyses are readily available. I accept that the flows below Salt End are substantially 2d and that more sophisticated sediment transport simulations will show areas of erosion and deposition not indicated by this model. However, in reality these changes are often short term and other long term mechanisms, not featured in the simulation, will result in very different and often steady state results. Such long term effects would even include maintenance dredging. I would therefore suggest that in such a case the use of a more "accurate" short term simulation might in fact act to reduce the overall quality of the model.

A similar problem arises in the selection of the numerous rate constants required to describe the various physical, chemical and biological processes described by this model. Direct measurement, in any practical sense, is usually impossible and thus reliance must be placed on published data backed up by adequate sensitivity analyses. Fortunately, WASP is strongly backed up by extensive research into such processes.

It must be acknowledged that the hydrodynamics and the metal transport and fate mechanisms covered by this model are complex and their treatment within this scheme has been substantially simplified. Broad averaging of the site geometry and of the input data has been employed partly out of the necessary reliance on published data for all model parameters and partly out of a desire to restrict the complexity of the study to that which is necessary to achieve the study's objectives. It should be realised that this study was not commissioned for the purposes of fundamental research but to answer the simple question of whether the proposed treatment plant is likely to have a significant effect on the environment or not. The degree of complexity that is justified for the study must therefore take into account the significance of the results being generated.

In this case, the model predictions are generally several orders of magnitude below existing levels or any EQS and therefore, even allowing for any local variations and "hot spots", negligible adverse impacts can confidently be predicted. Had the results of this model shown values approaching the EQS, then we would have had to have recommended a more lengthy and detailed programme of data acquisition and modelling in order to achieve adequate predictions.

REFERENCES

Ambrose, R. B., Woll, T. A., Connolly, J. P., Schanz, R. W., (1988), Model Theory, User's Manual, and Programmer's Guide for the WASP4 Hydrodynamic and Water Quality Model, Centre for Exposure Assessment Modelling, US EPA, Athens Georgia.

Environmental Assessment Services Ltd, Port of Hull Parliamentary Bill: Works 1-4, Environmental Statement (1988), Associated British Ports, Hull.

British Transport Docks Board: Report No. 290, Humber Estuary Sediment Flux, December 1981.

Humber Estuary Committee: Water Quality of the Humber Estuary, 1987.

25 Effects of tidal modulation on river plume spreading

K. Nakatsuji, K. Muraoka and S. Aburatani

ABSTRACT

A river–forced estuarine plume is studied three–dimensionally by using the Navier–Stokes and buoyancy–conservation equations with the hydrostatic relation. The study extends the earlier paper by authors to include the effects of the tidal flow and the earth's rotation. In case of a great flood, the buoyancy forcing and the earth's rotation are the major force to lead a strong geostrophic along–front jet. It propagates at significantly high speed to the right along the coast in the northern hemisphere. However, once the river plume is dispersed due to the advection effects by the tidal flow, a buoyancy–driven current becomes weakened and the spreading of river plume is dominated by tidally–interacting. Comparison with satellite and field data is given. It is also found that Osaka Bay is an estuary under the control of temporal and spatial variability of density fields.

1. Introduction

The buoyant outflow from rivers into coastal water exhibits itself as a surface plume with a well–defined frontal structure where the density change is rapid. On a larger scale in the order of the Rossby deformation radius, the earth's rotation makes a large surface plume deflect to the right in the northern hemisphere and the forward motion of its front relative to the ambient fluid appears. (Wang 1987, McClimans 1988)

The authors (1989) studied the Yodo River plume spreading in Osaka Bay in case of a flood of a maximum discharge 6,270 m^3/s by using a three–dimensional baroclinic numerical model in the previous paper. As a result, it was pointed out that the offshore spreading is suppressed and alongshore coastal jet is formed, when the horizontal length scale of the flood flow attains about 10 km in this case. The plume front was shown to propagate along the coast at a speed of about 0.6 m/s as much as 44 km away from the river mouth. Its numerical result well agreed with the surface temperature distribution in the infrared image taken by the satellite NOAA. However, the behaviour of river plume is noticeably different from that estimated from the tidal residual current chart for Osaka Bay. To complete this series of investigation, the effects of tidal forcing and wind–driven forcing are studied and analyzed. In particular, the tidal flows entering from the Straits of Akashi and Kitan play an important role on water quality transfer in Osaka Bay and they cause considerable distortion of the temporal evolution as well as spatial structure of the Yodo River plume. Therefore, the tidal current system cannot be ignored in the study of the Yodo River plume spreading.

In the present study, we adopted a three–dimensional baroclinic numerical model taking into consideration both the density current system and the tidal current system as the driving force for a river–forced estuarine plume. The aim of this study is to show how the tidal flow modulates the spreading of the Yodo River plume and how differences there are in its spreading behaviours between during a flood and during a normal river discharge.

2. The three–dimensional model

2.1 Basic equation

Under the Boussinesq and hydrostatic approximations, the equations governing the conservation of volume, momentum in the three directions and buoyancy are

$$\frac{\partial U_i}{\partial x_i} = 0 \tag{1}$$

$$\frac{\partial U}{\partial t} + U_i \frac{\partial U}{\partial x_i} - fV = -\frac{1}{\rho}\frac{\partial P}{\partial x} + \frac{\partial}{\partial x_i}(\varepsilon_i \frac{\partial U}{\partial x_i}) \tag{2}$$

$$\frac{\partial V}{\partial t} + U_i \frac{\partial V}{\partial x_i} + fU = -\frac{1}{\rho}\frac{\partial P}{\partial x} + \frac{\partial}{\partial x_i}(\varepsilon_i \frac{\partial V}{\partial x_i}) \tag{3}$$

$$0 = -g - \frac{1}{\rho}\frac{\partial P}{\partial z} \tag{4}$$

$$\frac{\partial \Delta\rho}{\partial t} + U_i \frac{\partial \Delta\rho}{\partial x_i} = \frac{\partial}{\partial x_i}(K_i \frac{\partial \Delta\rho}{\partial x_i}) \tag{5}$$

where (U, V, W) are velocities in the x, y and z directions, ζ is the water elevation, P is the pressure, $\Delta\rho(= \rho_a - \rho)$ is the density deviation against a reference sea water density ρ_a, ε_i and K_i are the eddy viscosity and diffusivity coefficients respectively in each direction, and f is the Coriolis parameter (0.8296×10^{-4} s^{-1}). The pressure is obtained by integrating the vertical momentum equation (4) from the water surface ($z = -\zeta$) to any depth. Therefore, the pressure gradient $\partial P/\partial x_i$ can be expressed as a sum of the water surface gradient $\partial \zeta/\partial x_i$ (barotropic mode) and the density gradient $\partial \Delta\rho/\partial x_i$ (baroclinic mode).

In case using constant eddy viscosity and diffusivity assumption, it is necessary to use empirical formulation for the reduction of mixing due to buoyancy effects on turbulence in stable situations. On the basis of the study of three–dimensional buoyant surface discharges

300

by Murota et al. (1988), the vertical eddy viscosity coefficient ε_z and eddy diffusivity K_z are used based on the Webb formula (1970) and Munk–Anderson formula (1948), respectively, with a neutral value of $\varepsilon_{zN} = 0.005$ m²/s. Both empirical formulae are a function of the gradient Richardson number. A very small horizontal eddy viscosity or diffusivity (20 m²/s) is used in all model runs. In the straits, however, turbulence energy increases in strength so significantly as to cause strong mixing. Hence, the values of 2,000 m²/s and 0.05 m²/s are used near the straits for horizontal and vertical eddy viscosity/diffusivity coefficients, respectively.

The numerical solution essentially follows the scheme of Murota et al. (1988). Equations and boundary conditions are finite–differenced in space to second order accuracies. In particular, the hybrid scheme by Spalding (1972) is adopted for the advection term in order to eliminate the numerical diffusion. For the time difference, the explicit scheme by the leapfrog method is basically used, and the implicit scheme only for the computation of water level ζ.

2.2 Model ocean

The computational domain of Osaka Bay is in the north of lat. 33°50' N. in the Kii Channel and in the east of long. 134°19' E. in the Sea of Harima as shown in Figure 1, which is designated in accordance with the environment assessment report by the Ministry of Transportation, MOT, Japan (1980). The domain is of a very wide range in order to take the water exchange through the Straits of Akashi and Kitan into account. The model's resolutions are $\Delta x = \Delta y = 2$ km, $\Delta t = 30$ s and $\Delta z = 2, 6, 12$ and 20 m from the sea surface. Since the tidal front is observed to form along 20 m in depth in Osaka Bay, the topography is reproduced by dividing into upper three layers to 20 m in depth and one lower layer.

Cosine waves having the amplitude of the sum of M_2 and S_2 tidal harmonic constants and phase lag of a M_2 tide are given for the tidal elevation at both open boundaries in the west end of the Sea of Harima and in the south end of the Kii channel, the values of which are determined according to the assessment report by MOT. That is, the amplitude in the former boundary is 40 cm, while that in the latter is 67 cm. The tidal cycle is set to 12 hours for convenience. Non–slip condition for the velocity is used for the coastline and the ocean bottom. A density flux of 0.0 is used for all boundaries except the river mouth of the Yodo River. The initial conditions are $U = V = W = 0$ and $\rho_a = 0.022$ kg/m³ for the ocean water. The density of river water is set to 0.0 kg/m³.

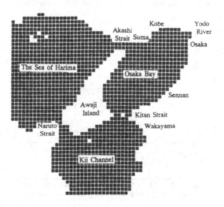

FIGURE 1 *Computational domain of Osaka Bay with two openings at the Straits of Akashi and Kitan and numerical grid*

301

2.3 Outline of computation

Model calibration Computation for the barotropic tidal flow is first performed in the model ocean taking no account of density difference in order to examine the model's potential for simulating real flow observed in Osaka Bay. When the computed values for the last two tides of 5 cycles run are compared, the reproducibility of the tidal flow is confirmed. Since the sea bottom topography is not correctly expressed, the conformity between the calculated and the observed values is slightly inferior to that of the computed results by the 2-level model generally used in the simulation of Osaka Bay. However, when comparing the observed values at the main observatory locations, these agree at a precision of within 7 $\%$ except -27 $\%$ at the southwest part of Osaka Bay and +12 $\%$ near the Akashi Strait in amplitude, and at a precision of within 3 degrees except 40 degrees in Akashi and 16 degrees in the northwest part of the Sea of Harima in phase lag. The model's potential could be judged to be very accurate in the simulation of the barotropic tidal flow.

Onset of river plume The development of river-forced estuarine plumes can be summarized by the following two experiments. In the first experiment, a flood outflow is examined. Figure 2 shows the hydrograph of Yodo River flood discharge, the maximum value of which is 6,270 m^3/s. It was observed at the observatory station in Hirakata. It rained ceaselessly occurred by the typhoon 10, which hit the Kansai district including Osaka, Kobe and Kyoto on August 1 to 2, 1982, and the succeeding passage of the low atmospheric pressure. The tidal elevations have been observed at Akashi Port near the Akashi Strait. Although the tide at Akashi shows a distorted diurnal tide type as well known, the tidal elevations and flows in other sea regions are the mixing type of a semidiurnal tide and a diurnal one. A cosine wave with a 12-hour period, therefore, is used in the computation. The high water time recorded in Akashi Port was found to coincide with the time when the maximum discharge was observed. The boundary conditions for tidal elevations at the open boundaries are, therefore, provided so that the computed maximum elevation occurs in Akashi Port just at the moment of the maximum river discharge. The computed tidal elevation at Akashi Port is also shown in Figure 2 corresponding to the hydrograph.

For the thickness of a river plume at the river mouth, 2.0 m is used as specified in the previous paper (1989). Since the horizontal grid spacing, Δx and Δy is 2,000 m against the width of the river mouth of 800 m, the velocity at the river mouth becomes 2/5 of the actual

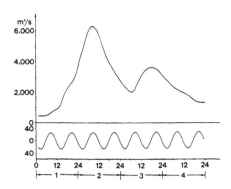

FIGURE 2 *Hydrograph of Yodo River flood on 1 to 4, August 1982 and presumed tidal elevation at Akashi Port*

velocity when the river discharge in Figure 2 is given. Under the such condition, the outflow becomes so subcritical that there appears quite different behaviours from observed river–forced estuarine plumes in a supercritical state. To perform the outflow under the hydrodynamically similar condition, it is necessary to allow the densimetric Froud numbers at the river mouth to coincide between model and natural phenomena. Therefore, we decided to give a discharge 5/2 times greater than the accurate discharge for the flood flow. The intruding Yodo River plumes generally consists of a strong but localized baroclinic motion.

In the second experiment, a river plume of a normal flow discharge, 500 m^3/s is examined. The tidal flow in this case is expected to have direct effects upon the river plume spreading. The same tidal conditions as in the first experiment are employed. In each experiment. the density difference $\Delta\rho$ between river and sea waters is 0.022 kg/m^3.

3. Numerical results and discussions

3.1 Development of Yodo River flood flow on 1–4 August 1982

Figure 3 shows the time variation of the horizontal flow fields at the surface layer averaged at 2 m after 10, 20, 30 and 35 hours from the beginning of Yodo River flood flow. Also shown are the density difference contours $\Delta\rho/\Delta\rho_a$ at 10 % intervals, in which $\Delta\rho_a$ represents the density difference between river and sea waters. At the first several hours when the rate of river discharge is small, the river water spreads out over the sea water in the radial directions so that it slides, and the isopycnal contours fan out with a concentric circle. The velocity vectors cross the isopycnal contours at right angles. It is a typical pattern of a surface plume. This tendency can be seen in the velocity and density fields after 10 hours. The isopycnal contours, however, show to be stretched out a little on Kobe side due to the effect of the earth's rotation. The radius of Rossby deformation, $\sqrt{(\Delta\rho/\rho_a)gh}/f$ is estimated approximately 10 km in the Yodo River plume, in which h is the layer thickness of the river plume. And, when the river water spreads to this scale, the buoyancy effect is equivalent to the effect of the earth's rotation. As a result, the river water spreading will be deviated toward the right through the geostrophic adjustment, and the offshore spreading will be suppressed to accelerate the alongshore flow.

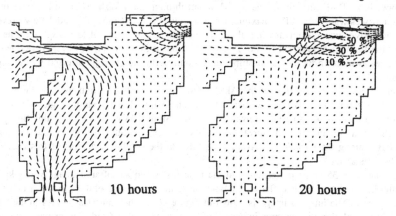

FIGURE 3 *Time variation of velocity vector and density fields at surface layer of 2 m illustrating the interaction of Yodo River flood flow with tidal flow: (a) 10 hours, and (b) 20 hours*

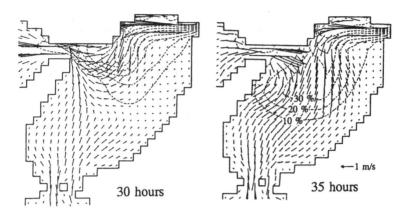

FIGURE 3 *Continued*: *(c) 30 hours and (d) 35 hours*

The velocity vectors and isopycnal contours after 20 hours correctly assume such an aspect. The Yodo River flood flows with spreading out to the offshore of Suma in parallel to Kobe coastline and maintaining the width of the Rossby deformation radius (about 10 km).

At the head of flood flow, the gradient of isopycnal contours becomes steep. It implies that the accumulated water due to the geostrophic adjustment forms an intruding current. This flow is called 'a coastal jet'. The propagation speed between 10 and 20 hours can be estimated 0.17 m/s from the change in isopycnal contours. It is a little smaller than 0.24 m/s obtained in the computation when no tidal flow was considered. (Nakatsuji, et al. 1991). The latter value almost agrees with the theoretical one for a rotational fluid, $1.4 \sqrt{(\Delta\rho/\rho_a)gh}$ given by Stern et al. (1982). The east–going tidal flow during 9 to 15 hours may has retarded the propagation.

At about 30 hours, the maximum westward flow occurs at the Akashi Strait. At two hours before that time the north–going flow from the Kitan Strait into Osaka Bay becomes a maximum. It branches two current; one flowing directly towards the Akashi Strait along the coastline of the Awaji Island, and the other counterclockwise flowing off Sennan coast towards the inner bay. Both currents merge and flow out through the Akashi Strait at the time of the maximum westward flow. The maximum of velocity attains 1.1 m/s. This tidal flow prevents the Yodo River water from flowing along the coastline, and causes it to transport to the Sea of Harima through the Akashi Strait. A major portion of flood water, however, is derived out for the sea away from the Suma coastline, and is transported to the right affected by the earth's rotation.

After 35 hours, the tidal flow at the Akashi Strait changes its direction towards the east. It is one hour before the maximum eastward flow. The major flow travels through a deep water area off the coast of Awaji Island towards the Kitan Strait. As it is clear from this figure, the Yodo River flood flow greatly spreads from the offshore of Suma to the central portion of Osaka Bay along the coastline of Awaji Island due to the combined actions of this tidal flow and the coastal jet.

The time after 35 hours corresponds to the time just when an infrared image was taken by the satellite NOAA, and also after only 6 hours from the occurrence of the maximum discharge of 6,270 m^3/s. The infrared image is shown in Figure 4(a). The numerical values in the figure represent the relative temperature in degree of the Celsius scale. Colder water was confirmed to correspond to river water on the basis of observation. The distribution of the sea surface temperature seen in the infrared image well agrees with the computation result of the density distribution of the surface layer shown in Figure 3(d).

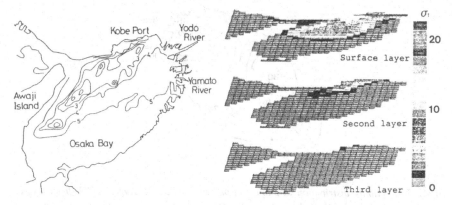

FIGURE 4 *Observed and computed density fields corresponding to 35 hours in Figure 3(d):*
(a) Infrared of surface temperature in Osaka Bay at 14 : 24 on 2 August 1982, and
(b) computed density fields in three layers of 1 m, 5 m, and 14 m depths

Figure 4(b) shows the entire area of Osaka Bay by slightly distorting the figure, and the horizontal density fields at depths of 1 m, 5 m and 14 m which are represented in a σ_t unit. These figure demonstrate a three dimensional structure of the flood flow. It is the tip of river water that is seen darkest in the figures. Taking notice of the density fields of second or third layers, water masses of σ_t larger than 2 are observed to locally exit at a depth of 5 m of the second layer. It can be seen that the river water spreading is about 5 m vertically.

It is concluded from the facts described above that, during a flood with enormous volume of river water as in this case, the Coriolis–deflected jet is characteristic of the Yodo River flood spreading despite the strong reciprocating tidal flow through the Akashi Strait: The Yodo River water flows westward offshore of Kobe, reaching the offshore of the eastern coast of Awaji Island.

Figure 5 shows the density fields of the surface layer after 50, 60 and 80 hours. Since the tidal time is different, it is difficult to separate the tidal flow system from the density current system for discussion. With the elapse of time, in other words, as the river water spreads out over the sea surface and the coastal jet–like characteristic becomes thinner, the advection due to the tidal flow becomes stronger. After 60 hours, the maximum eastward flow occurs in the

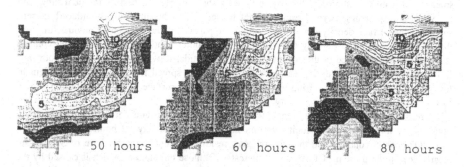

FIGURE 5 *Time variation of density fields at surface layer of 2 m*

305

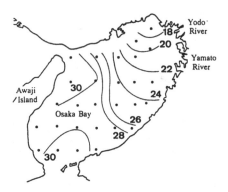

FIGURE 6 *Salinity distribution at 2 m depth observed on 5 to 6, August 1982 (from Kawana and Tanimoto (1984))*

Akashi Strait, and it is observed that the Yodo River water covers the entire area of Osaka Bay and also flows out from the Kitan Strait. After 80 hours – two hours after the maximum westward flow occurs –, the river water, which once flew out, flows from the Kitan Strait into Osaka Bay again. The density distribution is observed to be moved clockwise from the offshore of Sennan to the central portion of the bay. The distribution on the sea surface for salinity at 2 m below from the sea surface was drawn by Kawana and Tanimoto (1984) on the basis of the observation performed on 5 and 6, August after the flood. It is shown in Figure 6. The observed distribution qualitatively agrees with the computed density distribution after 80 days as shown in Figure 5(c).

3.2 Yodo River plume spreading of a normal flow rate

The second experiment was carried out based on simulation of the tidal oscillation with the summed amplitude of M_2 and S_2 tides with an inflow discharge of 500 m³/s from the rivers in the interior of the bay, which were collectively flown from the Yodo River. Since the total inflow flow rate increases with the computation, the steady solution cannot be obtained. So the computation results to be discussed are shown after 12th tidal cycle computation in which the variation of the density field against that of the preceding tide became small.

Figure 7 shows the velocity vectors and the density contours $\Delta\rho/\Delta\rho_a$ at 10 % intervals every 2 hours with the maximum eastward flow in the Akashi Strait as 0 time. The river water spreads out from the interior of the bay into the offshore of Sennan despite the tidal time, and the westward spreading beyond Suma cannot be seen. It is because tidally–induced current surpasses the surface density current in strength. In particular, the isopycnal contours for more than 20 % present almost steady distribution. On the other hand, the 10 % isopycnal contour is subject to the influence of the advection effect of the tidal flow, and tends to flow southward in the offshore of Sennan during eastern tidal flow and to spread from the offshore of Sennan toward the center of the bay during western tidal flow. The surface salinity distribution shown in Figure 6 is similar to these density fields.

Figure 8 shows features of the baroclinic and tidal residual currents for each layer which are obtained by integrating the velocity vectors for one cycle, namely 12 hours. In the surface layer, Okinose circulation and the eastern coast residual current in the offshore of Sennan are clearly seen to be present. Of particular interest is Okinose circulation, which is caused by the difference in paths of the reciprocating tidal flow through Akashi Strait as shown in Figure 7.

306

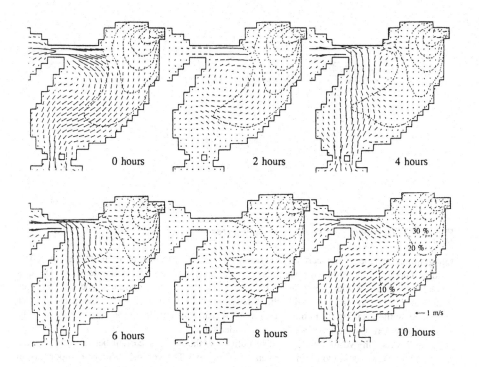

FIGURE 7 *Time variation of velocity vectors and density fields in surface layer every 2 hours in case of normal river flow rate*

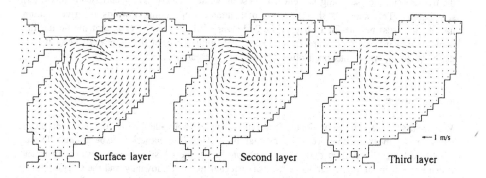

FIGURE 8 *Tidal residual and baloclinic circulation for surface, 2nd and 3rd layers*

The residual current off Sennan coast is probably associated with the Okinose circulation in addition to the density–induced residual current. The velocities of both currents are 45 cm/s and 30 cm/s at maximum respectively. As being pointed out by Fujiwara et al. (1989), the residual current at spring–tide in Osaka Bay is larger than expected. The other residual circulations off the coasts of Suma and Nishinomiya, whose existence has been pointed out by them from the observation, do not appear in this computation. This will be because the

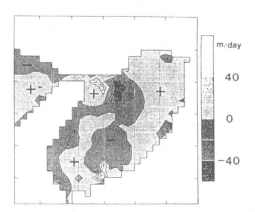

FIGURE 9 *Distribution of vertical residual current between surface and 2nd layers*

topographical shape is not correctly expressed in the model resolution of this study. The residual current in the 3rd layer, which is most probably not affected by the density current, is caused by the tidal barotropic system. By comparing the residual current in the surface layer with that in the 3rd layer or with the tidally–induced residual flow in the surface layer computed without taking into consideration the density current, the importance of the density current system in Osaka Bay can be easily understood.

Figure 9 shows the vertical velocity distribution for the residual current between the surface layer and 2nd layer in contours in intervals of 20 m/day. The vertical velocity components in this three–dimensional model are computed at the center of the upper surface of each control volume using the equation of continuity. Positive flows are directed upward. In the eastern sea region where the river water spreading can be seen with a depth of 20 m in Osaka Bay as the border line, the ascending current can be seen, while the descending current in the western sea region. This suggests the existence of a undercurrent going towards the river mouth, namely towards the interior of bay in the 2nd layer. It is of great interest that the border line between upward and downward flows agrees well with the front line appeared along the sea depth of 20 m, which has been observed by many researchers.

4. Conclusion

We discussed the effects of tidally–interacting and Coriolis–deflecting on the river plume spreading by means of the three–dimensional baroclinic flow numerical experiments. During the Yodo River flood flow due to the typhoon 8,210, the density current system surpassed. It can be seen in the computation result that, affected by both buoyancy and the Coriolis force, the river water flows to the right along the coastline of Awaji Island from the offshore of Kobe while forming a coastal jet. Computed density fields agrees well with the infrared image taken by the satellite NOAA. However, as the result of spreading of the river water over the entire area of Osaka Bay due to the advection effect by the tidal flow hereafter, the coastal jet–like characteristic becomes thinner, and it could be confirmed that the river water spreading is controlled by the tidal flow system. This result well agrees with the salinity distribution measured on August 5 to 6 after the flood.

Next, the baroclinic and tidal residual current was examined by performing the three dimensional density current and tidal flow computation for providing the inflow of normal flow discharge, Okinose circulation and the eastern coast residual current offshore Sennan are noticeably observed, whose existence have been pointed out from the observation. The river plume is transported by the residual current, and spreads out from the interior of the bay into the offshore of Sennan. Also the existence of a front could be predicted from the vertical velocity distribution for the baloclinic and tidal residual current. From the above, it turned out that the density current has a significant influence on the movement and spreading of substances in the Osaka Bay. In other words, the numerical experiment result suggests that Osaka Bay is an estuary under the control of temporal and spatial variability of the density current.

ACKNOWLEDGEMENT

This work is supported by the Grant–in–Aid for Scientific Research, The Ministry of Education and Science under Grants 03650423 and 02302067. We are indebted to Honorary Professor A. Murota for providing some useful discussion and Mr. N. Yamamoto and Mr. T. Sato for their programming assistance.

REFERENCES

[1] Fujiwara, T., Higo, T. and Takasugi, Y., (1989), Residual Current, Tidal Flow and Vortex Observed in Osaka Bay, *Proc. Coastal Eng. JSCE*, vol. **36**, pp. 209–213. (in Japanese)

[2] Kawana, K. and Tanimoto, T., (1984), Measured Result of Suspended Matter Distribution in Osaka Bay After Heavy Rain, *Repts. The Government Industrial Res. Institute, Chugoku*, No. **22**, pp. 67–74. (in Japanese)

[3] McClimans, T. A., (1988), Estuarine Fronts and River Plumes, *Physical Processes in Estuaries*, Ed. by Dronker, J. & Leussen, W. V., Spring–Verlag, pp. 55–69.

[4] Munk, W. H. and Anderson, E. R., (1948), Notes on a Theory of the Thermocline, *Jour. Marine Res.*, vol. **7**, pp. 276–295.

[5] Murota, A., Nakatsuji, K. and Huh, T. Y., (1988), A Numerical Study of Three-Dimensional Buoyant Surface Jet, *Proc. 6th Congress Asian and Pacific Regional Division of IAHR*, vol. **3**, pp. 33–40.

[6] Nakatsuji, K., Huh, J. Y. and Murota, A., (1989), Effects of the Earth's Rotation on the Behaviour of River Plumes, *Hydraulic and Environmental Modelling of Coastal, Estuarine and River Waters*, Ed. by Falconer, R. A. et al., Gower Technical, pp. 310–321.

[7] Nakatsuji, K., Yamamoto, N. and Muraoka, K., (1990), Outflow and Three–Dimensional Spreading of River Water in Enclosed Bay, *Marine Pollution Bulletin*, vol. **23**, Pergamon Press, pp. 551–559.

[8] Spalding, D. B., (1972), A Novel Finite Difference Formulation for Differential Expressions Involving Both First and Second Derivations, *Int. Jour. Numerical Methods in Engineering*, vol. **4**, pp. 551–559

[9] Stern, M. E., Whitehead, J. A. and Hua, B. L., (1982), The Intrusion of a Density Current Along the Coast of a Rotating Fluid, *Jour. Fluid Mech.*, vol. **123**, pp. 237–265.

[10] The Third Construction Bureau of the Ministry of Transport, (1980), Environmental Assessment Report on the Kansai International Airport Island Project. (in Japanese)

[11] Wang, D. P., (1987), The Strait Surface Outflow, *Jour. Geophys. Res.*, vol. **92**, No. C10, pp. 10,807–10,825.

[12] Webb, W. K., (1970), Profile Relationships, the Log–Linear Range and Extension to Strong Stability, *Quart. Jour. Royal Met. Soc.*, vol. **96**, pp. 67–90.

26 A delay-diffusion description for two-dimensional contaminant dispersion

Y. Liu

ABSTRACT

In the earlier stage of dispersion process, the rate of dispersion associated with earlier discharge in relatively large because the memory term extends further back in time. In order to investigate the early properties of two-dimensional dispersion, we propose the delay–diffusion equation as following:

$$\frac{\partial \bar{C}}{\partial t} + \bar{u}\frac{\partial \bar{C}}{\partial x} + \bar{V}\frac{\partial \bar{C}}{\partial y} - \bar{K}_{xx}\frac{\partial^2 \bar{C}}{\partial x^2} - \bar{K}_{yy}\frac{\partial^2 \bar{C}}{\partial y^2} - \int^{\infty}\left\{\left(\frac{\partial D_{xx}}{\partial \tau}\frac{\partial^2}{\partial x^2} + \left(\frac{\partial D_{xy}}{\partial \tau}\right.\right.\right.$$

$$\left.\left. + \frac{\partial D_{yx}}{\partial \tau}\right)\frac{\partial^2}{\partial x \partial y} + \frac{\partial D_{yy}}{\partial \tau}\frac{\partial^2}{\partial y^2}\right\}\bar{C}(x-\int^{\tau}\tilde{u},\ y-\int^{\tau}\tilde{v},\ t-\tau)d\tau = \bar{q}$$

which is based on the ansatz derived by Smith (1981). The D_{xx}, D_{yx}, D_{xy}, D_{yy} and $\tilde{u}$, $\tilde{v}$ had been obtained.

1 Introduction

Since G. I. Taylor (1953, 1954) [1], [2] showed that the dispersion process play an important role in contaminant diffusion, many author ([1]–[9]) devoted their wisdoms to investigate the dispersion process. But Taylor pointed out

311

[3] that the large diffusivity of contraminant dispersion could be gained only after a sufficient length of time had elasped. In order to remedy this shortcoming, Gill & Sankarasubramanian (1970) have advocated that a variable coefficient diffusion equation:

$$\frac{\partial \bar{C}}{\partial t} + \bar{u}(t)\frac{\partial \bar{C}}{\partial x} - [K + D(t)]\frac{\partial^2 \bar{C}}{\partial x^2} = 0 \tag{1.1}$$

be used to model all stages of the dispersion process. Eq.(1.1) has exact the area, centroid and variance of the bulk concentration distribution $\bar{C}(x,t)$. However, it is difficult to reconcide the equations with underlying physical process as Taylor (1959) concluded 'It seems therefore that no physical meaning can be attached to the use of equations in which the coefficient of diffusion varies with the time of diffusion, eventhough the formulae produced by their use do represent adequately the concentrations in particular cases.'

It is Smith (1981) that first posed a new ansatz:

$$C - \bar{C} = \sum_{j=1}^{\infty} \int_0^{\infty} l_j(y,z,\tau)\frac{\partial^j}{\partial x^j}\bar{C}(x-\int_0^{\tau} \tilde{u}d\tau', \ t-\tau)d\tau \tag{1.2}$$

to analyse the early longitudinal dispersion. Here, $\tilde{u}$ is transport velocity, l_j weighed function, C concentration, $\bar{C}$ sectionally-averaged concentration. Eq.(1.2) has the same remarkable properties as Eq.(1.1), but Eq.(1.2) overcomes the shortcoming of Eq.(1.1).

We extended the Eq.(1.2) to two-dimensional form in order to deal with early dispersion properties in bays, lakes, estuaries and shallow water where the horizontal mixing of contaminant is dominant. The coefficients of shear dispersion and transport velocities for two-dimensional flows had been obtained and discussed.

2   Horizontal and Vertical Dispersion Equations

As the starting point for our mathematical analysis we take the full equation to have the form:

$$\begin{cases} \dfrac{\partial C}{\partial t} + u\dfrac{\partial C}{\partial x} + v\dfrac{\partial C}{\partial y} - K_{xx}\dfrac{\partial^2 C}{\partial x^2} - K_{yy}\dfrac{\partial^2 C}{\partial y^2} - \dfrac{\partial}{\partial z}K_{zz}\dfrac{\partial C}{\partial z} = q \\[2mm] K_{zz}\dfrac{\partial C}{\partial z}\bigg|_{\partial A} = 0 \end{cases} \tag{2.1}$$

here, ∂A is impermeable boundary, K_{xx}, K_{yy}, K_{zz} diffusivities, q discharge rate, u, v horizontal velocity component. For two-dimensional dispersion, we propose the following anstaz:

$$
\begin{cases}
C - \bar{C} = \\[2mm]
\displaystyle\sum_{m=1}^{\infty} \int_0^{\infty} l_{ij}^m (z,\tau) \frac{\partial^m}{\partial x^i \partial y^j} \bar{C}(x-\int_0^\tau \tilde{u},\; y-\int_0^\tau \tilde{v},\; t-\tau)d\tau \\[4mm]
i + j = m, \quad i,\; j \geq 0
\end{cases}
\tag{2.2}
$$

If the ansatz (2.2) is substituted into the vertically averaged form of dispersion (2.1), we arrive at the integro-differential equation:

$$
\frac{\partial \bar{C}}{\partial t} + \bar{u}\frac{\partial \bar{C}}{\partial x} + \bar{v}\frac{\partial \bar{C}}{\partial y} - \bar{K}_{xx}\frac{\partial^2 \bar{C}}{\partial x^2} - \bar{K}_{yy}\frac{\partial^2 \bar{C}}{\partial y^2}
$$

$$
+ \sum_{m=1}^{\infty}\int_0^{\infty}\left(l_{ij}^m (u-\bar{u})\frac{\partial^{m+1}}{\partial x^{i+1}\partial y^j} + l_{ij}^m (v-\bar{v})\frac{\partial^{m+1}}{\partial x^i \partial y^{j+1}}\right.
$$

$$
\left. - l_{ij}^m (K_{xx}-\bar{K}_{xx})\frac{\partial^{m+2}}{\partial x^{i+2}\partial y^j} - l_{ij}^m (K_{yy}-\bar{K}_{yy})\frac{\partial^{m+2}}{\partial x^i \partial y^{j+2}}\right)
$$

$$
\cdot \bar{C}(x-\int^\tau \tilde{u},\; y-\int^\tau \tilde{v},\; t-\tau)d\tau = \bar{q}(x,\; y,\; t)
\tag{2.3}
$$

The essential mathematical properties of Eq.(2.3) is determined by the $\partial/\partial t + \bar{u}\partial/\partial x + \bar{v}\partial/\partial y - \bar{K}_{xx}\partial^2/\partial x^2 - \bar{K}_{yy}\partial^2/\partial y^2$, and it is acceptable to truncate the l_{ij}^m series at any level. At the lowest-order truncation ($m = 1$) of Eq.(2.3), we obtained the delay-diffusion equation for two-dimensional dispersion:

$$
\frac{\partial \bar{C}}{\partial t} + \bar{u}\frac{\partial \bar{C}}{\partial x} + \bar{v}\frac{\partial \bar{C}}{\partial y} - \bar{K}_{xx}\frac{\partial^2 \bar{C}}{\partial x^2} - \bar{K}_{yy}\frac{\partial^2 \bar{C}}{\partial y^2} - \int_0^{\infty}[\frac{\partial D_{xx}}{\partial \tau}\frac{\partial^2}{\partial x^2}
$$

$$
+ \frac{\partial D_{xy}}{\partial \tau}\frac{\partial^2}{\partial x \partial y} + \frac{\partial D_{yx}}{\partial \tau}\frac{\partial^2}{\partial y \partial x} + \frac{\partial D_{yy}}{\partial \tau}\frac{\partial^2}{\partial y^2}]
$$

$$
\bar{C}(x-\int^\tau \tilde{u},\; y-\int^\tau \tilde{v},\; t-\tau)d\tau = \bar{q}
\tag{2.4}
$$

In order to get l_{ij}^m, $\tilde{u}$, $\tilde{v}$, we must analyse the vertical dispersion equations. Substituting (2.2) into (2.1) and using Eq.(2.3) and boundary conditions, we obtained:

313

$$\left\{
\begin{aligned}
&[I^1_{10} + (u-\bar{u})]\big|_{\tau=0}\, \frac{\partial \bar{C}}{\partial x} + [I^1_{01} + (v-\bar{v})]\big|_{\tau=0}\, \frac{\partial \bar{C}}{\partial y} \\[4pt]
&+ [I^2_{20}(z,0) + \bar{K}_{xx} - K_{xx}]\big|_{\tau=0}\, \frac{\partial^2 \bar{C}}{\partial x^2} + [I^2_{02}(z,0) \\[4pt]
&+ \bar{K}_{yy} - K_{yy}]\big|_{\tau=0}\, \frac{\partial^2 \bar{C}}{\partial y^2} + \sum_{m=1}^{\infty}\int_0^{\infty} \left(\frac{\partial I^m_{ij}}{\partial \tau} - \tilde{u} I^{m-1}_{i-1\,j} \right. \\[4pt]
&- \tilde{v} I^{m-1}_{i\,j-1} - \overline{I^{m-1}_{i-1\,j}\,u} - \overline{I^{m-1}_{i\,j-1}\,v} + u I^{m-1}_{i-1\,j} \\[4pt]
&+ v I^{m-1}_{i\,j-1} + \overline{I^{m-2}_{i-2\,j}K_{xx}} + \overline{I^{m-2}_{i\,j-2}K_{yy}} - I^{m-2}_{i-2\,j}K_{xx} \\[4pt]
&- I^{m-2}_{i\,j-2}K_{yy} - \frac{\partial}{\partial z}K_{zz}\frac{\partial I^m_{ij}}{\partial z} \left. \right) \frac{\partial^m}{\partial x^i \partial y^j} \\[4pt]
&\cdot \bar{C}\left(x-\int^{\tau}\tilde{u},\ y-\int^{\tau}\tilde{v},\ t-\tau\right)d\tau = 0 \\[6pt]
&\sum_{m=1}^{\infty}\int_0^{\infty} K_{zz}\frac{\partial I^m_{ij}}{\partial z}\frac{\partial^m}{\partial x^i \partial y^j}\bar{C}\,d\tau \bigg|_{\partial A} = 0
\end{aligned}
\right. \tag{2.5}$$

Note:

$$I^m_{ij} = 0, \qquad \text{if } i+j \le 0$$

Since the solution procedure must be valid for any $\bar{C}$ profile, it is natural to require that the individual equations corresponding to each $\partial^m \bar{C}/\partial x^i \partial y^j$ coefficient and integrand be satisfied separately. This prescription leads to the sequence of initial-boundary-value problems (to second order):

$$\left\{
\begin{aligned}
&\frac{\partial}{\partial \tau}I^1_{10} - \frac{\partial}{\partial z}K_{zz}\frac{\partial I^1_{10}}{\partial z} = 0 \\[4pt]
&K_{zz}\frac{\partial I^1_{10}}{\partial z}\bigg|_{\partial A} = 0, \qquad I^1_{10}\big|_{\tau=0} = \bar{u} - u
\end{aligned}
\right. \tag{2.6}$$

$$\left\{
\begin{aligned}
&\frac{\partial}{\partial \tau}I^1_{01} - \frac{\partial}{\partial z}K_{zz}\frac{\partial I^1_{01}}{\partial z} = 0 \\[4pt]
&K_{zz}\frac{\partial I^1_{01}}{\partial z}\bigg|_{\partial A} = 0, \qquad I^1_{01}\big|_{\tau=0} = \bar{v} - v
\end{aligned}
\right. \tag{2.7}$$

314

$$\begin{cases} \dfrac{\partial}{\partial \tau} I_{20}^2 - \dfrac{\partial}{\partial z} K_{zz} \dfrac{\partial I_{20}^2}{\partial z} = \tilde{u} l_{10}^1 + \overline{u l_{10}^1} - u l_{10}^1 \\[4mm] K_{zz} \dfrac{\partial I_{20}^2}{\partial z}\bigg|_{\partial A} = 0, \quad I_{20}^2\big|_{\tau=0} = K_{xx} - \bar{K}_{xx} \end{cases} \qquad (2.8)$$

$$\begin{cases} \dfrac{\partial}{\partial \tau} I_{02}^2 - \dfrac{\partial}{\partial z} K_{zz} \dfrac{\partial I_{02}^2}{\partial z} = \tilde{v} l_{01}^1 + \overline{v l_{01}^1} - v l_{01}^1 \\[4mm] K_{zz} \dfrac{\partial I_{02}^2}{\partial z}\bigg|_{\partial A} = 0, \quad I_{02}^2\big|_{\tau=0} = K_{yy} - \bar{K}_{yy} \end{cases} \qquad (2.9)$$

$$\begin{cases} \dfrac{\partial}{\partial \tau} I_{11}^2 - \dfrac{\partial}{\partial z} K_{zz} \dfrac{\partial I_{11}^2}{\partial z} = \tilde{u} l_{01}^1 + \tilde{v} l_{10}^1 \\[4mm] \qquad\qquad + (\overline{u l_{01}^1} - u l_{01}^1) + (\overline{v l_{10}^1} - v l_{10}^1) \\[4mm] K_{zz} \dfrac{\partial I_{11}^2}{\partial z}\bigg|_{\partial A} = 0, \quad I_{11}^2\big|_{\tau=0} = 0 \end{cases} \qquad (2.10)$$

These equations describe the vertical variation of concentration and the intial values and forcing terms for these equations are analogues of the generation mechanism of non-uniform advection.

3 Delay-dispersion Function

The eigenfunctions Ψ_m for the decay of vertical concentration variation provide a convenient means of solving equations (2.6) – (2.10). Eigenfunctions for Eq.(2.6) – (2.10) satisfy the eigenvalue problem:

$$\begin{cases} \dfrac{d}{dz} K_{zz} \dfrac{d\Psi_m}{dz} + \lambda_m \Psi_m = 0 \\[4mm] K_{zz} \dfrac{d\Psi_m}{dz}\bigg|_{\partial A} = 0 \end{cases} \qquad (3.1)$$

The lowest mode $\Psi_0 = 1$ has $\lambda_0 = 0$ and corresponds to the steady state of a cross-sectionally uniform concentration distribution. To represent the velocity profiles u, v, we introduce the coefficients:

$$\begin{cases} u_m = \overline{(u - \bar{u})\Psi_m} \; / \; (\overline{\Psi_m^2})^{1/2} \\[4mm] v_m = \overline{(v - \bar{v})\Psi_m} \; / \; (\overline{\Psi_m^2})^{1/2} \end{cases} \qquad (3.2)$$

315

From the eigenfunction representation of the initial conditions, it is clear that solutions of Eq.(2.6) - (2.7) can be written

$$
\begin{cases}
l^1_{10} = - \sum_{m=1}^{\infty} u_m \exp(-\lambda_m \tau) \Psi_m(z) / (\overline{\psi_m^2})^{1/2} \\
l^1_{01} = - \sum_{m=1}^{\infty} v_m \exp(-\lambda_m \tau) \Psi_m(z) / (\overline{\psi_m^2})^{1/2}
\end{cases}
\tag{3.3}
$$

In terms of u_m, v_m, the delay-dispersion functions in the lowest-order truncation are given by

$$
\begin{cases}
\dfrac{\partial}{\partial \tau} D_{xx} = \overline{l^1_{01}(\bar{u} - u)} = \sum_{m=1}^{\infty} u_m^2 \exp(-\lambda_m \tau) \\
\dfrac{\partial}{\partial \tau} D_{yy} = \overline{l^1_{01}(\bar{v} - v)} = \sum_{m=1}^{\infty} v_m^2 \exp(-\lambda_m \tau) \\
\dfrac{\partial}{\partial \tau} D_{xy} = \dfrac{\partial}{\partial \tau} D_{yx} = \sum_{m=1}^{\infty} u_m v_m \exp(-\lambda_m \tau)
\end{cases}
\tag{3.4}
$$

4 Determination of Transport Velocities

Proceeding to the non-homogeneous problem (2.8) - (2.10) for l^2_{20}, l^2_{02} and l^2_{11}, we introduce further coefficients

$$
\begin{cases}
u_{mn} = \overline{(u\Psi_m - u\Psi_m)\Psi_n} / (\overline{\psi_m^2})^{1/2} (\overline{\psi_n^2})^{1/2} \\
v_{mn} = \overline{(v\Psi_m - v\Psi_m)\Psi_n} / (\overline{\psi_m^2})^{1/2} (\overline{\psi_n^2})^{1/2} \\
K_{xm} = \overline{(K_{xx} - \bar{K}_{xx})\Psi_m} / (\overline{\psi_m^2})^{1/2} \\
K_{ym} = \overline{(K_{yy} - \bar{K}_{yy})\Psi_m} / (\overline{\psi_m^2})^{1/2}
\end{cases}
\tag{4.1}
$$

Thus if we represent l^2_{20}, l^2_{02} and l^2_{11} by the eigenfunction expansions:

$$
\begin{cases}
l^2_{20} = \sum_{m=1}^{\infty} a_m(\tau) \exp(-\lambda_m \tau)(\Psi_m / (\overline{\psi_m^2})^{1/2}) \\
l^2_{02} = \sum_{m=1}^{\infty} b_m(\tau) \exp(-\lambda_m \tau)(\Psi_m / (\overline{\psi_m^2})^{1/2}) \\
l^2_{11} = \sum_{m=1}^{\infty} c_m(\tau) \exp(-\lambda_m \tau)(\Psi_m / (\overline{\psi_m^2})^{1/2})
\end{cases}
\tag{4.2}
$$

316

then the amptitude factors $a_m(\tau)$, $b_m(\tau)$ and $c_m(\tau)$ satisfies the ordinary differential equations:

$$
\begin{cases}
\dfrac{da_m}{d\tau} = \displaystyle\sum_{n=1}^{\infty} u_{mn} u_n \exp((\lambda_m - \lambda_n)\tau) - \tilde{u}(\tau)u_m, \\[2mm]
a_m(0) = K_{xm} \\[4mm]
\dfrac{db_m}{d\tau} = \displaystyle\sum_{n=1}^{\infty} v_{mn} v_n \exp((\lambda_m - \lambda_n)\tau) - \tilde{v}(\tau)v_m, \\[2mm]
b_m(0) = K_{ym} \\[4mm]
\dfrac{dc_m}{d\tau} = \displaystyle\sum_{n=1}^{\infty} [u_{mn}v_n + v_{mn}u_n]\exp((\lambda_m - \lambda_n)\tau) \\[2mm]
\qquad\qquad - (\tilde{u}(\tau)v_m + \tilde{v}(\tau)u_m), \qquad c_m(0) = 0
\end{cases}
\tag{4.3}
$$

The solutions can be written as

$$
\begin{cases}
a_m = K_{xm} + u_{mm}(u_m\tau - \int_0^\tau \tilde{u}d\tau \\[2mm]
\qquad + \displaystyle\sum_{n\neq m} u_{mn}u_n [\exp[(\lambda_m-\lambda_n)\tau]-1]/(\lambda_m-\lambda_n)\} \\[4mm]
b_m = K_{ym} + v_{mm}(v_m\tau - \int_0^\tau \tilde{v}d\tau \\[2mm]
\qquad + \displaystyle\sum_{n\neq m} v_{mn}v_n [\exp[(\lambda_m-\lambda_n)\tau]-1]/(\lambda_m-\lambda_n)\} \\[4mm]
c_m = v_m(-\tau u_{mm} + \int_0^\tau \tilde{u}d\tau) + u_m(-\tau v_{mm} + \int_0^\tau \tilde{v}d\tau) \\[2mm]
\qquad + \displaystyle\sum_{n\neq m} [u_{mn}v_n + v_{mn}u_n [\exp[(\lambda_m-\lambda_n)\tau]-1]/(\lambda_m-\lambda_n)\}
\end{cases}
\tag{4.4}
$$

and are explicitly dependent upon the chosen value of the velocity $\tilde{u}$, $\tilde{v}$.

What we ideally would wish to achive with our choice for $\tilde{u}(\tau)$, $\tilde{v}(\tau)$ is that the solutions of (2.4) should be as close as possible to the solutions of the exact (2.1). Therefore, we consider the higher dispersion equations. From the equation (2.5), we find that the integrands for the $\partial^3\bar{C}/\partial x^3$ and $\partial^3\bar{C}/\partial y^3$ contribution to the dispersion are given by

$$
\overline{l^1_{10}(\bar{K}_{xx}-K_{xx})} + \overline{l^2_{20}(u-\bar{u})} = 2\sum_{m=1}^{\infty} K_{xm} u_m \exp(-\lambda_m\tau)
$$

317

$$+ \sum_{m=1}^{\infty} (u_{mm} \tau - \int_0^\tau \tilde{u} d\tau) u_m^2 exp(-\lambda_m \tau)$$

$$+ \sum_{m=1}^{\infty} \sum_{n \neq m} u_n u_m u_{mn} \left\{ \frac{exp(-\lambda_n \tau) - exp(-\lambda_m \tau)}{\lambda_m - \lambda_n} \right\} \qquad (4.5)$$

$$\overline{I_{01}^1 (\bar{K}_{yy} - K_{yy})} + \overline{I_{02}^2 (v - \bar{v})} = 2 \sum_{m=1}^{\infty} K_{ym} v_m exp(-\lambda_m \tau)$$

$$+ \sum_{m=1}^{\infty} (v_{mm} \tau - \int_0^\tau \tilde{v} d\tau) v_m^2 exp(-\lambda_m \tau)$$

$$+ \sum_{m=1}^{\infty} \sum_{n \neq m} v_n v_m v_{mn} \left\{ \frac{exp(-\lambda_n \tau) - exp(-\lambda_m \tau)}{\lambda_m - \lambda_n} \right\} \qquad (4.6)$$

The first terms on the right-hand side show the effect of lateral variations in diffusivities, and the next terms represent the higher-approximation to the shear effects than are given by the delay-dispersion function (3.4). Clearly, optimal choice for $\tilde{u}$, $\tilde{v}$ is to make the correction for shear effects be identically zero:

$$\begin{cases}
\dfrac{\partial D_{xx}}{\partial \tau} \int_0^\tau (u - \bar{u}) d\tau = \tau \sum_{m=1}^{\infty} (u_{mm} - \bar{u}) u_m^2 exp(-\lambda_m \tau) \\[2mm]
\quad + \sum_{m=1}^{\infty} \sum_{n \neq m} u_m u_n u_{mn} \dfrac{exp(-\lambda_n \tau) - exp(-\lambda_m \tau)}{\lambda_m - \lambda_n} \\[4mm]
\dfrac{\partial D_{yy}}{\partial \tau} \int_0^\tau (v - \bar{v}) d\tau = \tau \sum_{m=1}^{\infty} (v_{mm} - \bar{v}) v_m^2 exp(-\lambda_m \tau) \\[2mm]
\quad + \sum_{m=1}^{\infty} \sum_{n \neq m} v_m v_n v_{mn} \dfrac{exp(-\lambda_n \tau) - exp(-\lambda_m \tau)}{\lambda_m - \lambda_n}
\end{cases} \qquad (4.7)$$

Then, transport velocity may be expressed as

318

$$\ddot{u}(\tau) = [\sum_{m=1}^{\infty} u_m^2 exp(-\lambda_m \tau)]^{-2} \{\sum_{m=1}^{\infty} u_m^2 exp(-\lambda_m \tau)$$

$$\cdot [\sum_{m=1}^{\infty} u_m^2 u_{mm} exp(-\lambda_m \tau) - \tau \sum_{m=1}^{\infty} \lambda_m u_{mm} u_m^2 exp(-\lambda_m \tau)$$

$$+ \sum_{m=1}^{\infty} \sum_{n \neq m} u_n u_m u_{mn} \frac{\lambda_m exp(-\lambda_m \tau) - \lambda_n exp(-\lambda_n \tau)}{\lambda_m - \lambda_n}]$$

$$+ \sum_{m=1}^{\infty} u_m^2 \lambda_m exp(-\lambda_m \tau)[\tau \sum_{m=1}^{\infty} u_{mm} u_m^2 exp(-\lambda_m \tau)$$

$$+ \sum_{m=1}^{\infty} \sum_{n \neq m} u_m u_n u_{mn} \frac{exp(-\lambda_n \tau) - exp(-\lambda_m \tau)}{\lambda_m - \lambda_n}]\}$$

$$\ddot{v}(\tau) = [\sum_{m=1}^{\infty} v_m^2 exp(-\lambda_m \tau)]^{-2} \{\sum_{m=1}^{\infty} v_m^2 exp(-\lambda_m \tau)$$

(4.8)

$$\cdot [\sum_{m=1}^{\infty} v_m^2 v_{mm} exp(-\lambda_m \tau) - \tau \sum_{m=1}^{\infty} \lambda_m v_{mm} v_m^2 exp(-\lambda_m \tau)$$

$$+ \sum_{m=1}^{\infty} \sum_{n \neq m} v_n v_m v_{mn} \frac{\lambda_m exp(-\lambda_m \tau) - \lambda_n exp(-\lambda_n \tau)}{\lambda_m - \lambda_n}]$$

$$+ \sum_{m=1}^{\infty} v_m^2 \lambda_m exp(-\lambda_m \tau)[\tau \sum_{m=1}^{\infty} v_{mm} v_m^2 exp(-\lambda_m \tau)$$

$$+ \sum_{m=1}^{\infty} \sum_{n \neq m} v_m v_n v_{mn} \frac{exp(-\lambda_n \tau) - exp(-\lambda_m \tau)}{\lambda_m - \lambda_n}]\}$$

In the limits of large and small times, we have

$$\ddot{u}(0) = \bar{u} + \overline{(u-\bar{u})^3} / \overline{(u-\bar{u})^2}, \quad \ddot{u}(\infty) = u_{11}$$
$$\ddot{v}(0) = \bar{v} + \overline{(v-\bar{v})^3} / \overline{(v-\bar{v})^2}, \quad \ddot{v}(\infty) = v_{11}$$

(4.9)

In fact, the extent of the skewness of concentration field depends on the difference between u, v and $\bar{u}$, $\bar{v}$. When $K_{xx} =$ const., $K_{yy} =$ const., the centroid, area and variance determined from Eq.(2.4) are exact, which can be verified by (2.3) timing $x^i y^j$ and integrating partly.

Also, the solution conditions for D_{ij} may be given by

$$\begin{cases} D_{ij}(x) = \int u_i^1 \, dz \int \frac{dz}{K_{zz}} \int u_j^1 \, dz \\[2mm] u_i^1 = u_i - \bar{u}_i \end{cases} \tag{4.10}$$

or expressed by Chatwin's form function:

$$\begin{cases} D_{ij} = \overline{u_i f_j} \\[2mm] \dfrac{d}{dz} K_{zz} \dfrac{df_x}{dz} = \bar{u} - u, \quad \bar{f}_x = 0, \quad K_{zz} \dfrac{\partial f_x}{\partial z} \bigg|_{\partial A} = 0 \qquad (4.11) \\[2mm] \dfrac{d}{dz} K_{zz} \dfrac{df_y}{dz} = \bar{v} - v, \quad \bar{f}_y = 0, \quad K_{zz} \dfrac{\partial f_y}{\partial z} \bigg|_{\partial A} = 0 \end{cases}$$

5 One Kind of Shallow Water Flow

Fisher's shallow water flow model is shown in Fig.5.1. From (4.10), we got:

$$D_{ij}(x) = \frac{d^2}{K_{zz}} \begin{bmatrix} \dfrac{u_0^2}{120} & \dfrac{5u_0 v_0}{192} \\[3mm] \dfrac{5u_0 v_0}{192} & \dfrac{v_0^2}{120} \end{bmatrix} \tag{5.1}$$

Eigenvalue and eigenfunction are

$$\lambda_m = m^2 \pi^2 K_{zz} / d^2, \quad \Psi_m = \cos m\pi \frac{y}{d} \tag{5.2}$$

Then the coefficients may be written as:

$$\begin{cases} U_m = \dfrac{u_0 \sqrt{2}(1-(-1)^m)}{m^2 \pi}, \quad V_m = (-1)^{m+1} \dfrac{2\sqrt{2}\, v_0}{m\pi} \\[3mm] u_{mm} = v_{mm} = 0 \\[3mm] u_{mn} = \dfrac{2u_0(m^2+n^2)}{\pi^2(m^2+n^2)}[1-(-1)^{m+n}](-1)^{[\frac{m+n}{2}]+1} - \dfrac{2m}{m^2-n^2} \qquad (5.3) \\[3mm] v_{mn} = \dfrac{2v_0}{\pi^2}[-\dfrac{\sin\frac{m+n}{2}\pi}{m+n} + \dfrac{\sin\frac{m-n}{2}\pi}{m-n}] \end{cases}$$

The delay dispersion coefficients are given by

$$
\begin{cases}
D_{xx} = \dfrac{1}{120}\,\dfrac{u_0 d^2}{K_{zz}} \\[4mm]
\qquad + \displaystyle\sum_{m=1}^{\infty} \dfrac{u_0 d^2}{K_{zz}}\,\dfrac{2[1-(-1)^{m+1}]}{\pi^4 m^6}\,exp(-\lambda_m \tau) \\[6mm]
D_{xy} = D_{yx} = \dfrac{5}{192}\,\dfrac{v_0 u_0 d^2}{K_{zz}} \\[4mm]
\qquad + \displaystyle\sum_{m=1}^{\infty} \dfrac{u_0 v_0 d^2}{K_{zz}}\,\dfrac{4[(-1)^{m+1}+1]}{\pi^4 m^5}\,exp(-\lambda_m \tau) \\[6mm]
D_{yy} = \dfrac{1}{12}\,\dfrac{v_0 d^2}{K_{zz}} + \displaystyle\sum_{m=1}^{\infty} \dfrac{v_0 d^2}{K_{zz}}\,\dfrac{8}{\pi^4 m^4}\,exp(-\lambda_m \tau)
\end{cases}
\qquad (5.4)
$$

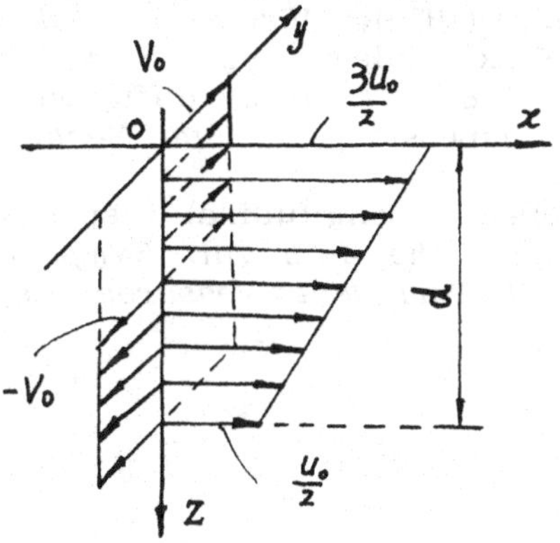

Fig. 5.1.

6 Conclusion and Discussion

To derive the delay–diffusion equation (2.4) for a given flow situation, the order of procedure is as follows:

1. Determine the velocity field and eigenfunction and eigenvalue;

2. Calaulate the delay-dispersion functions (3.4);

3. Evaluate the coefficients u_{mn}, v_{mn}, u_m, v_m, and solve Eq.(4.8) for $\bar{u}$, $\bar{v}$.

Generally, we can discuss effects of the memory on area, centroid and variance by the sign of $u-\bar{u}$ and $v-\bar{v}$. Of course, the Thacker's diffusion-telegraph equation may be extend to two-dimensional one directly ([7], [8]).

321

REFERENCES

[1] Taylor, G.I., (1953), Dispersion of Soluble Matter in Solvent Flowing Slowly through a Tube, *Proc. Roy. Soc.* A219, pp 186-206.

[2] Taylor, G.I., (1954), The Dispersion of Matter in Turbulent Flow through a Tube, *Proc. Roy. Soc.* A223, pp 446-468.

[3] Taylor, G.I., (1959), The Present Position in the Theory of Turbulent Diffusion, *Adv. Geophys*, b, pp 101-111.

[4] Chatwin, P.C., (1970), The Approach to Normality of the Concentration Distribution of a Soluble in Solvent Flowing along a Pipe, *J. Fluid Mech.*, 43, pp321-362.

[5] Gill, W.M., and Sankarasubramanian, R., (1970), Exact Analysis of Unsteady Convective Diffusion, *Proc. Roy. Soc.*, A316, pp 341-350.

[6] Thacker, W.C., (1976), A Solvable Model of Shear Dispersion, *J. Phys. Oceanog.*, 6, pp 66-75.

[7] Fisher, H.B., (1979), *Mixing in Inland and Coastal Water*, New York Academic, pp 125-148.

[8] Smith, R., (1981), A Delay-diffusion Deseription for Contaminant Dispersion, *J. Fluid Mech.*, Vol. 105, No. 9, pp 469-486.

[9] Liu, Y.L., (1990), Longitudinal Dispersion in Unsteady Flows, *Proc. of the 4-th Int. Symp. on Refined Flow Modelling and Turbulent Measurements*, Wuhan, China, pp 196-204.

27 Finite element simulation of flow and pollution transport applied to a part of the River Rhine

R. Feldhaus, J. Hottges, T. Brockhaus and G. Rouve

ABSTRACT

In an 25 km^2 area in the southern part of the river Rhine possible effects of a multitude of flood control measures on flow properties had to be analyzed. A two-dimensional mathematical numerical model based on the Finite-Element-Technique was used to assess detailed information of high spatial resolution. With the aid of the computed data for water levels and stream velocities a differentiated estimation of flood damage risk for man and nature is facilitated. Moreover the pollution transport was modelled for various cases of instantaneous injection of non-buoyant solutes. Based on the calculated flow field and empirically computed mixing coefficents the effect of pollutant injections on dead zones is demonstrated.

1   Introduction

Two-dimensional mathematical numerical simulations were carried out for an approximately 14 km long section of the river Rhine near Karlsruhe [1]. The investigations were subjected to the following objectives:

- flow simulations for quantitative assessment of flood control measures including ecological aspects,
- simulation of pollution transport.

2   Description of the investigated area

The investigated area, in the following paragraphs called "Rastatter Rheinniederung" covers an approximately 25 km^2 area of the Rhine plain. This area is limited from the surrounding by dykes [fig. 1]. On the left bank of the Rhine, that is on the French side, the Rhine dykes are orientated close to the stream whereas on the German side natural flood plain is still available. The "Rastatter Rheinniederung" is devided in two sections by the river Murg falling into the Rhine at km 344.5. South of the confluence of Rhine and Murg the former uncultivated Rhine was splitted up in individual watercourses. Today these watercourses determine clearly the topographical structure of the Rhine flood plain. Due to the smaller slope north of the Murg, the Rhine began to meet in a distinct main branch. Here the structure of topography is determined by gravel extraction of dredge lakes with considerable depth.

By this strong surface heterogeneity as well as the high variability of the vegetation, high demands on the capacity and stability of a mathematical numerical model are requested.

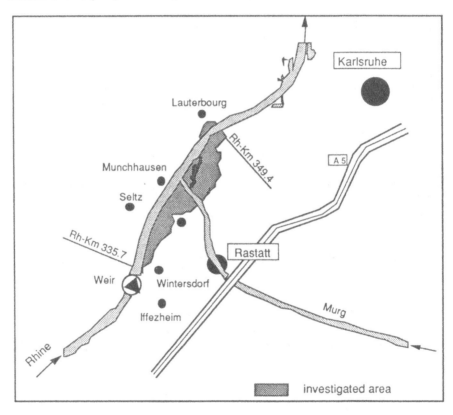

FIGURE 1   *Map of the investigated area*

3   Mathematical-numerical models for flow and pollution transport

Because of the complexity of the regional structure described above, the application of the Finite-Element-Method is convenient. Its theoretical bases are described in [2], [3] and [9]. The Finite-Element-Method permits a very realistic reproduction of the area by local refinements of the finite element mesh. Profiting the Front-Solver-Technology [7], a special order for the solution of differential equations later to be demonstrated allows the use of the simulation model even on powerful personal computers.

3.1   Simulation of flow

The flow properties of fluids can be described exactly by the formulation of balance equations of mass and momentum. In spite of the high capacity of the most progressive computers available at the moment, the compact solution of the equation system is not possible in its whole complexity at present. A usual simplification is represented by reduction of the three-dimensional to a one- or two-dimensional view. This procedure limits the validity of the equation system to those flow conditions being dominant in the examined dimensions. The effects in the dimension not considered have to be realized by averaged model assumptions.

It is profitable to carry out a two-dimensional approach expanded by depth averaging. This approach can be derived from the integration of the universally valid three-dimensional differential equation system by time and depth of water. By use of the tensor notation the so-called shallow-water-equations [8] can be written as:

$$\frac{\partial h}{\partial t} + \frac{\partial (u_i\, h)}{\partial x_i} = 0 \tag{1}$$

$$\frac{\partial u_i}{\partial t} + u_j \frac{\partial u_i}{\partial x_j} = -g \frac{\partial}{\partial x_i}\left(h + z_{so}\right) - \frac{1}{\rho} \frac{\partial\left(T_{ij} + D_{ij}\right)}{\partial x_j} + S_i \tag{2}$$

with:

h	:	averaged flow-depth
t	:	time
$u_{i,j}$	:	averaged flow-velocity in the direction of x and y
g	:	Gravitation
z_{so}	:	bottom level
ρ	:	density of fluid
T_{ij}	:	diffusive momentum transport
D_{ij}	:	dispersive momentum transport
S_i	:	external forces

Further simplification is made by considering only stationary flow:

$$\frac{\partial h}{\partial t} = 0 \quad ; \quad \frac{\partial u_i}{\partial t} = 0 \tag{3}$$

By the external forces S_i all exterior forces on the examined water body are understood. In the modelled form applied here these forces consist of the shear stresses at wall and bottom, the wind-shear acting on the water surface and the Coriolis force.

The terms resulting from time- and depth-averaging have to be expressed by averaged variables in order to make the equation system solvable. Stresses due to the viscosity of the fluid as well as the so-called Reynolds-stresses are combined in the stress tensor T_{ij}. Several turbulence closure models are used for solution of T_{ij}.

The differential momentum transport expressed by the tensor D_{ij} originates from the deviations of the real vertical distribution of a variable from its averaged value of time and depth. This term is neglected by most of the users. However, because of its considerable influence on results, it certainly represents an item of research [10].

3.2 Simulation of pollution transport

For the assessment of the influence of pollutants on the flood plain and the individual watercourses, e.g. injected as a result of an accident, pollutant transport was simulated for steady flow conditions. The depth-averaged advection-diffusion equation for this problem is:

$$\frac{\partial c}{\partial t} + u_i \cdot \frac{\partial c}{\partial x_i} = \frac{1}{h} \cdot \frac{\partial}{\partial x_i}\left(h \cdot E_{ij} \cdot \frac{\partial c}{\partial x_j}\right) \tag{4}$$

with:

c	:	depth-averaged concentration
E_{ij}	:	mixing coefficient tensor

The mixing coefficient is assumed similar to [6] as composed of longitudinal ε_ξ and transverse ε_η components. They find the mixing tensor E_{ij} with the angle ϕ (positive counter-clockwise) of the flow vector from the x-axis:

$$E_{xx} = e_\xi \cdot \cos^2 \phi + e_\eta \cdot \sin^2 \phi \tag{5}$$

$$E_{xy} = E_{yx} = (e_\xi - e_\eta) \cdot \sin \phi \cdot \cos \phi \tag{6}$$

$$E_{yy} = e_\xi \cdot \sin^2 \phi + e_\eta \cdot \cos^2 \phi \tag{7}$$

Elder [4] relates the components to the local waterdepth h and shear velocity u_* by:

$$e_\xi = e_L \cdot u_* \cdot h = 5.93 \cdot u_* \cdot h \tag{8}$$

$$e_\eta = e_T \cdot u_* \cdot h = 0.23 \cdot u_* \cdot h \tag{9}$$

e_L denotes the longitudinal-, e_T the transversal mixing coefficient.

For the numerical solution of the equations a full Finite-Element approach in all three coordinates is used. Because the differential equation behaves in a parabolic manner in time, it is possible to compute the whole process devided in single timesteps. For every timestep the concentration field vector C is used as a boundary condition additional to the concentration values in the inlet, which must be defined for every timestep. This yields a number of unknowns which is similar to other methods, i.e. the number of nodes in the computational area minus these on the inlet.

For discretisation purposes the numerical model uses prismatic elements with triangular and quadrangular footing where the magnitude of the prism is the length of a time step. The shape-function is trilinear and each element has six or eight nodes.

The equations system is inverted once instead of solving for every timestep, and for every timestep only a matrix multiplication is necessary.

The matrix calculations are carried out in band matrix form to reduce the number of operations. For the matrix inversion it is analytically not guaranteed that the inversion of a band matrix forms another band matrix, but it seems likely, because at low Courant numbers (eqn. 10) there is no influence from one node to another, when they are far from each other (the solute particles are not that fast). Numerical experiments proved that the bandwidth increases by a factor of about 2 - 5. Anyhow, the timesteps have to make sure that the element Courant Cu number is distinctly smaller than 1.

$$Cu = \frac{v \cdot \Delta t}{\Delta x} \tag{10}$$

with:

Δt	:	timestep
Δx	:	minimum of element length
v	:	viscosity

4   Assessment of flood control measures

In the biologically hypersensitive area of "Rastatter Rheinniederung" situated right above the weir of Iffezheim (cf. fig. 1), a prediction of flow conditions is necessary for each of the considered modifications of topography in order to take flood control measures that meet ecological concerns. The range of discharge encloses an area of 2000 m^3/s (mean water) to 5000 m^3/s (high water) respectively. Studying a large number of possible measures mathematical numerical modelling was given preference over a traditional experimental research due to reasons of economy and flexibility.

The following paragraphs illustrate the use of the Finite-Element simulation model in order to assess economical and ecological aspects of flood control measures.

4.1   Discretization

The area of "Rastatter Rheinniederung" is topographically represented by approximately 8000 nodes arranged in 3000 elements [fig. 2]. The structure of the pictured finite element mesh is orientated close to topography and distribution of vegetation. A major part of nodes is situated in hectometre stations of the Rhine in which topography was mesured in detail.

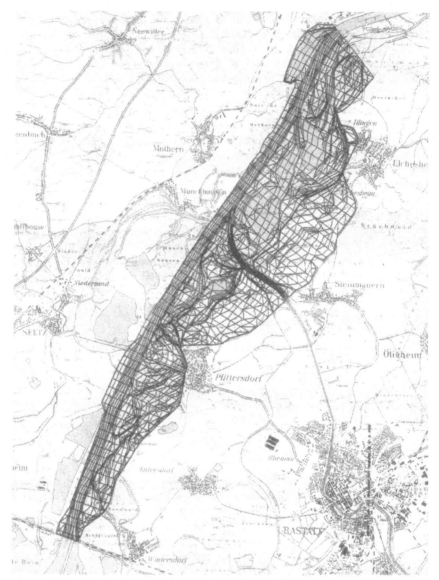

FIGURE 2   *Finite element mesh of Rastatter Rheinniederung*

4.2   Calibration and Verification

The empirical parameters of the model have to be adapted to the characteristics of the investigated area. This is done in a model calibration process by the comparison of calculated water
levels and velocities with values measured in nature for known discharges. Subsequently the
determined model parameters are to be verified by comparing them to another discharge configuration with known flow data.

Fig. 3 shows the comparison between measured and calculated water levels along the German dykes of the river Rhine used for model verification. According to this, one can proceed on the assumption that the tolerance of accuracy forecasting water levels averaged is less than 7 cm.

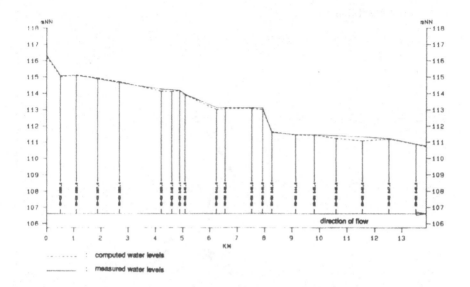

: computed water levels
: measured water levels

FIGURE 3   *Comparision of measured / computed water levels along the dykes of the River Rhine*

4.3   Results

In the whole investigated area water levels and stream velocities were calculated for about 60 variants of discharge and topography. To illustrate one of the manifold possibilities of using the calculated water levels, a flood risk analysis for the German dykes of the Rhine will be demonstrated. The suggested removal of the actually still existing flow obstacles in the investigated area leads to a considerable decrease of damage risk at dams [fig. 4].
Furthermore streamvelocities [fig. 5] are reliably prognosticated. In this paper only an example can be presented. The velocities directly can be used by ecologists and engineers of water management for quantitative description e.g. of possible entry of oxygen, position of deadwater zones, regions in danger of erosion etc..

329

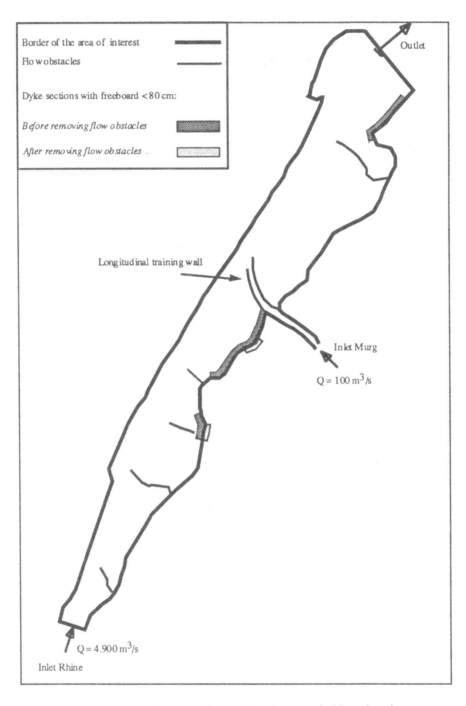

Border of the area of interest

Flow obstacles

Dyke sections with freeboard < 80 cm:

Before removing flow obstacles

After removing flow obstacles .

Outlet

Longitudinal training wall

Inlet Murg

$Q = 100 \, m^3/s$

$Q = 4.900 \, m^3/s$

Inlet Rhine

FIGURE 4   *Decrease of damage risk at the dykes by removal of flow obstacles*

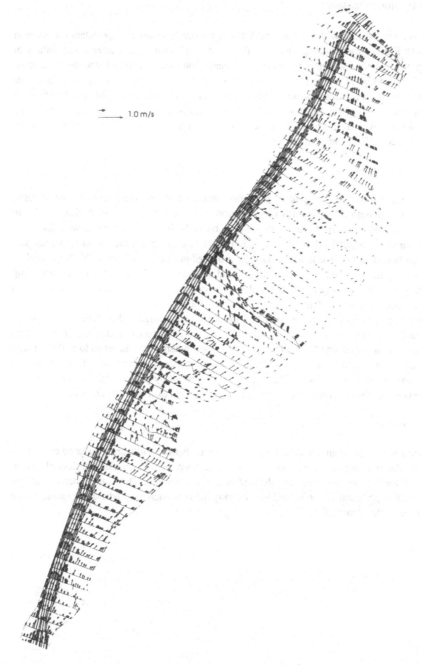

FIGURE 5   *Evaluated stream velocities (Q$_{total}$ = 5.000 m^3/s)*

5   Pollutant transport

First computation runs with the numerical model for the description of pollutant transport in the river Rhine produced high numerical oscillations. That means alternating values at contiguous nodes and additionally negative concentration values at some locations. In contrast to the flow simulation where the functions for velocities are approximated quadratically over one element, the simulation of pollution transport uses only linear interpolation functions to minimize computing time. Because of this, the steep gradients in water depth and velocity data could not be resolved by the finite element mesh outlined in figure 2. The solution of the problem was: Refinement of the mesh.

5.1   Discretization

Because mesh modification by hand is a time-consuming and boring business, sophisticated meshing and remeshing procedures were developed. As very often in river hydraulics also in this case the nodes which had to be combined to build the finite element mesh were locations of measured water depth. A mesh generation procedure combines these nodes to triangles, based on boundary nodes which need only be provided manually in the case that the boundary is very complicated. In this case it could be found automatically, because there were no strong indentations in the boundary. The triangulation procedure computer time only increases with less than $O(n^{1.5})$ where O denotes the order and n is the number of nodes.

The triangles can be generated adapted to the flow, which means that their smaller width preferably lies rectangular to the velocity vector. Additional nodes can be inserted automatically to avoid small elements. After that, the elements are combined to build as many quadrilaterals as possible. During the whole meshing procedure one can limit the bandwidth of the matrix for the linear Finite-Element equation system by preventing the connection of elements where the differences between their node numbers exceed a certain limit.

5.2   Results

Test runs with refined and modified mesh gave much better results and produced only small wiggles. An example can be seen in fig. 6. To investigate the influence on the floodplains, an injection for 140 seconds has been defined at the inlet section. The main statement of this figure is, that pollution is transported very quickly in the mainchannel whereas it remains for a long time on the floodplains.

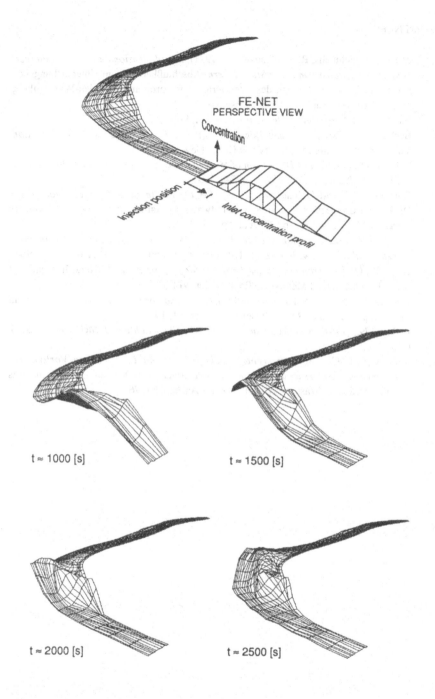

FIGURE 6   *Spreading of pollution*

REFERENCES

[1]   Arnold, U., Feldhaus, R. and Rouvé, G., (1991), *Zweidimensionale Strömungsberech-nung im Retentionsraum Murg/Steinmauern*, Abschlußbericht zum Untersuchungsvor-haben des IWW im Rahmen des Integrierten Rheinprogramms Baden-Württemberg, Institut für Wasserbau und Wasserwirtschaft, RWTH Aachen.

[2]   Bathe, K.J., (1986), *Finite-Elemente-Methoden*, Springer Verlag.

[3]   Becker, E.B., Carey, G.F. and Oden, J.T., (1981), *Finite Elements, An Introduction*, Vol. 1, Prentice Hall, Inc., Inglewood Cliffs, New Jersey.

[4]   Elder, J. W., (1959), *The Dispersion of Marked Fluid in Turbulent Shear Flow*, Journal of Fluid Mechanics, Vol. 5, No. 4, pp 544-560.

[5]   Gambolati, G. and Galeati, G., (1989), *On the Finite Element Integration of the Dispersion-Convection Equation*, Groundwater Contamination: *Use of Models in Decision-Making* ed. Jousma, G. et al, pp 231 - 241.

[6]   Holly, F.M. and Usseglio-Polatera, J.-M., (1984), *Dispersion Simulation in Two-Dimensional tidal Flow*, Journal of Hydraulic Engineering, Vol. 110, No. 7, July 1984.

[7]   Hood, P., (1976), *Frontal Solution Program for Unsymmetric Matrices*, International Journal for Numerical Methods in Engineering, Vol. 10.

[8]   Kuipers, J. and Vreugdenhil, C.B., (1973), *Calculation of Two-Dimensional Horizontal Flow*, Report on Basic Research Delft, p 163, part 1, 19.

[9]   Reddy, J.N., (1984), *An Introduction to the Finite Element Method*, McGraw-Hill Book Company.

[10]   Stein, C.J., (1990), *Mäandrierende Fließgewässer mit überströmten Vorländern-Experimentelle Untersuchung und numerische Simulation*, Mitteilungen des Instituts für Wasserbau und Wasserwirtschaft, RWTH Aachen, Nr. 76.

28 Analysis of the balance of point and diffuse source cadmium loading in the Rhine River basin

A. Almassy, G. Jolankai, I. Biro, W. van Deursen and J. Kwadijk

ABSTRACT

A Geographical Information System (GIS) has been coupled with a steady state, network type hydrological and material transport model for assessing the balance of point and diffuse source Cadmium loadings in the Rhine River basin. Highly varying spatial information, stored in thematic maps, has been manipulated and used in order to estimate the non-point source contribution to the overall Cd loadings. Longitudinal concentration profiles have been calculated throughout the Rhine river system, and calibrated against measurements.

1. Introduction

The present study, first phase of a larger study due for completion in April 1992, was carried out jointly by the Water Resources Research Centre, Budapest, Hungary (VITUKI) and the International Institute for Applied Systems Analysis, IIASA, Laxenburg, Austria (IIASA) with the assistance of the University of Utrecht, the Netherlands, for the National Institute of Public Health and Environment Hygiene (RIVM) of the Netherlands. This paper presents the results of the first phase of the study, which was completed in January 1991.

335

The final goal of the study was to provide a better understanding of the contribution of point and diffuse sources from the different regions of the Rhine river basin to the overall Cadmium loadings, to help identify where to concentrate efforts to reduce heavy metal pollution sources.

The ultimate sink of environmentally hazardous substances discharged by rivers are the estuaries, seas and oceans. The North Sea is surrounded by one of the most industrialized areas of the world. As a consequence it receives large contaminant loads of which perhaps the most hazardous are heavy metals. A significant part of these loads originates from the Rhine River basin and ends up in the sediments of the downstream part of the river or in the sea. Five heavy metals were identified as contaminants of interest to the RIVM, but in the first phase of the study only one, Cadmium, has been used for testing and demonstrating capabilities of the modelling approach. Similarly, three time slices of interest have been identified, but in the first phase of the study, data from the year 1975 have been used.

The Cd loadings of the Rhine originate from both point and diffuse sources. If the inputs (pollutant emissions from industrial or other point sources and diffuse sources) are known, the in-stream concentrations are usually modelled by one dimensional, network type dynamic or steady state models. Point source pollutant loads may be measured, or available data are used to estimate their contribution to the total loads. To assess the non point source part of the pollutant loads, either extensive and time consuming manual methods, or the application of new techniques is required. In the course of this study, the assessment of the diffuse side cf the Cd loadings has been made by applying GIS operations on thematic maps, whilst for the calculation of the in-stream concentrations a steady state network type model (SENSMOD) has been applied. In the following sections both SENSMOD and the applied GIS will be briefly presented.

2. The SENSMOD model as adopted to the present study

The model SENSMOD (based on the ideas of Dr G. Jolánkai) has originally been developed in the Water Resources Research Centre (VITUKI), Budapest, Hungary, for supporting regional water management planning decisions involving water quality aspects. It was applied to several river networks in Hungary for assessing the water quality impacts of planned changes in the river system, and for supporting decisions on discharge consents. Its original version contains modules for describing any contaminant with first order decay, and modules for describing coupled oxygen processes and nitrification.

336

2.1 A short description of SENSMOD

The version of SENSMOD used in the present study predicts
longitudinal discharge and Cd concentration profiles for the
defined streamflow network of the Rhine river basin. This is
achieved by accumulating the specific runoffs of the
subcatchment areas of the reach concerned, and solving the
one dimensional transport equation with first order reaction
kinetics for each reach of the river network. The equations
of the reach-by-reach calculations are the following.
  For the calculation of the discharge profile:

$$Q(x) = Q_o + R \cdot A \cdot \frac{x}{L} \qquad\qquad (1)$$

where x is the distance from the upstream end of the reach,
$Q(x)$ is the discharge at the point x, Q_o is the discharge at
the upstream end of the reach, R is the average areal
specific runoff for the catchment area of the actual reach
(supposed to be distributed uniformly along the reach), A is
the catchment area of the reach and L is the reach length.
  For calculating the concentration profile the explicit
solution of the following differential equation has been
used:

$$\frac{\partial}{\partial x}\Big[Q(x) \cdot C(x)\Big] = \frac{R \cdot A}{L} \cdot C_d - K \cdot \frac{Q(x)}{v} \cdot C(x) \qquad\qquad (2)$$

where $C(x)$ is the concentration at the point x, C_d is the
concentration of the runoff (contains the diffuse load), K is
the decay rate of the contaminant in the river reach and v is
the average flow velocity of the reach.
  The explicit solution of the differential equation (2) can
be expressed as follows:

$$C(x) = C_o \cdot F(x)^{\beta(x)} + \frac{C_d}{\beta(x)} \cdot \Big(1 - F(x)^{\beta(x)} \Big) \qquad\qquad (3)$$

where C_o is the concentration at the upstream end of the
reach, $F(x)$ is the dilution factor at the point x and $\beta(x)$ is
the decay term.
  The dilution factor can be expressed as follows:

$$F(x) = \frac{Q_o}{Q_o + \frac{Q_1 - Q_o}{L} \cdot x} \qquad\qquad (4)$$

where Q_1 is the discharge at the downstream end of the reach.

The decay term can be expressed as:

$$\beta(x) = 1 + \frac{K}{v} \cdot \left[\frac{Q_o}{Q_1 - Q_o} \cdot L + x \right] \tag{5}$$

The concentration of the land runoff can be written as

$$C_d = \frac{\sum\limits_{u} L_u \cdot e^{-k_u \cdot \bar{t}_u}}{Q_1 - Q_o} \tag{6}$$

where L_u is the total contaminant load coming from the u-th land use form in the catchment area of the reach, k_u is the decay rate of the contaminant travelling on the terrain and exported from the u-th land use form and $\bar{t}_u$ is the average time of travel of the contaminant exported from the u-th land use form to the recipient.

This average travel time can be expressed by means of an empirical formula as

$$\bar{t}_u = \alpha_1 \left[\frac{\ell}{\sqrt{\frac{\Delta H}{\ell}}} \right]^{\alpha_2} \tag{7}$$

where ℓ is the average length of the runoff pathways from the actual cell to the recipient, averaged for the u-th land use form, ΔH is the average elevation difference between the two above mentioned points, averaged for the u-th land use form and α_1, α_2 are empirical coefficients describing the roughness conditions of the runoff pathway in function of the different land use forms.

The total contaminant load coming from the u-th land use form L_u can be calculated as

$$L_u = A_u \cdot Y_u \tag{8}$$

where A_u is the area of the u-th land use form within the catchment area of the reach in concern and Y_u is the export rate of the contaminant in the u-th land use form.

This latter can be expressed according to the unit area loading concept:

$$Y_u = \sum\limits_{k} I_k \cdot a \cdot R^* \tag{9}$$

338

where the sum represents the total contaminant input (coming from k different pollution origins) to the u-th land use form, a is the transport availability factor [3], the fraction of total contaminant input that can be removed by a unit value of runoff and R* is a transformed value of runoff conditions which can be expressed as follows:

$$R^* = R_{max} \cdot \frac{R}{R_{0.5} + R} \tag{10}$$

where R_{max} is the maximum runoff from the area, R is the actual runoff from the area and $R_{0.5}$ is the runoff value at which the half of the maximum potential export rate is achieved.

For Cadmium four major source of diffuse contaminant input have been taken into consideration:

$$\sum_k I_k = I_p + I_s + I_a + I_c \tag{11}$$

where I_p is the diffuse Cd input from phosphorus fertilizer use, I_s is the diffuse Cd input from sewage sludge use, I_a is the diffuse Cd input from atmospheric deposition and I_c is the diffuse Cd input from communal wastewaters.

In the reach-by-reach calculation point sources are also taken into consideration in the following way. For the discharge profile the point sources (concentrated water discharges) are simply added to the actual discharges; in the case of the concentration an immediate and complete mixing is assumed and the resultant concentration is calculated on the basis of the mass balance.

Once the listed variables are known, it is possible to substitute into equations (1) and (3) and repeat the calculation reach by reach from the most upstream points of the network to the downstream boundary. In this way the first approximations of the discharge and concentration profiles can be calculated until a measurement point. At each gauging station a calibration algorithm applies a common correction factor to the parameters to be calibrated for the sub-tree bordered by gauging sections. This correction factor is calculated in function of the discrepancies between the calculated and measured data. The above mentioned parameters used for calibration were: the specific runoff (for the discharge profile) and the decay rates (for the concentration profile). It should be stressed that the term "decay rate" means here a rate coefficient that combines the effects of all processes that affect the fate of the contaminants concerned, while they travel downstream.

3. The GIS used in the study

There are many commercially available GIS systems, and their use is able to facilitate the solution of almost any problem where the user has to handle large spatially distributed data sets. However, in some special cases, as in the modelling of water movement governed processes, the usual set of GIS manipulations may not be sufficient to provide a complete solution of the problem. This has been the reason for the development of a special, hydrology and water management oriented GIS at the University of Utrecht, Institute of Geographical Research for PC-s, which includes some special water movement oriented operations as well as most of the classical GIS tasks.

3.1 The Local Drainage Direction concept

Perhaps the most important advantage of this system is that in addition to the usual set of GIS operations (arithmetical and logical operations performed on thematic maps) it contains several special operations, designed for hydrological modelling applications. When modelling hydrology related processes by GIS tools, the basic and primary information is contained in the topography of the catchment basin, since the water movement is regulated by a simple principle: "water flows downstream". All the spatial units traditionally used in hydrology (catchment areas of given streamflow sections or reaches etc.) depend on the topographical conditions, namely on the magnitudes and directions of the local elementary slopes. The GIS used in this study is able to handle these concepts.

The basis of these special facilities is the LDD (Local Drainage Direction) concept. An LDD map is prepared on the basis of the original elevation map; it contains an integer code for each raster cell denoting the elementary direction of the local slope. In essence the LDD map shows, which of the eight neighbouring cells the actual cell drains into. This is achieved by selecting the greatest elevation difference between the elevation of the cell under consideration and the neighbouring ones. The information stored in the LDD maps can be considered as a elementary network of the superficial water movement.

The LDD map enables the contents of individual raster cells to be accumulated along the elementary flow path network. The user can select the downstream limit of this path and calculations, such as average values, can then be readily performed for the catchment to the user selected points. The inverse of this operation (spreading values from user selected points to their subcatchments) is also included into the system.

340

3.2 Other features

In addition to this LDD concept, specially designed for
hydrological modelling purposes, the system contains most of
the usual GIS facilities, i.e. point type - raster type
interface utilities for loading point type data into raster
maps and extracting raster attributes to point type data
bases, or presentation utilities for controlling and
visualizing interim or final result maps.

4. Data preparation and calculations

SENSMOD, being a network type model, requires data aggregated
to its own elementary units, i.e. to river reaches. Therefore, after defining the schematic network of the river
system and the reach bordering points, all the data necessary
for substitution into the equations in section 2.1, were
transformed from thematic maps into average values representing subcatchment areas of the defined river reaches.

4.1 Miscellaneous maps

The LDD map was prepared on the basis of the digitized
elevation map (see Figure 1) of the Rhine River basin [2]
(bounded by the Dutch-German border on the North). A fragment
of the LDD map overlayed on the elevation map is shown in
Figure 2. Subsequently the schematic network of the river
system was defined together with the user defined river reach
bordering points; then the reach lengths and the catchment
areas were calculated.

A land use map has also been digitized from the Rhine
Monograph [2].

The three areal specific runoff maps, for equation (10),
were prepared by the spreading operation (described earlier)
on the basis of the characteristic discharge values calculated from daily time series of 21 discharge gauging stations
in the area. Then, the average areal specific runoff values
for the defined river reaches were extracted from the map.

4.2 Calculation of the diffuse Cadmium loadings

According to equation (11) four main diffuse Cd input sources
have been identified; from phosphorus fertilizer
application, from sewage sludge use, from atmospheric
deposition and from general communal diffuse sources. For the
first two inputs spatially distributed data were available
from a German data base containing data on a "kreise"
(counties of Germany) basis. Consequently after digitizing
the kreise border locations the respective maps could be

341

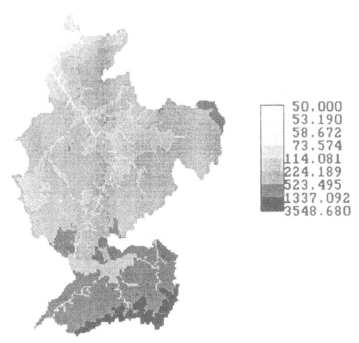

FIGURE 1 *Elevation map of the Rhine River basin [m]*

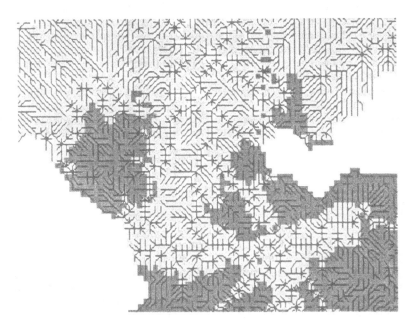

FIGURE 2 *LDD map overlayed on the elevation map*

342

prepared. As an example, the diffuse Cd input from phosphorus fertilizer application is shown in Figure 3. For the atmospheric deposition circulation modelling results of the Transboundary Air Pollution Project of the IIASA were used. Finally, the Cd input from general communal diffuse sources were assessed on the basis of population density maps, consequently the map corresponding to equation (11) could be prepared.

Substituting into equation (9) the Cd export map could be created (Figure 4.).

For substituting into equation (6) the map corresponding to equation (7) is needed. By means of the operations briefly described in section 3.1 the map containing in each raster cell the travel times necessary for the water particle to arrive at the nearest point of a defined river reach after following the elementary runoff pathways (the LDD directions) could be prepared (Figure 5.). Then, extracting from this the average travel time values corresponding to the different river reaches and substituting into equation (6) the diffuse Cadmium concentrations could be calculated, and accordingly the equations (1) and (3) could be substituted reach by reach.

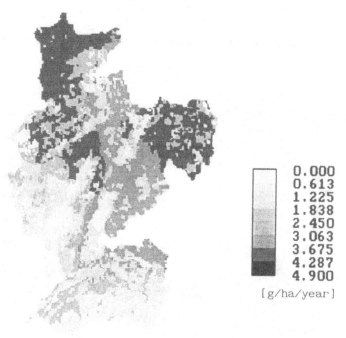

0.000
0.613
1.225
1.838
2.450
3.063
3.675
4.287
4.900
[g/ha/year]

FIGURE 3 *Map of diffuse Cd inputs from P fertilizer use*

343

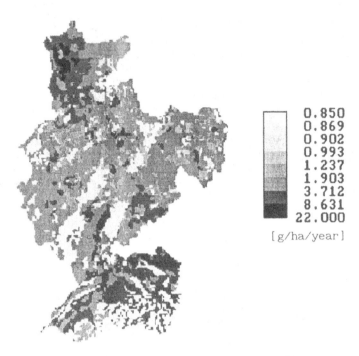

	0.850
	0.869
	0.902
	0.993
	1.237
	1.903
	3.712
	8.631
	22.000

[g/ha/year]

FIGURE 4 *Cd land runoff export map from all diffuse sources*

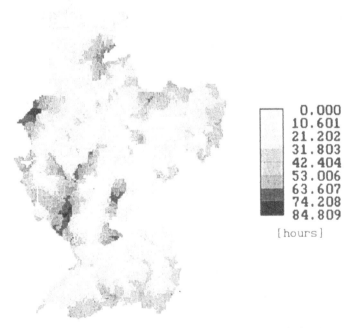

	0.000
	10.601
	21.202
	31.803
	42.404
	53.006
	63.607
	74.208
	84.809

[hours]

FIGURE 5 *Travel times from grid cell to the recipient*

344

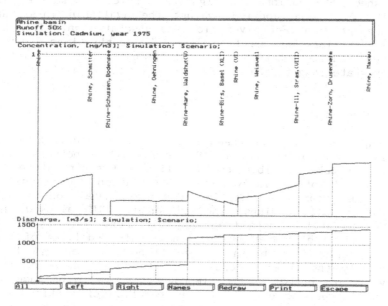

Figure 6.a *Longitudinal discharge and concentration profiles*

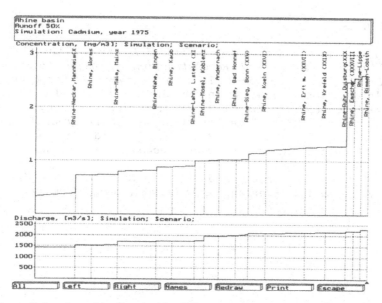

Figure 6.b *Longitudinal discharge and concentration profiles*

Finally, after the inclusion of the Cadmium loads from the significant industrial and municipal point sources in the watershed, the calibration algorithm, briefly described in section *2.1*, were applied against the available in-stream measurement data. For river flow the discharge data from 21 gauging stations were used, and for Cd concentrations 5 water quality gauging stations were available. The calculated steady state longitudinal profiles are presented in Figure 6.

5. Conclusion

The paper presents an example of how the coupling of a special, hydrology-oriented GIS system with a network type model can make possible the handling and manipulation of large amounts of spatially varying information, for assessing the diffuse source contribution of a heavy metal contaminant to the overall loadings. The use of the GIS allowed the authors to utilize data from very different and incoherent sources; it facilitated the regional-scale modelling of the international river basin of the Rhine. However, in this paper only the interim results of the first phase of the study are presented. Further improvements are to be made on several issues, as e.g. in the preparation of continuously varying areal specific runoff maps. Similar analyses for other heavy metal components and for other time slices are also to be carried out.

REFERENCES

[1] Ayers R.U., Ayers L.W., Tarr J.A. and Widgrey R.C. (1985), *A Historical Reconstruction of Major Pollutant Levels in the Hudson-Raritan Basin*. Variflex Corporation
[2] Internationale Kommission für die Hydrologie des Rheingebietes (1989), *Das Rheingebiet*, Koblenz, Germany
[3] Lum K.R. (1982), "The potential availability of P, Al, Cd, Co, Cr, Cu, Fe, Mn, Ni, Pb and Zn in urban particulate matter" *Environ. Technol. Letters* No.3. pp 37-62.
[4] Salomons W., Förstner, Ü. (1984), *Metals in the Hydrocycle*, Springer-Verlag, Berlin, Heidleberg, Germany

29 Numerical simulation of water flow and water quality in urban river networks

S. Q. Liu and B. Li

Abstract

A one dimensional mathematical model of water flow and water quality in tidal river networks is presented. The equations are numerically solved using a two-stage finite differencing scheme for simulating the time-dependent variations of the hydrodynamic and solute transport processes. The model was applied to an urban river network in Fuzhou City, China in order to predict flooding and water quality. The results were used for environmental planning and management of water resources in the city area.

Introduction

It is a common phenomenon that rivers are connected to form a network with a branched or looped structure. The hydrodynamic behaviour in such networked rivers is more complex than that in a single river reach. For the purposes of project planning and management of water resources, flooding control and water quality control, computer simulation has been widely applied as an efficient tool. It has also been proved that the one-dimensional model, comprising the Saint Venant equations and the advective-diffusion equation, has been successfully used to describe the water flow pattern and solute transport processes in rivers. As the number of rivers in a river network increases and the complexity of the water system becomes higher, some specific techniques for numerical computation are required for simplifying the problems. Based on the Saint Venant equations and the advective-diffusion equation, confluence equations, which describe water flow movement and solute mixing at the river junctions in a network, are introduced to be a part of the mathematical model, Tucci et al.[4]. The finite differencing approach is used to formulate linear differencing equations, which are solved by a Gaussian reduction procedure in order to obtain water levels, water flows and solute concentrations at every cross-section within the river system. To simplify the calculation procedure and to reduce the computer memory requirements, a two-stage finite differencing scheme, modified from Zhang[6] and Ye[5], was applied. At the first stage, equations, called the river-reach equations, which cover every river reach in the network and are formed by using the known values of water flow and water quality at the cross-sections. The unknown values of water flow and solute concentrations at both the upstream and downstream ends of each river reach can be obtained by solving these equations. At the second stage, equations, called the section equations, which cover every section in each river reach are formed. The section equations are solved by using the results from the first stage as boundary conditions for the river reach. The modelling results provide information of water flow, water surface elevation and solute concentration for simulating and predicting the hydrodynamic process and water quality variations in the networked river system. An urban river network in Fuzhou City of China, which has been historically functioning for flooding control, wastewater collection, navigation and public recreation, was simulated as a case study and model test. The river network system is dominated by tidal movement of the Minjiang River, an estuary going through the city to the East China Sea. Water quality of the rivers has been contaminated following industrial development and population increase in recent years. The study was implemented to find out the critical factors for optimal planning of an engineering project in order to maximise use of the environmental capacity of the river system for flood control and wastewater discharge.

Mathematical Model

Hydraulic equations

The one dimensional unsteady flow in rivers can be represented by the Saint Venant equations as follows:-

$$\frac{\partial Q}{\partial x} + B\frac{\partial H}{\partial t} = q_e \tag{1}$$

$$\frac{\partial u}{\partial t} + u\frac{\partial u}{\partial x} + g\frac{\partial H}{\partial x} + g\frac{u|u|}{C^2 R} = \frac{q_e v_e}{A} \tag{2}$$

where, Q-- the flow discharge of the river, x -- distance along the river reach, B-- width of water surface, H-- elevation of water surface, t -- time, q_e-- the lateral inflow to the river, u -- mean velocity of flow, g-- gravity acceleration, C -- chezy roughness coefficient, R -- the hydraulic radius, v_e -- mean velocity of the lateral inflow, A --cross-sectional area of the river.

Figure 1 shows a simple example of a river network, with six river reaches. For each single river reach, the Saint Venant equations can be applied to simulate the water flow patterns in it.

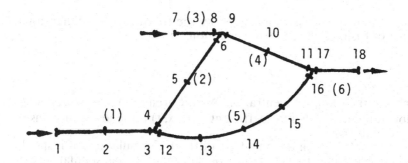

Figure 1   *Sketch of an example of river network*

(1) to (6) -- code numbers of river reaches; 1 to 18 -- code numbers of cross-sections

Since the water flows in the river reaches affect each other through the confluences, where two or more river reaches join together, the confluence equations, which describe the flow characteristics, were introduced as follows (see Tucci [4]):-
the mass conservation equation

$$\sum_{i \in j} Q_i = 0 \tag{3}$$

and the energy conservation equation

$$H_k + \frac{u_k^2}{2g} = H_l + \alpha_{k-l}\frac{u_l^2}{2g} + h_{f(k-l)} \tag{4}$$

in which, j -- code number of confluence, i -- code number of the cross-sections most closely connected to j, k and l -- code numbers of cross-sections at the upstream and downstream of the confluence j, α_{k-l} -- correction factor of energy loss in the section from k to l, $h_{f(k-l)}$-- energy loss between cross-sections k and l.
Applying the equations (1) and (2) to each river reach and the confluence

equations (3) and (4) to each river junction, a set of simultaneous partial equations were formed to represent hydraulic movement of the flows in the whole networked river system.

Mass transport equations

The river water quality is represented by the following one-dimensional advective-diffusion and dispersion equation,

$$\frac{\partial(AC)}{\partial t} + \frac{\partial(QC)}{\partial x} - \frac{\partial}{\partial x}\left(E_i A \frac{\partial C}{\partial x}\right) = \sum S(C, x, t) \tag{5}$$

in which, C-- the solute concentration, A-- the area of the cross-section, E_i -- the longitudinal dispersion coefficient, S(C, x, t) -- the losses or gains of the solute in the water.

For the term S(C, x, t) in the equation, different formulations of it should be applied according to the nature of substance being simulated. For example, a refined form of Streeter-Phelps equations for biochemical oxygen demand of 5 days (abbreviated as BOD in this paper) and dissolved oxygen (DO), see McBride [2] and Rinaldi [3], are as follows, respectively,

$$S_{BOD}(C, x, t) = -(k_1 + k_3) AC_{BOD} + AL_a + q_e C_{eBOD} \tag{6}$$

and

$$S_{DO}(C, x, t) = -k_1 AC_{BOD} + k_2 A(C_s - C_{DO}) - D_b A + q_{eDO} \tag{7}$$

in which, $S_{BOD}(C, x, t)$ -- the loss or gain of BOD, k_1-- the BOD carbonaceous reaction rate, (day^{-1}), k_3 -- the rate of removal of BOD by reaction, absorption and sedimentation, etc.(day^{-1}), C_{bod}-- concentration of BOD, L_a -- the rate of addition of BOD (mg/l per day), C_{ebod} -- the concentration of BOD in the lateral inflow(mg/l), k_2 -- the reaction constant(day^{-1}), C_s -- saturation concentration of DO(mg/l), C_{do} -- DO concentration(mg/l), D_b -- the removal of oxygen by benthal deposits and plant respiration, etc. (mg/l), and C_{edo}-- the DO concentration in the lateral inflow(mg/l).

Also, many other constituents can be included for numerical modelling by using similar functional representations of mass transport described by the EPA of the United States, see Bowie [1].

At a river junction, it is assumed that the constituents are conservative and well mixed. The mass balance equation can be written as:-

$$C_j = \frac{\sum(Q_i C_i)}{\sum Q_i} \tag{8}$$

where, C_j -- the solute concentration at the confluence j, Q_i -- inflows from the cross-sections at the upstream of the confluence, C_i -- the solute concentrations in the inflows.

The concentrations at all the junctions are calculated at every time step and are used as boundary conditions in the simulating processes.

Two-stage Finite Differencing Techniques

In order to solve the equations (1) and (2) for obtaining water surface elevation and water flow at each cross-section of the rivers, a finite differencing scheme, taking the mean velocity at a cross-section as $u = Q/A$, is applied as following,

$$\frac{\partial Q}{\partial x} = \frac{1}{\Delta x_j}(Q_{j+1}^{t+1} - Q_j^{t+1})$$

$$\frac{\partial H}{\partial t} = \frac{1}{2\Delta t}(H_j^{t+1} + H_{j+1}^{t+1} - H_j^t - H_{j+1}^t)$$

$$B = \frac{1}{2}(B_j^t + B_{j+1}^t)$$

$$\frac{\partial\left(\frac{Q}{A}\right)}{\partial x} = \frac{1}{\Delta x_j}\left(\left(\frac{Q}{A}\right)_{j+1}^{t+1} - \left(\frac{Q}{A}\right)_j^{t+1}\right)$$

$$\frac{\partial\left(\frac{Q}{A}\right)}{\partial t} = \frac{1}{2\Delta t}\left(\left(\frac{Q}{A}\right)_j^{t+1} + \left(\frac{Q}{A}\right)_{j+1}^{t+1} - \left(\frac{Q}{A}\right)_j^t - \left(\frac{Q}{A}\right)_{j+1}^t\right)$$

$$\frac{Q}{A} = \frac{1}{2}\left(\left(\frac{Q}{A}\right)_j^t + \left(\frac{Q}{A}\right)_{j+1}^t\right)$$

$$\frac{\partial H}{\partial x} = \frac{1}{\Delta x_j}(H_{j+1}^{t+1} - H_j^{t+1})$$

$$\frac{\frac{Q}{A}\left|\frac{Q}{A}\right|}{C^2 R} = \frac{1}{2}\left(\left(\frac{\frac{Q}{A}\left|\frac{Q}{A}\right|}{C^2 R}\right)_j^t + \left(\frac{\frac{Q}{A}\left|\frac{Q}{A}\right|}{C^2 R}\right)_{j+1}^t\right)$$

Substituting the finite differencing representations into equations (1) and (2), two finite differencing equations, which describe water flows and water elevations at the upstream cross-section j and the downstream cross-section j+1, can be formulated as follows:-

$$P_{(2j,1)}H_j^{t+1} + P_{(2j,2)}Q_j^{t+1} + P_{(2j,3)}H_{j+1}^{t+1} + P_{(2j,4)}Q_{j+1}^{t+1} = P_{(2j,5)} \qquad (9)$$

$$P_{(2j+1,1)}H_j^{t+1} + P_{(2j+1,2)}Q_j^{t+1} + P_{(2j+1,3)}H_{j+1}^{t+1} + P_{(2j+1,4)}Q_{j+1}^{t+1} = P_{(2j+1,5)} \qquad (10)$$

in which,

$$P_{(2j,1)} = \frac{\Delta x_j}{4\Delta t}(B_j^t + B_{j+1}^t)$$

$$P_{(2j,2)} = -1$$

$$P_{(2j,3)} = P_{(2j,1)}$$

$$P_{(2j,4)} = -P_{(2j,2)}$$

$$P_{(2j,5)} = \frac{\Delta x_j}{4\Delta t}(B_j^t + B_{j+1}^t)(H_j^t + H_{j+1}^t)$$

$$P_{(2j+1,1)} = -2g\frac{\Delta t}{\Delta x_j}$$

$$P_{(2j+1,2)} = \left\{1 - \frac{\Delta t}{\Delta x_j}\left(\left(\frac{Q}{A}\right)_j^t + \left(\frac{Q}{A}\right)_{j+1}^t\right)\right\}/A_j^t$$

$$P_{(2j+1,3)} = -P_{(2j+1,1)}$$

$$P_{(2j+1,4)} = \left\{1 + \frac{\Delta t}{\Delta x_j}\left(\left(\frac{Q}{A}\right)_j^t + \left(\frac{Q}{A}\right)_{j+1}^t\right)\right\}/A_j^t$$

$$P_{(2j+1,5)} = \left(\frac{Q}{A}\right)_j^t + \left(\frac{Q}{A}\right)_{j+1}^t - g\Delta t\left\{\left(\frac{\frac{Q}{A}\left|\frac{Q}{A}\right|}{C^2 R}\right)_j^t + \left(\frac{\frac{Q}{A}\left|\frac{Q}{A}\right|}{C^2 R}\right)_{j+1}^t\right\}$$

Obviously, equations (9) and (10) can be simplified, respectively, as

$$H_j + \alpha_{2j}H_{j+1} + \beta_{2j}Q_{j+1} = \gamma_{2j} \tag{11}$$
$$Q_j + \alpha_{2j+1}H_{j+1} + \beta_{2j+1}Q_{j+1} = \gamma_{2j+1} \tag{12}$$

For each cross-section, the water flow and elevation of the water surface will be defined by the two equations. Therefore, there will be 2n equations in the whole river network system with n cross-sections, including the boundary values at the upstream cross-section of each river reach. By rewriting (11) and (12) to matrix form as,

$$Y_j + M_j Y_{j+1} = N_j \tag{13}$$

i.e.

$$Y_j = N_j - M_j Y_{j+1} \tag{14}$$

where,

$$Y_j = \begin{pmatrix} H_j \\ Q_j \end{pmatrix}$$

352

$$Y_{j+1} = \begin{pmatrix} H_{j+1} \\ Q_{j+1} \end{pmatrix}$$

$$M_j = \begin{pmatrix} \alpha_{2j} & \beta_{2j} \\ \alpha_{2j+1} & \beta_{2j+1} \end{pmatrix}$$

$$N_j = \begin{pmatrix} \gamma_{2j} \\ \gamma_{2j+1} \end{pmatrix}$$

An equation for each river reach, called the river reach equation, can be derived by a substitution process on equation (14), from $j=s$ until $j=d$, as

$$Y_s + M'_{d-1} Y_d = N'_{d-1}$$

in which, s and d -- code numbers of the first and last cross-section of a river reach respectively, Y_s- the unknown H_s and Q_s at the first cross-section of the river reach, Y_d-- the unknown H_d and Q_d at the last cross-section of the river reach, M'_{d-1} and N'_{d-1}-- constants obtained by substituting Y_j into equation(14), from $j=s$ until $j=d$.

Solving the simultaneous linear equations (15), together with the confluence equations (3) and (4), the unknown values of water flow and water elevation at both ends of each river reach can be obtained.

Using the results from the river reach equations as boundary conditions, the water flow characteristics at each cross-section can be solved from the section equations (11) and (12) in each river reach.

Similarly, an implicit finite differencing equation of equation (5) for BOD concentration at cross-section j can be written as

$$\frac{C_j^{t+1} - C_j^t}{\Delta t} + u \frac{C_j^t - C_{j-1}^t}{\Delta x_j} = E \frac{C_{j+1}^{t+1} - 2C_j^{t+1} + C_{j-1}^{t+1}}{\Delta x_j^2} - \frac{1}{2} k_1 (C_j^{t+1} + C_{j-1}^t) \quad (16)$$

or

$$\alpha_j C_{j-1}^{t+1} + \beta_j C_j^{t+1} + \gamma_{j+1} C_{j+1}^{t+1} = \delta_j \quad (17)$$

in which,

$$\alpha_j = -E / \Delta x_j$$

$$\beta_j = 1/\Delta t + 2E/\Delta x_j^2 + k_1/2$$

$$\gamma_j = C_j^t (1/\Delta x_j - u_j/\Delta x_j) + C_{j-1}^t (u_j/\Delta x_j - k_1/2)$$

From equation (17), the relationship between the first cross-section and the last one in a river reach can be derived from the following procedure.

Figure 2   *A river reach with n+1 cross-sections*

353

Figure 2 shows a river reach with n + 1 cross-sections. Equation (16) for the last cross-section is

$$\alpha_n C_{n-1}^{t+1} - \beta_n C_n^{t+1} = \gamma_n \tag{18}$$

i.e.

$$C_n^{t+1} = \frac{\gamma_n}{\beta_n} - \frac{\alpha_n}{\beta_n} C_{n-1}^{t+1}$$

or,

$$C_n^{t+1} = \Psi_n + \Phi_n C_{n-1}^{t+1} \tag{19}$$

where,

$$\Psi_n = \gamma_n / \beta_n$$
$$\Phi_n = -\alpha_n / \beta_n$$

Substituting (19) into equation (17) For the cross-section j = n-1, we can have

$$C_{n-1}^{t+1} = \frac{\delta_{n-1} - \gamma_{n-1} \Psi_n}{\beta_{n-1} + \delta_{n-1} \Phi_n} - \frac{\alpha_{n-1}}{\beta_{n-1} + \gamma_{n-1} \Phi_n} C_{n-2}^{t+1}$$

or,

$$C_{n-1}^{t+1} = \Psi_{n-1} - \Phi_{n-1} C_{n-2}^{t+1} \tag{20}$$

in which,

$$\Psi_{n-1} = \frac{\delta_{n-1} - \gamma_{n-1} \Psi_n}{\beta_{n-1} + \delta_{n-1} \Phi_n}$$

$$\Phi_{n-1} = -\frac{\alpha_{n-1}}{\beta_{n-1} + \gamma_{n-1} \Phi_n}$$

This can be continued by the above recurrence procedure until j = 1, as

$$C_1^{t+1} = \Psi_1 - \Phi_1 C_0^{t+1}$$

where, C_0 is the known value at the boundary of the river reach.
Following a back-substitution process, a river-reach equation of each reach can be derived as,

$$C_n^{t+1} = L_n + S_n C_0^{n+1} \tag{21}$$

in which, L_n and S_n are constants and are obtained by the back-substitution calculation.

Combining equations (21) and (8), a set of simultaneous linear equations are formed and can be solved for obtaining concentrations at both ends of each river reach.

At the second stage of the water quality simulation, the results from the river reach equations are used as boundary conditions for equation (17). The solute concentrations at each cross-section in each river reach can be obtained.

Case Study

The mathematical model was applied to study the urban river network in Fuzhou City of China to assist the local environmental agency for environmental management and engineering project planning of the city. As shown in Figure 3, the urban river system is a complex looped network with a total length of about 60 kilometres. The water flow in the river system is dominated by semidiurnal tide with 5 hours of flood tide and 7.25 hours of ebb tide from the Minjiang River. In the river network, there are 28 river reaches and 24 confluences. Five confluences are connected directly to Minjiang River, and the tidal levels of them were measured. The river network has been historically functioning in the city for flooding control, wastewater discharge from domestic houses and industries, navigation and public recreation. Also, it affects the climate significantly in the urban area. With the development of industry and the increase in population, the water quality of the river system has been seriously declining. Engineering projects have been continuously implementing for water environment protection. The results of the study indicated that the water environmental capacity for receiving waste discharges has been exceeded, which has brought critical pollution problems in the rivers. Figure 4 shows that the reach 5 always brings fresh water from the Minjiang River, which flushes the river network and benefits the water quality in the rivers. However, the fresh water flow is very small and should be increased as much as possible. An engineering approach has been proposed to dredge this reach for diverting more fresh water into the river network. Figure 5 gives the concentrations of BOD and DO in reaches 11, 18 and 20. The BOD concentrations were about 4 mg/l and did not change significantly with time. Reaches 18 and 20 are located at the upstream of the network and receive fresh water from the sources at points 1 and 2. The DO levels were maintained to be positive, but their values were only about 40% of the saturation level. Reach 11, on the other hand, is at the downstream of the network, and it is the main waterway of outflow from the river network. A big part of the solutes in the inner rivers is flushed out through it. The water quality of reach 11 changes with the tidal level and the zero DO level appears at the low tide. The water quality in the downstream area behaves in similar fashion to that of reach 11. Although there is shortage of field data available and the model has not

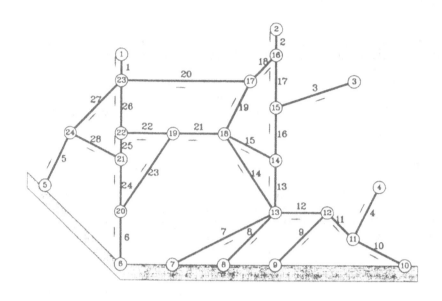

Figure 3   *Sketch Map of River Network in Fuzhou City*

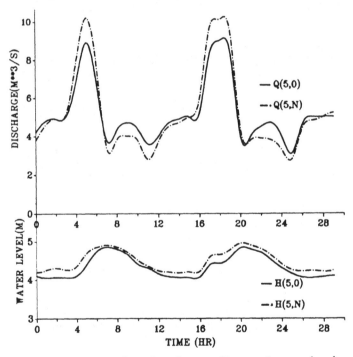

Figure 4   *Simulated water flow and water level at the inlet and outlet of reach 5*

356

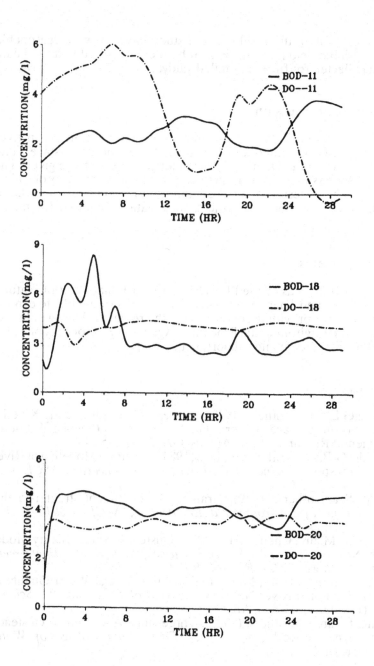

Figure 5   *Simulated variations of BOD and DO concentrations*
in the reaches 11, 18 and 20

357

been fully calibrated, the results of the study have shown a reasonable similarity with the field investigations. It has been required that field data should be collected for further detailed study.

Conclusion and Discussion

Mathematical model of water flow and water quality in an urban river network has been presented and solved by the use of a two-stage finite differencing procedure. Details of the numerical approach are given. The model was used to simulate an urban river network in Fuzhou City of China, and it has produced results of a close similarity to field investigations. For further detailed study on this project, field data is required for model calibration.

Acknowledgements

The research study was funded by The National Foundation of Natural Science of China and Tongji University. The authors would like to express thanks to them for providing conditions for the research. They would also like to thank the Environmental Protection Agency of Fuzhou City of China for their encouragement and support on this study.

References

[1] Bowie,G.L. and others,(1985), Rates, Constants, and Kinetics Formulations in Surface Water Quality Modeling (Second Edition), Environmental Research Lab., Athens, GA, PB85 245 314.
[2] McBride, G.B. and Rutherford, J.C., (1984), Accurate Modeling of River Pollutant Transport, *Journal of Environmental Engineering, ASCE, Vol. 110, No. 4, pp808-827*
[3] Rinaldi, S. and others, (1979), Parameter Estimation of Streeter-Phelps Models, *Journal of Environmental Engineering, ASCE, Vol.105, EE1, pp75-88*
[4] Tucci, C.E.M. and Chen, Y.H., (1981), Unsteady Water Quality Model For River Network, *Journal of Water Resources Planning and Management Division, ASCE, Vol.107, WR2, pp477-493*
[5] Ye, Changming, (1986), Mathematical Modelling of Water Quality in non-fixed directional River Networks, *Journal of Environmental Science* (China), (in Chinese), *Vol.6, No.3*
[6] Zhang, Erjun, and others, (1982), Three-stage Agorithm of Unsteady Flow in River Networks, *Journal of East China College of Water Conservancy,* (in Chinese), *Vol.1*

30 Dispersion in compound open channel flow

P. Prinos

ABSTRACT

Dispersion in compound open channel flow is studied with the use of numerical model which solves the parabolic three-dimensional Navier-Stokes equations in conjunction with a k-ε model of turbulence and a transport equation for concentration. Computed velocity contours, concentrations and turbulent diffusivities indicate the effect of the interaction mechanism on the transport of pollutants.

For weak interacting conditions (high relative depts) spreading is more intense in the vertical direction while for strong interacting conditions (low relative depths) spreading and maximum concentration occurs away from the interface and towards the flood plain side walls. The dimensionless diffusion coefficient D_z is calculated through the model and is found to vary within the mixing region with values ranging from 0.1 to 0.35 for the experimental conditions considered.

1 Introduction

Measurements and computations of mean flow and turbulence characteristics in compound open channels have been performed by many investigators [1,5,6,7,10,12,13,14,17,19]. The effects of the strong shear layer, developed in the junction region of the compound cross section, on the mean velocity, boundary shear stress and overall capacity have been determined for various geometric and hydraulic characteristics.

However, dispersion studies in compound open channels are rather limited with various shortcomings.

Wood and Liang [18] have studied computationaly and experimentally the dispersion in compound open channels. They have computed the depth-averaged concentration of a passive effluent at various cross-sections along the channel using a diffusion equation in which the lateral turbulent diffusivity D_z was estimated by the relationship $D_z = 0.130 \, U_* \, h$. The effluent was released from various locations in the main channel or in the flood plain and certain conclusions were derived with regard to the position of release.

The main shortcomings of the study were (a) the distance from the inlet at which the effluent was released. This distance was 3.5m from the inlet and hence the flow was not fully developed at this section and (b) the relatively high relative depth used in both computations and experiments. The relative depth ($y_r = y_{fp} / y_{tot}$) was approximately 0.5 for which the interaction mechanism between main channel and flood plain flow is known to be weak.

Djordjevic et al [2] have performed a similar study in which the above shortcomings were overcomed. The effluent was released at a distance 10.5m from the entrance in which the flow was fully developed and the relative depth was varied from 0.1 to 0.25. They used the depth-averaged concentration equation in which the coefficient D_z was set equal to $0.2 U_* h$. This value was found to give best agreement between the computed and measured concentration distribution at various cross sections.

James [4] has presented a numerical model for simulating the transfer of suspended sediments from a channel to an adjacent flood plain for various sediment diameters (75, 150, 300 μm) and relative flow depths (0.20, 0.30). He assumed that the sediment diffusivity is equal to momentum diffusivity and solved the concentration equation with a transverse diffusivity D_z estimated by an equation proposed by Rajaratram and Ahmadi [15] in the interaction region and also by a similar equation proposed by Lau and Krishnappan [8] beyond the interaction zone.

Arnold et al [1] have presented some laboratory results for the dimensionless transverse diffusivity e_z^* and the turbulent Schmidt-number σ_t. The measured results of e_z^* varied in the range of 0.27 to 0.85 with a mean of 0.45, corresponding with field measurements rather than with laboratory values. Also, the σ_t was varied from 0.5 (smooth flood plains and low interaction effects) to 1.0 (medium flood plain roughness and high interaction effects).

Finally Guymer et al [3] have reported some initial results from a dispersion study in a compound open channel using various injection points in the main channel and in the flood plain. They also used a random walk model for describing the measurement solute distributions.

In this study, dispersion of "passive" contaminants in compound open channels is studied numerically using the three-dimensional Reynolds-averaged Navier-Stokes equations together with a transport equation for concentration and the k-ε model of turbulence. The equations used, are parabolic in the flow direction i.e. downstream events do not influence the flow and hence an efficient and economic marching forward procedure is applied for solving them [11].

With this flow modelling the diffusivity coefficients in the vertical and transverse directions are not input into the concentration equation, as in the previous studies, but are calculated directly as part of the solution procedure. Hence computed coefficients can be compared with those used in other studies and those reported in experimental studies. The k-ε model of turbulence is used for "closing" the system of equations, although the model does not produce any turbulence-driven secondary currents [16]. For the conditions of Wood and Liang [18] study the flow is not fully developed, such a secondary motion has not been developed and hence it is thought that such a turbulence model performs satisfactorily. However, an algebraic model of turbulence can be incorpated into the above numerical model as it has been done in a previous study of turbulence modelling in compound channel flow [14].

The model has been applied for the experimental conditions of [2,18] in which the effluent release was in the main channel and in the flood plain [18] and also at the main channel/flood plain interface [2]. Computed isovels and isoconcentrations are presented at characteristic stations downstream of the release position while D_z coefficients are compared against those used in the original studies.

2 The Governing Equations

The three-dimensional Reynolds averaged Navier-Stokes equations, parabolic in the longitudinal direction are used for the prediction of the mean velocity in conjunction with a transport equation for predicting the concentration field.

Continuity equation:

$$\frac{\partial U}{\partial x} + \frac{\partial V}{\partial y} + \frac{\partial W}{\partial z} = 0 \qquad (1)$$

x- Momentum :

$$U\frac{\partial U}{\partial x}+V\frac{\partial U}{\partial y}+W\frac{\partial U}{\partial z}=-g\frac{\partial y_f}{\partial x}+gS_o+\frac{\partial}{\partial y}\left(v_t\left(\frac{\partial U}{\partial y}+\frac{\partial V}{\partial x}\right)\right)+$$

$$\frac{\partial}{\partial z}\left(v_t\left(\frac{\partial U}{\partial z}+\frac{\partial W}{\partial x}\right)\right)$$

(2)

y- Momentum :

$$U\frac{\partial V}{\partial x}+V\frac{\partial V}{\partial y}+W\frac{\partial V}{\partial z}=-\frac{1}{\varrho}\frac{\partial p}{\partial y}+2\frac{\partial}{\partial y}\left(v_t\frac{\partial V}{\partial y}\right)+\frac{\partial}{\partial z}(v_t(\frac{\partial V}{\partial z}+\frac{\partial W}{\partial y}))$$

(3)

z- Momentum :

$$U\frac{\partial W}{\partial x}+V\frac{\partial W}{\partial y}+W\frac{\partial W}{\partial z}=-\frac{1}{\varrho}\frac{\partial p}{\partial z}+\frac{\partial}{\partial y}\left(v_t(\frac{\partial W}{\partial y}+\frac{\partial V}{\partial z})\right)+2\frac{\partial}{\partial z}(v_t\frac{\partial W}{\partial z})$$

(4)

Concentration equ.:

$$U\frac{\partial C}{\partial x}+V\frac{\partial C}{\partial y}+W\frac{\partial C}{\partial z}=\frac{\partial}{\partial y}(\frac{v_t}{\sigma_C}\frac{\partial C}{\partial y})+\frac{\partial}{\partial z}(\frac{v_t}{\sigma_C}\frac{\partial C}{\partial z})$$

(5)

where U, V, W = mean velocity components in the x- (longitudinal) y- (vertical) and z- (transverse) directions respectively; y_f = flow depth; S_o = channel slope; g = acceleration due to gravity, v_t = eddy viscosity; ϱ = fluid density; C = mean concentration; σ_C = turbulent Schmidt number for the pollutant concentration.

In the above equations the eddy-viscosity hypothesis is assumed and the isotropic eddy viscosity v_t is calculated through the k-ε model of turbulence as

$$v_t=c_\mu\frac{k^2}{\varepsilon}$$

(6)

where c_μ = constant (= 0.09), k = turbulence kinetic energy and ε = its rate of dissipation.

The distribution of k and ε is obtained by the solution of the following modelled transport equations for k and ε:

k-equation:

$$U\frac{\partial k}{\partial x}+V\frac{\partial k}{\partial y}+W\frac{\partial k}{\partial z}=\frac{\partial}{\partial y}(\frac{v_t}{\sigma_k}\frac{\partial k}{\partial y})+\frac{\partial}{\partial z}(\frac{v_t}{\sigma_k}\frac{\partial k}{\partial z})+P-\varepsilon \qquad (7)$$

ε-equation:

$$U\frac{\partial \varepsilon}{\partial x}+V\frac{\partial \varepsilon}{\partial y}+W\frac{\partial \varepsilon}{\partial z}=\frac{\partial}{\partial y}(\frac{v_t}{\sigma_\varepsilon}\frac{\partial \varepsilon}{\partial y})+\frac{\partial}{\partial z}(\frac{v_t}{\sigma_\varepsilon}\frac{\partial \varepsilon}{\partial z})+c_{\varepsilon 1}\frac{\varepsilon}{k}P-c_{\varepsilon 2}\frac{\varepsilon^2}{k} \qquad (8)$$

where σ_κ , σ_ε , $c_{\varepsilon 1}$, $c_{\varepsilon 2}$ = constants (= 1.0, 1.3, 1.44, 1.92 respectively) and P = production of k. The latter is also calculated through the eddy-viscosity hypothesis from the following relationship.

$$P=v_t\left\{2[(\frac{\partial W}{\partial z})^2+(\frac{\partial V}{\partial y})^2]+(\frac{\partial W}{\partial y}+\frac{\partial V}{\partial z})^2+(\frac{\partial U}{\partial y})^2+(\frac{\partial U}{\partial z})^2\right\} \qquad (9)$$

The first term in the r.h.s of equ. (2) is equal to zero for uniform flow considered here and the coefficient σ_C, appearing in equ. (5), represents the ratio of eddy viscosity v_t to turbulent diffusivity D and is set equal to 1.0 suitable for dispersion calculations in compound channels and especially in the interaction region [1].

Equations (1) to (9) form a set of equations which is solved simultaneously using appropriate boundary conditions. The solution procedure and the boundary conditions employed in this study are presented in the following section.

3 Numerical Procedure

All the differential equations from (1) to (9) introduced previously are parabolic in the longitudinal direction and they can be expressed in a common form as follows:

$$U\frac{\partial \phi}{\partial x}+V\frac{\partial \phi}{\partial y}+W\frac{\partial \phi}{\partial z}=\frac{\partial}{\partial y}(\Gamma_\phi\frac{\partial \phi}{\partial y})+\frac{\partial}{\partial z}(\Gamma_\phi\frac{\partial \phi}{\partial z})+S_\phi \qquad (10)$$

where ϕ = unknown variable (ϕ = 1 (continuity), U, V, W (momentum),C (concentration), k, ε (k-ε model)), Γ_ϕ = effective diffusion coefficient and S_ϕ = source term.

363

An efficient forward-marching solution procedure can be employed, as described in [11]. With this procedure the calculation domain is covered only once without iteration, starting from given initial conditions and two-dimensional storage of the variables is required at the grid-points located in one cross-section. In the present study the numerical scheme proposed by Patankar and Spalding [11] was employed.

At the inlet compound cross-section a uniform distribution of all variables was prescribed. The secondary velocities V and W were set equal to zero and k and ε had such small values that the eddy viscosity was approximately 10-15 times the kinematic viscosity ν. The concentration was set equal to 1 within the control volumes surrounding the grid nodes at the position of effluent's release, and zero everywhere else in the compound cross section. The effluent was released through a small tube downstream of the inlet cross section as in the experiments of Wood and Liang [18] while , in the case of Djordjevic et al [2] the effluent was released through the whole interface flow depth. In the first case the dimensions of the effluent control volume are approximately five times greater than the actual tube opening but this does not introduce any significant errors.

Boundary conditions are also specified at solid walls and free surface for all variables U, V, W, k, ε, C. At solid walls, the wall function technique proposed by Launder and Spalding [9] was used, by which the boundary conditions are specified at a grid point which lies outside the laminar sublayer. The boundary conditions for k and ε are also specified at the same point.
The boundary conditions at the free surface are specified following the approach of Lau and Krishnappan [7] which considers the free surface as a symmetry plane. Hence normal gradients of U, W, k were set to zero. Also the value of ε at the free surface was calculated by a relationship given in [7].The flux of pollutants at the solid walls and the free surface is zero and hence the concentration gradients normal to walls and free surface are prescribed as zero.

The step-by-step integration was carried out until the flow became fully developed (no change in velocity distribution in the longitudinal direction) and for a distance to cover the experimental length. At each step the y- and z-momentum equations were solved with a guessed pressure field (the longitudinal momentum equation had a known pressure gradient equal to gS_o) with the upstream values taken as guesses. Pressure and velocity fields were corrected subsequently to satisfy the continuity equation. Finally the equations of k, ε, and C were solved.

The results were stored at particular stations downstream of the effluent's position where experimental results are available. The solutions were obtained with a grid (60x30) (60 in the horizontal and 30 in the vertical direction) distributed non-uniformly over the compound cross-section. A finer grid was also used in the vertical direction (up to 55 grid lines) but the results were almost identical with the coarser grid. Hence the coarser grid was used in the subsequent calculations. The longitudinal step was taken approximately (1/40) of the total flow depth for most of the test cases considered.

4 Computational Results

Initially the model was applied for the experimental conditions of Wood and Liang [18] which can be summarized as follows: Channel Slope = 0.00047, Discharge = $0.0047 \, m^3/s$, main channel flow depth = 0.051m, flood plain depth = 0.025m, main channel width = 0.371m and flood plain width = 0.189m.

The source of pollutant was located (a) in the main channel (30mm from the channel bed and 40mm from the interface) and (b) in the flood plain (5mm from the flood plain bed and 40mm from the interface). For both cases the distance of the source from the inlet was 3.5m. For the above flow conditions the relative depth y_r is relatively high ($y_r = 0.5$) and hence the interaction mechanism generated near the interface is rather weak.

Computed results were obtained at various distances along the channel for which experimental results are available (4.5, 5.0, 6.0, 7.0 and 8.0m). However, for the sake of brevity, computed flow characteristics and pollutant concentrations are presented at characteristic stations x = 4.5 m and x = 8.0 m.

Fig. 1 (a,b,c) shows the velocity contours, isoconcentrations and dimensionless diffusion coefficient respectively at x = 4.5m (1.0 from the source) for the source located in the main channel.The latter coefficient has been calculated as D_z/U_*h (where U_* is the local shear velocity and h = the local flow depth) for direct comparison against the constant depth-averaged value of 0.13 which has been used in [18]. This coefficient is in essence equal to the dimensionless eddy viscosity coefficient v_t/U_*h since the coefficient σ_C has been set equal to 1.

The velocity contours indicate that for such high relative depth the interaction mechanism is weak with the interface plane of zero apparent shear stress being almost vertical and weak velocity gradients in the transverse direction.

The isoconcentration lines indicate the intensive dispersion of pollutants in the vertical rather than in the transverse direction due to greater velocity gradients in this direction. The isodiffusivity lines indicate that the value of the coefficient increases with the distance from the channel bed, as the eddy viscosity v_t, and the average value used by Wood and Liang [18] is rather low and a larger value should be used. This also has been suggested in [18] for quantitative agreement between experimental and theoretical results.

Fig. 2 shows the isoconcentration lines at a distance x = 8.0m (4.5 from the source) for the same conditions which indicate the more spreading of the pollutants in the vertical rather than in the transverse direction.

When the source was moved in the flood plain the dispersion of pollutants is more pronounced in the vertical rather than in the transverse direction (fig.3) due to weak interaction mechanism in the interfacial region.

The model was also applied for the test B of Djordjevic et al [2] which has the following flow and geometric characteristics: Channel Slope = 1.5% Discharge = 12.9 l/s, main channel flow depth = 0.25m, flood plain depth = 0.05, main channel width = 0.35m and flood plain width = 0.35m.

The dye was released from the whole flow depth at the main channel/flood plain interface, at a distance 10.5m from the channel inlet. Concentrations were measured and calculated at distances from the source equal to 6.0, 12.0 and 21.0m. The above relative depth y_r, equal to 0.2, indicates the rather intense interaction mechanism which is expected to have a strong effect on the dispersion of pollutants.

The computed flow characteristics at distance $x = 16.5m$ (6m from the due source) are shown in fig.4 (a,b,c). The velocity contours (fig.4(a)) indicate the rather strong interaction mechanism with the plane of the zero apparent shear stress being almost horizontal. Horizontal planes of zero shear stress have also been proposed by Wormleaton et al [19] for calculating the discharge in compound channels with strong interaction mechanism. The isoconcentration lines indicate the dispersion of pollutants away from the interface with the maximum concentration located towards the flood plain side walls.

The values of the dimensionless D_z (or v_t) coefficient are shown to vary within the interaction region and have a maximum value inthere.

Djordjevic et al [2] have used a value of 0.2 for the depth-averaged D_z which may be appropriate for the main channel but slightly lower than those computed in the flood plain.

Similar conclusions can be derived from the computed flow characteristics at a distance $x = 32m$ (22m from the source). The velocity contours have similar pattern as before (fig.5 (a)) while the maximum concentration has moved further towards the side wall, indicating the strong effect of the transverse shear layer on the dispersion of pollutants (fig.5(b)). Finally, D_z lines indicate similar behaviour with that computed previously with a maximum value of 0.35 in the flood plain and 0.25 in the main channel.

5 Conclusions

The dispersion of pollutants in compound open channels has been studied numerically by solving the parabolic, three-dimensional Navier-Stokes equations in conjunction with the k-ε model of turbulence and a transport equation for concentration. The model has been applied for the experimental conditions of [2] and [18] and the following conclusions can be derived:

a)   For high relative depths (weak transverse shear layer) the dispersion of pollutants is more pronounced in the vertical rather than in the transverse direction for both source locations (in the main channel and the flood plain.

b)   For strong interacting conditions (low relative depth) the effect of the shear layer on the dispersion of pollutants is very significant and has as a result the spreading and the maximum concentration to move away from the interface and towards the flood plain side walls.

c)   The diffusion coefficient D_z is calculated as part of the computational technique, based on the hydrodynamic model, and is not used as an input to the concentration equation. The value of the dimensionless coefficient is shown to vary in the interaction region and a constant depth-averaged value of 0.134 or 0.20 used by other investigators may not be appropriate for such conditions. The computed local value of the dimensionless D_z coefficient was found to vary from 0.1 to 0.35 in most part of the compound cross section.

REFERENCES

[1] Arnold U, Hottges J. and Rouve G.,(1989), "Turbulence and mixing mechanisms in compound open channel flow", *Proc. of XXIII Congress of IAHR*, pp. A-133-A140.
[2] Djordjevic S., Petrovic I., Maksimovic C. and Radojkovic M., (1989), "Experimental tracer investigations in a compound
laboratory channel",*Proc. of HYDROCOMP '89*, pp. 269-278.
[3] Guymer I., Brockie N.J.W. and Allen C.M., (1990), "Towards Random Walk models in a large scale laboratory facility" *Proc. of Int. Conf. on Transport and Dispersion*, pp. 12.B.1-12.B.6
[4] James C.S.,(1985) "Sediment transfer to overbank sections", *J. of Hydraulic Research*, vol. 23, no 1.
[5] Kawahara Y. and Tamai N.,(1989) "Mechanism of lateral
momentum transfer in compound channel flows", *Proc. of XXIII Congress of IAHR*, pp. B463-B470.
[6] Knight D.W. and Demetriou J.D.,(1983), "Flood Plain and Main Channel Flow Interaction", *J. of Hydraulic Eng.*, vol. 109(8), pp.1073-1092.
[7] Law U.L. and Krishnappan B.G.,(1986), "Turbulence Modelling of Flood

Plain Flows", *J. of Hydraulic Eng.*, vol. 112(4), pp.251-267.

[8] Lau Y.L. and Krishnappan B.G., (1977), "Transverse Dispersion in Rectangular channels" *J. of Hydraulics Div.*, ASCE, vol.103, pp. 1173-1189.

[9] Launder B.E. and Spalding D.B.,(1974), "The Numerical computation of Turbulent Flow", *Computer methods in Applied Mechanics and Eng.*, vol.3, pp. 269-289.

[10] McKeogh E.J. and Kiely G.K.,(1989), "Experimental study of the mechanisms of flood flow in meandering channels", *Proc. of XXIII Congress of IAHR*, pp. B491-B498.

[11] Patankar S.V. and Spalding B.D.,(1972), "A Calculation procedure for Heat, Mass and Momentum transfer in the Three-Dimensional Parabolic Flows", *Int. Journal at Heat and Mass transfer*, vol. 15, pp. 1787-1806.

[12] Prinos P., Townsend R.D. and Tavoularis S.,(1985), "Structure of Turbulence in Compound channel Flows", *Journal of Hydraulic Eng.*, vol. 111(9), pp. 1246-1261.

[13] Prinos P.,(1989), "Experiments and Numerical Modelling in Compound Open Channels and Duct Flows" ,*Proc. of "HYDROCOMP 89"*, pp. 215-225.

[14] Prinos P.,(1990), "Turbulence Modelling of main channel - flood plain flows with an algebraic stress model", *Proc. of the Int. Conf. on River Flood Hydraulics*, pp. 173-185.

[15] Rajaratnam N. and Ahmadi R., (1981), "Hydraulics of channels with flood plains" ,*J. of Hydraulic Res.*, vol. 19, no 1, pp. 43-60.

[16] Rodi W.,(1982), Turbulence models and their Application in Hydraulics, IAHR Publication .

[17] Tominaga A., Nezu I. and Ezaki K.,(1989), "Experimental study on secondary currents in compound open-channel Flows" *Proc. of XXIII Congress of IAHR*, pp. A15-A22.

[18] Wood I.R. and Liang T., (1989), "Dispersion in an open channel with a step in the cross-section", *J. of Hydraulic Res.*, vol. 27, no 5, pp. 587-601.

[19] Wormleaton P.R., Allen J. and Hadjipanos P.,(1982), "Discharge Assesment in Compound Channel Flow", *J. of Hydraulics*, ASCE, vol. 108, pp. 975-994.

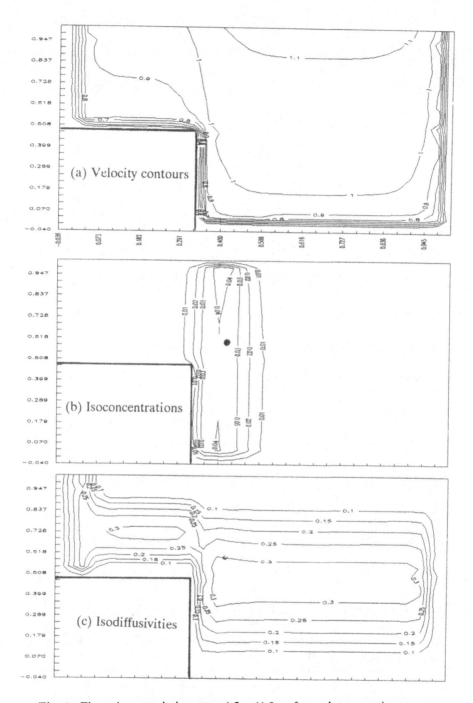

Fig. 1: Flow characteristics at x=4.5m (1.0 m from the source)
(a) Velocity contours,(b) Isoconcentrations, (c) Isodiffusivities.

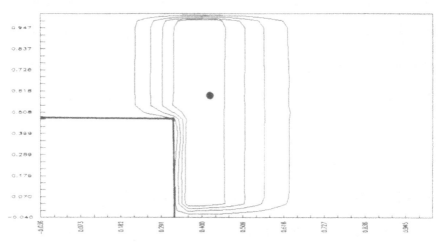

Fig. 2: Isoconcentrations at x=8.0 (4.5m from the source)

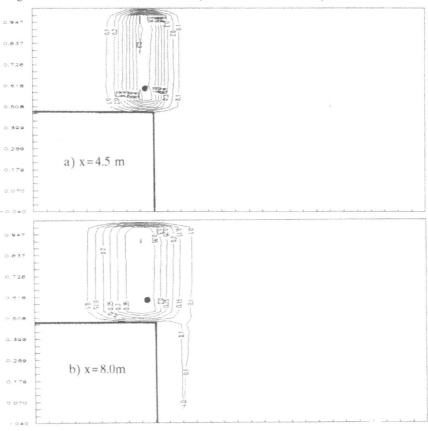

a) x=4.5 m

b) x=8.0m

Fig. 3: Isoconcentrations with the source on the flood plain
    a) x=4.5 m, b) x=8.0m

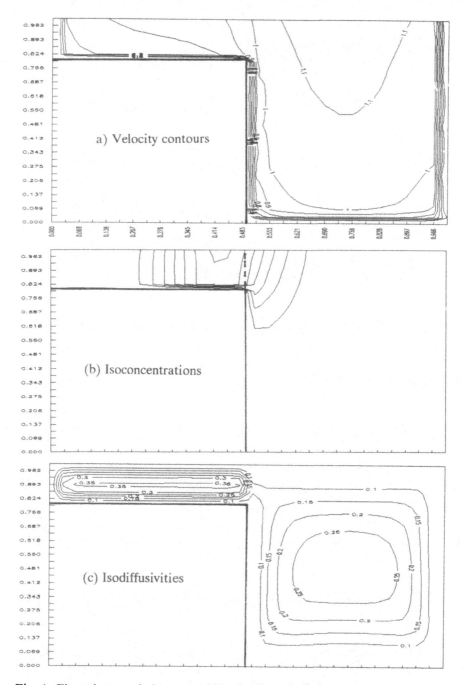

Fig. 4: Flow characteristics at x = 16.5 m (6.0m from the source)
a) Velocity contours,(b) Isoconcentrations, (c) Isodiffusivities

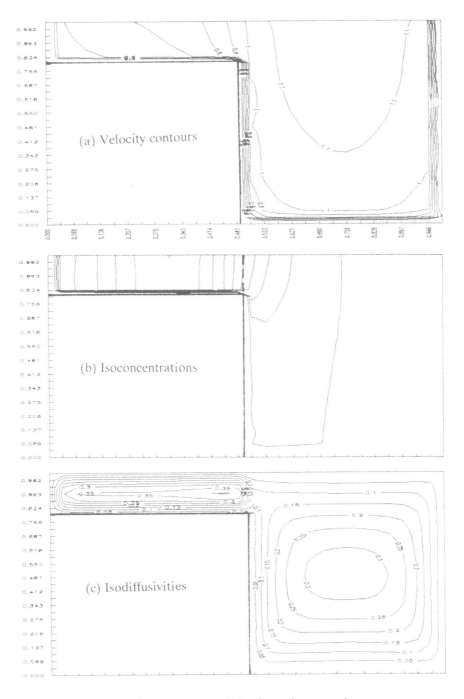

Fig. 5: Flow characteristics at x = 32.0m (22m from the source)
(a) Velocity contours,(b) Isoconcentrations, (c) Isodiffusivities

31 Prediction of pollutant transport in compound channel flows

B. Lin and K. Shiono

ABSTRACT

Three dimensional computational models have recently been developed for solving the momentum equation in conjunction with the refined Algebraic Stress Model (ASM+) and the non-linear k- ϵ model in a compound channel. Using the turbulent parameters predicted by the models, a prediction of pollutant transport rates has been attempted. The computed results are discussed in terms of the depth mean concentration of tracer, and the turbulent diffusion coefficients are also investigated through the Prandtl number. The results show a clear difference in the 3-D structure of the tracer concentration between cases with secondary flow present and those without. The numerical results agree well with the experimental results in the 3-D flow area although the flow condition is not quite uniform.

Introduction

Experimental investigations on tracer concentration in a compound channel have been studied by [1] and more recently at the SERC flood Channel Facility. The distribution of the depth mean tracer concentration [1] shows a noticeable change of the concentration near the main channel/flood plain (M/F) junction where flow becomes of a highly three dimensional nature, in particular the secondary flow becomes strong. This phenomenon has been observed in [2],[10]. The predicted depth mean concentration of tracer using the 2-D mathematical model [1] showed under-estimation in the flood plain and over-estimation in the main channel near the M/F junction. Therefore a 3-D mathematical model may be required to improve the accuracy for prediction of the concentration. In this study, the refined Algebraic Stress Model(ASM+) [8] and non-linear k-ε model [13] are adopted to generate the secondary flow, and the linear k-ε model is used to predict the non-secondary flow field for comparison purposes. An investigation of pollutant transport rates for dye injected continuously in a vertical line source in steady uniform flow is undertaken by uncoupling between the hydrodynamic and contaminant transport models.

Hydrodynamic models

The three-dimensional mean velocity components in a steady uniform open channel may be obtained by solving the continuity and momentum equations. The primary shear stress terms, $\overline{uv}$ and $\overline{uw}$ in the momentum equation are determined with the aid of the isotropic eddy viscosity hypothesis :

$$-\overline{uv} = v_t \frac{\partial \overline{U}}{\partial y}, \quad -\overline{uw} = v_t \frac{\partial \overline{U}}{\partial z} \tag{1}$$

where x,y and z are longitudinal, lateral and vertical directions and u,v and w are turbulent fluctuations respectively. The streamwise gradients of the turbulent quantities have been neglected by assuming uniform flow. The isotropic eddy viscosity, v_t, is calculated as in the conventional $k-\epsilon$ model with $v_t = C_\mu k^2 \epsilon^{-1}$ where $C_\mu = 0.09$. The empirical constants are $C_{\epsilon 1} = 1.44, C_{\epsilon 2} = 1.92$ and $\sigma_t = \sigma_c = 1.225$. This model generates a non-secondary flow field.

The source of secondary flow is non-zero values of the normal Reynolds stress terms $\overline{ww} - \overline{vv}$ and the cross Reynolds stress term, $\overline{vw}$ in the streamwise vorticity equation. The normal Reynolds stresses and the cross Reynolds stress are expressed in terms of the solvable quantities with some coefficients [7],[8],[13], the so called ASM, ASM+ and non-linear k-ε models respectively. These models have been applied to a compound channel [12] and the results have been agreed reasonably well with the experimental data. Examples of the predicted secondary flow and the experimental data [14] are shown in Fig.1. The ASM+ and non-linear k-ε models are selected in this paper to provide the 3 components of velocity and eddy viscosities for the diffusion equation.

Turbulent transport of pollutant

374

Turbulence transports passive contaminants such as chemical species, salinity, heat and particles. Assuming the analogy to Ficks' law to be valid and the scale of the random motion of turbulence much greater than that of the molecular motion, the diffusion equation of a passive contaminant for steady state conditions can be written as :

$$\overline{U}\frac{\partial \overline{C}}{\partial x}+\overline{V}\frac{\partial \overline{C}}{\partial y}+\overline{W}\frac{\partial \overline{C}}{\partial z}=\frac{\partial}{\partial x}\left(D_x\frac{\partial \overline{C}}{\partial x}\right)+\frac{\partial}{\partial y}\left(D_y\frac{\partial \overline{C}}{\partial y}\right)+\frac{\partial}{\partial z}\left(D_z\frac{\partial \overline{C}}{\partial z}\right) \qquad (2)$$

where D_x, D_y and D_z = the eddy diffusivity, and the time averaged $\overline{cu}$, $\overline{cv}$ and $\overline{cw}$ are the mean rates of transport due to turbulence of the contaminant, whose contaminant ,c, is across a unit area normal to the component of velocity. The diffusivities are normally associated with the eddy viscosity, v_t. The ratio of the eddy viscosity to the eddy diffusivity known as the turbulent Prandtl number, Pr, has been obtained from experiments and field work and found to vary between 0.5 and 1.0.

Boundary conditions

1) Solid wall ; the wall function [6] is adopted that is a resultant streamwise velocity expressed by a local friction velocity at the first grid point of the computational domain adjacent to the wall.
2) Free surface ; the velocity component and turbulent fluctuation normal to the free surface are zero, while the gradient normal to the surface is taken as zero for all other variables except for the rate of the turbulent energy dissipation. The expression of the dissipation rate at a first grid point below the free surface is given as :

$$\epsilon = \frac{C_\mu^{3/4} k^{3/2}}{\kappa}\left(\frac{1}{Y'}+\frac{1}{0.07H}\right) \qquad (3)$$

where H is depth of water and Y' is distance from the solid side wall.

Numerical procedure

The SIMPLER method[9] has been adopted to solve all the partial differential equations, which were assumed to be parabolic in the streamwise direction but elliptic in the cross-stream planes. The solution was obtained by marching the 2-D cross-section solutions along the streamwise direction until the flow became a fully developed turbulent flow over the distance , x/R=500, where R=Hydraulic Radius. A 'staggered grid' mesh was used, in which the secondary velocity components were placed at positions different from all the other variables. The mesh sizes were set uniformly except near the boundaries where the mesh sizes were adjusted by the application of the wall function. An appropriate initial value of each variable was assumed and the velocity components and pressure were computed first, using the momentum and continuity equations, then the turbulent kinetic energy and dissipation rate were computed, finally the Reynolds stresses were estimated. For the pollutant transport rate computation, the flow field and turbulent quantities associated with the diffusion equation were assumed to be unchanged in a cross section after a flow was fully developed. Therefore the pollutant diffusion equation was solved using the computed velocity components and turbulent quantities.

Results and comments

The secondary flow is computed using the ASM+ and non-linear k-ε models and the results are shown in Fig. 1. The patterns of the secondary flow are both similar and the ASM+ model produces stronger secondary flow than the non-linear k-ε model. It is noted that vectors are drawn to different scales in different figures. The pattern of the results is similar to the experimental results and the details of the analysis for the turbulent quantities can be seen in [12].

The distribution of the depth mean velocity is shown in Fig.2 for the three models. It can be seen from Fig.2 that the depth mean velocity is little changed between the models except near the M/F interface where the flow has a strong 3-D nature. It can be noted that the lateral momentum is transferred from the main channel to the flood plain near the M/F junction on the flood plain for the k-ε model (no secondary flow case), but the other models show it transfering the opposite direction. This phenomenon corresponds well to the bulging flow pattern near the M/F interface recognised experimentally [10].

The bed shear stress plotted in Fig.3 is distributed in a similar manner to the depth mean velocity. The maximum bed shear stress occurs near y/H=1.5 where the secondary flow direction becomes downwards(see Fig.1) and is consistent with the experimental data[14]. The magnitude of the depth eddy viscosity is attenuated as the strength of the secondary flow increases at the main channel and near the M/F interface in the flood plain, (see Fig.4(a)).

The non-dimensional eddy viscosity defined as ,$\lambda = \epsilon / (u \cdot H)$ has also been computed and is shown in Fig.4(b). The values of λ for the k-ε, ASM+ and non-linear k-ε model are 0.095, 0.09 and 0.07 at the centre of the main channel respectively, which are similar to the experimental values [11]. The value on the flood plain becomes 0.22 except near the wall regions. Using the empirical formula [11] for the eddy viscosity ratio between the main channel and flood plain over the shallow range of flood plain depths will give the same order of non-dimensional eddy viscosity magnitude as in the main channel if the formula is extended in deep water. The value may be higher but is similar to the value of 0.2 used for the 2-D mathematical model [1].

An investigation of pollutant behaviour was carried out injecting tracer as a line source at the M/F interface where a flow was assumed to be well developed. The turbulent Prandtl number of 0.5 was used to compute the tracer concentration distribution at the downstream section, 25H (= 2m) from the injection point shown in Fig5. The 3-D concentration distribution shows a clear distinction between the non-secondary flow (k- ϵ model) and the secondary flow (ASM+) cases. The lateral position of the peak concentration coincides at the water surface and the bed in the flood plain for the non-secondary flow case, but is skewed more into the flood plain at the water surface than at the bed for the secondary flow case. This is consistent with the secondary flow pattern (see Fig.1). The distribution of the concentration becomes undulated in the top half layer of the main channel where strong secondary flow is present.

The depth mean concentration of tracer is also calculated at the 1,2,4 and 6 meter downstream sections for the three models and is shown in Fig.6. The peak concentration moves into the flood plain and is reduced as the strength of the secondary flow magnitude increases, (ie. ASM+ model generates the largest secondary flow, second for the non-linear k- ϵ model and no secondary flow for the k- ϵ model). The concentration is also attenuated with the increase of secondary flow in the M/F interface region of the main channel. This clearly indicates that the secondary flow contributes to the mixing significantly. At 6m downstream, the distribution does not vary significantly laterally and the concentration approaches a well mixed condition. Therefore it may be important to consider the secondary flow effect near field of an injection point but may it not be important further downstream.

An investigation of the turbulent Prandtl number ,Pr, was carried out to understand the turbulent diffusion processes and the results are shown in Fig.7. It can be seen from Fig.7(a) that the peak of the concentration is shifted towards the flood plain as the Prandtl number for the transverse component increases while Pr was kept constant at 0.5 for the vertical component. An increase of the Prandtl number means a reduction of the turbulent diffusion so that the influence of the secondary flow becomes predominant. The distribution of the concentration in the main channel becomes an undulated shape near the M/F interface in the main channel as Pr increases.

377

For a change of Pr for the vertical component while Pr was kept constant at 0.5 for the lateral component, the peak of the concentration is attenuated at a faster rate than that of the lateral component as the Pr value decreases (see Fig. 7(b)). The concentration also tends to become well mixed at a faster rate. Therefore it may be concluded that the vertical diffusion coefficient is dominant when considering mixing for the whole area but insensitive to secondary flow mixing. On the other hand the lateral diffusion coefficient may be sensitive to the influence of secondary flow so that the mixing occurs locally. Fig. 8 shows the sensitivity of the peak concentration to the diffusion coefficient.

An application of the turbulent diffusion models has been made with the aim of comparing with the existing experimental data [1]. The results are shown in Fig.9 using a Prandtl number of 0.5 for both components. The transverse profile of the concentration has a similar characteristic to the previous results, (see Fig. 6) for an asymmetric compound channel. In particular the results of the ASM+ model agree well with the data in the M/F interface region although the flow condition is not quite uniform. It may be conjectured that the secondary flow is significant in the M/F interface region as has been seen in the other experimental results.

Conclusions

By solving the 3-D turbulent diffusion equation using the predicted velocity field and turbulent parameters, the tracer concentration has been predicted for an asymmetric compound channel. The influence of the secondary flow is recognised as significant in the transport behaviour near the tracer injection point but not for further downstream. The study of the Prandtl number shows that the mixing process is dominated by vertical turbulent diffusion. Lateral turbulent diffusion is sensitive to the secondary flow magnitude. The ASM+ model predicts the experimental results well despite non-uniform flow conditions. If there exists a strong 3-D flow in a channel, the 3-D mathematical model can improve the prediction capability. The k-ε model is capable of reasonable prediction for a weak secondary flow case.

References

1. Djordevic,S., Petrovic,J., Maksimovic,C. And Radojkovic,M. (1989), "Experimental tracer investigations in a compound laboratory channel", Proc. Of the Hydrocomp 89', Dubrovnik, Yugoslavia, pp269-278.

2. Imamoto,H., Ishigaki,T and Shiono,K. (1992), "On hydraulic characteristics in an open compound channel", Annual Bulletin of the Disaster Prevention Institute, Kyoto University, 6.

3. Kawahara,Y and Tamai,N. (1988), "Numerical calculation of turbulent flows in compound channel with an albebraic stress turbulence model", Proc. the 3rd Intl. Symp. on refined flow modelling and turbulence measurements, (Editors by Y.Iwasa, N.Tamai and A.Wada), Tokyo, Japan, pp.9-16.

4. Knight, D.W. And Shiono, K. (1990), "Turbulent measurements in a shear layer region of a compound channel", J. Hyd. Res. IAHR, Vol. 28, NO. 2, pp.175-196.

5. Larson,R. (1988), " Numerical simulation of flow in compound channels", Proc. of the 3rd Intl. Symp. on refined flow modelling and turbulence measurements, (Editors by Y.Iwasa, N.Tamai and A.Wada), Tokyo, Japan, pp. 537-544.

6. Launder,B.E. And Spalding,D.B. (1974), "The numerical computation of turbulent flow", Comp. Meth. In Appl. Mech. And Eng., 3. pp.269-289.

7. Launder,B.E. And Ying,W.M. (1973), "Prediction of flow and heat tranfer in ducts of square cross-section", Proc. Instn. Mech. Engrs, Vol. 187, pp.455-461.

8. Noat,D. And Rodi,W. (1982), " Calculation of secondary currents in channel flow", J. ASCE, HD, Vol.108, HY8, August, pp.948-968.

9. Partanker,SV. (1980), "Numerical heat transfer and fluid flow",Hemisphere Publ. Corp., Washington.

10. Shiono,K and Knight,DW (1989), "Vertical and transverse measurements of reynolds stress in a shear region of a compound channel", Proc. Of 7th Intl. Symp. On Turbulent Shear Flows, Stanford, USA, pp.28.1.1-28.16.

11. Shiono,K and Knight,DW (1991), "Turbulent open channel flows with variable depth across the channel", J. Fluid Mech., 222, pp617-646.

12. Shiono,K. And Lin,B. (1992), " Three dimensional numerical models for two stage open channel flows", Hydrocomp 92', VITUKI, Hungary, May.

13. Speziale,C.G. (1987), " On nonlinear k-l and $k - \epsilon$ models of turbulence", J. Fluid. Mech., Vol.178, pp459-475.

14. Tomainage,A. And Nezu,I.(1991), "Turbulent structure in compound open channel flows", J. Hyd. Eng. ASCE, Vol.117, No.1, pp.21-41.

379

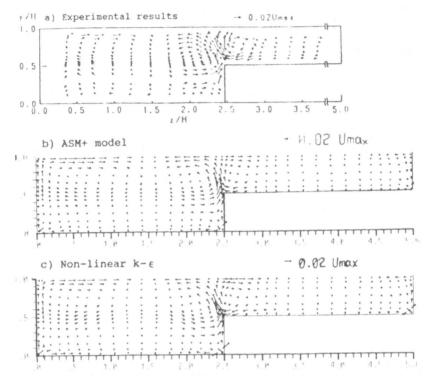

Figure 1 Computed secondary flow : a) Experimental date [14],
b) ASM+ model and c) non-linear k-ε model.

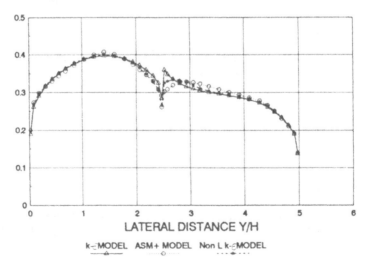

Figuer 2 Computed depth mean velocity.

380

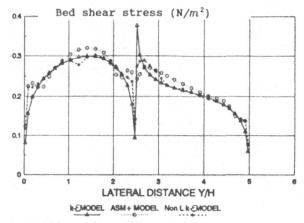

Figure 3 Bed shear stress.

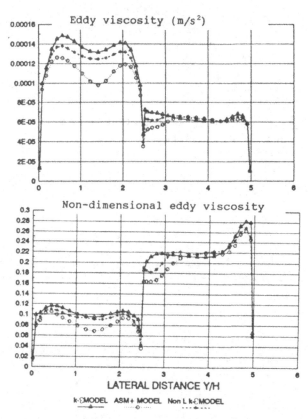

Figure 4 Eddy viscosity and non-dimensional eddy viscosity.

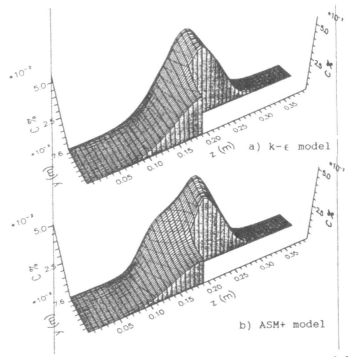

Figure 5 3-D concentration distribution : a) k- ε model
and b) ASM+ model.

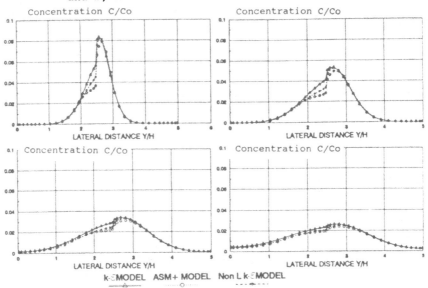

Figure 6 Depth mean concentration: a) 1m, b) 2m, c) 4m
and d) 6m downstream.

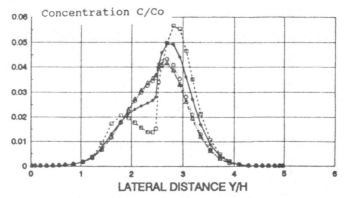

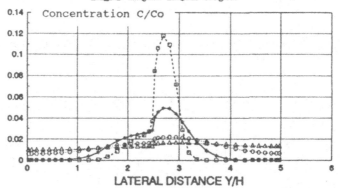

Pr=0.05 Pr=0.1 Pr=0.5   Pr=5.

Figure 7 Investigation of the Prandtl number : a) Lateral component
and b) Vertical component.

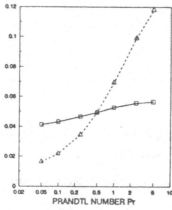

Figure 8   Sensitivity test of the peak concentration by
the Prandtl number.

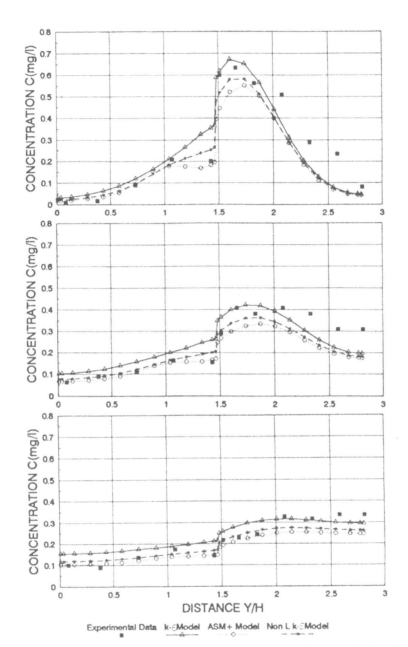

Figure 9 Prediction of depth mean concentration : a) 6m, b) 12m and c) 21m downstream.

32 Numerical model of three-dimensional flow processes and pollution transport in the rivers

A. L. Mironovsky

ABSTRACT

The problem about calculation of 3-dimensional fields of the velocity and of the pollution concentration in turbulent flow in the straight channels with complex cross-section is solved. The mathematical model, which allows to calculate the secondary flows (vortexs with longitudinal axis) , which influence greatly on the dynamic of the flow and on the mass transport processs is used. The calculation results are compared with experiments and are in qualitative agreement. The recomendation about method of pollution effluent which leads to more rapidly pollution delution are given.

1   INTRODUCTION

During two last decades a great attention is paid to mathematical modeling and numerical calculation of the turbulent flows and pollution spreading in the rivers. Moreover, the models for depth-averaged values or standart "K-E" turbulence modelare used in most cases [1]. However, this approach does not allow to correspond the main characteristics of the flow and of pollution spreading exactly enough and detailed while solving many problems. Therefore it is necessary to use mathematical models which take into consideration the anisotropy of turbulence and 3-dimentional turbulence structure. The numerical realization of such models is sufficiently difficult therefore nowadays they are used only for flows in straight channels.3-dimentional effects in these cases take place in the channels with the non-round cross-section. The secondary currents are generated in these channels because of the anisotropy of turbulence.

The corresponding calculations for the first time were carried out by Launder, Ying and Reece [2,3] for ducts and open channels with square cross-section. The calculation results for channels with trapezoidal and compound cross-section are shown in [4,5]. However, the authors of these papers did not investigate the influence of the secondary currents on pollution spreading.

The survey of the published papers testifyes, that it is necessary to elaborate calculation method which will allow to calculate turbulent flow and pollution spreading in the straight channels with a complex cross-section which accounts the secondary currents.

2   MATHEMATICAL MODEL

2.1   *Equations*

The turbulent flow of the liquid under the effect of the gravitation and pollution transport from sourses in straight channel with complex cross-section have been examined. The movement of the liquid is considered to be steady and completely developed along longitudinal co-ordinate X (Fig.1).

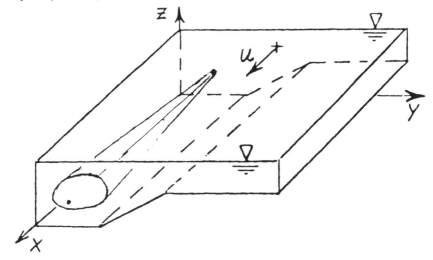

FIGURE 1   *Scheme of the flow and pollution spreading*

It is necessary to calculate 3-dimentional longitudinally-homogeneous velocity field (taking in consideration the secondary currents) and 3-dimentional unsteady field of the pollution concentration. The secondary flows appear because of the anisotropy of turbulence intensity. While calcaulating the velocity field Reynolds equations with algebraic correlations for the Reynolds stress [7] are used. The secondary currents equations are written in the terms of " vortex - streamline ". It allows to pick out obviously the term of the generation of the secondary currents vorticity and simplifies the numerical realization of the ma-

386

thematical model. The system of the equations which corresponds the liquid movement consists of the equation for the primary mean velocity, two equations for the secondary currents, equations for the kinetic energy of turbulence and rate of its dissipation and closing correlations:

$$V\frac{\partial u}{\partial y} + W\frac{\partial u}{\partial z} = \frac{\partial}{\partial y}\left(\nu_t \frac{\partial u}{\partial y}\right) + \frac{\partial u}{\partial z}\left(\nu_t \frac{\partial u}{\partial z}\right) + gi \qquad (1)$$

$$V\frac{\partial \zeta}{\partial y} + W\frac{\partial \zeta}{\partial z} = C_0\left[\frac{\partial^2}{\partial y \partial z}\frac{k^2}{\varepsilon^2}\left(\left(\frac{\partial u}{\partial y}\right)^2 - \left(\frac{\partial u}{\partial z}\right)^2\right) + \right.$$
$$\left. + \left(\frac{\partial^2}{\partial z^2} - \frac{\partial^2}{\partial y^2}\right)\frac{k^2}{\varepsilon^2}\frac{\partial u}{\partial y}\frac{\partial u}{\partial z}\right] \qquad (2)$$

$$\Delta \psi = \zeta \qquad (3)$$

$$V\frac{\partial k}{\partial y} + W\frac{\partial k}{\partial z} = \frac{\partial}{\partial y}\left(\frac{\nu_t}{G_k}\frac{\partial k}{\partial y}\right) + \frac{\partial}{\partial z}\left(\frac{\nu_t}{G_k}\frac{\partial k}{\partial z}\right) + P - \varepsilon \qquad (4)$$

$$V\frac{\partial \varepsilon}{\partial y} + W\frac{\partial \varepsilon}{\partial z} = \frac{\partial}{\partial y}\left(\frac{\nu_t}{G_\varepsilon}\frac{\partial \varepsilon}{\partial y}\right) + \frac{\partial}{\partial z}\left(\frac{\nu_t}{G_\varepsilon}\frac{\partial \varepsilon}{\partial z}\right) + \frac{\varepsilon}{k}(C_1 P - C_2 \varepsilon) \qquad (5)$$

where $\zeta = \frac{\partial v}{\partial z} - \frac{\partial w}{\partial y}$; $v = \frac{\partial \psi}{\partial z}$; $w = -\frac{\partial \psi}{\partial y}$;

$\nu_t = C_\mu \frac{k^2}{\varepsilon}$; $P = \nu_t \left(\left(\frac{\partial u}{\partial y}\right)^2 - \left(\frac{\partial u}{\partial z}\right)^2\right)$;

where U - primary mean velocity; V,W - secondary currents; X - longitudinal coordinate; Y,Z - diametrical co-ordinates; Vt - kinematic koefficient of the turbulent viscosity; g - gravitational acceleration; i - channel slope; ζ - vorticity of the secondary currents; ψ - streamline function of the secondary currents; K - kinetic energy of the turbulence; E - rate of the kinetic energy dissipation; P - generation of the kinetic energy; Gk,Ge - Prandtl turbulent numbers; C0,C1,C2,Cm - constants; according to [1,2] : Gk=1.0 Ge=1.3 C0=0.0033
C1=1.44 C2=1.92 Cm=0.09.
When calculating unsteady 3-dimentional field of the pollution concentration the equation of the passive admixture transfer is solved. The admixture have been transferred by primary mean velocity along the flow and by secondary currents and turbulent diffusivity across the flow. Besides that, admixture may come to the free surface or sink. The intensity of the admixture sourses can be unsteady and sourses themselfs can spread both in the cross-section and along the flow. The equation which corresponds the distribution of the admixture concentration Φ is:

$$\frac{\partial \Phi}{\partial t} + u\frac{\partial \Phi}{\partial x} + V\frac{\partial \Phi}{\partial y} + W\frac{\partial \Phi}{\partial z} = \frac{\partial}{\partial y}\left(\frac{\nu_t}{\sigma_\Phi}\frac{\partial \Phi}{\partial y}\right)$$

$$+ \frac{\partial}{\partial z}\left(\frac{\nu_t}{\sigma_\Phi}\frac{\partial \Phi}{\partial z}\right) - W_\Phi \frac{\partial \Phi}{\partial z} + F\left(x, y, z, t\right) \qquad (6)$$

where t - time; Gf - Prandtl turbulent number (Gf=0.5÷1.0 for different typyes of the admixture); WΦ - velocity of the coming to the free surface or siking; F - admixture sourses.

2.2 Boundary conditions

Boundary conditions to the equation (1) The improved method of near-wall functions which takes into consideration the peculiarities of the cross-section shape [8] is used. The nearest to the solid surface (bottom or wall) points of grid have been disposed in the region of the

logarithmic velocity profile. These points are marked with index 'P'. The nearest to the free surface points of grid are marked with index 'S'.
Shear stress along the perimeter of the cross-section is:

$$\begin{cases} T_p = \chi^2 u_p |u_p| / (ln(En_p^+))^2 \\ T_s = 0 \end{cases} \quad (7)$$

where χ - von Karman's constant (χ=0.41); E - constant; n_p^+ - non-dimensional distance from solid surface; Δ - equivalent rough value; u_* - dynamic velocity ($u_* = \sqrt{T_p/\rho}$); ν-kinematic viscosity.
E=30 and $n_p^+ = n_p/\Delta$ for raugh surface,
E= 9 and $n_p^+ = u_*/\nu$ for smooth surfase.
If T_p and T_s are known then Up and Us can be calaulated by solving the discret analog of the equation (1) for points P and S because diffusivity transfer of the velocity U from surface is proportional T.

Boundary conditions to the equations (2,3) The type of the boundary conditions to the equations (2,3) is discussed in detail in [9,10]. For our problem the next condition is advisable:

$$\begin{cases} \dfrac{\partial \zeta}{\partial n} = 0 & (8) \\ \psi = 0 & (9) \end{cases}$$

where n - normal to the surface.

Boundary conditions to the equations (4,5) The diffusivity transfer of the turbulence kinetic energy from the solid surface is equal to zero. Therefore Kp can be calaulated by solving the discret analog of the equation (4) for points P. The boundary condition at the free surface is:

$$\left. \dfrac{\partial k}{\partial z} \right|_s = 0 \quad (10)$$

The value of the rate of kinetic energy dissipation can be calaulated [1] from correlations:

$$\varepsilon_p = \dfrac{C_\mu k_p^{3/2}}{\chi n_p} \quad (11)$$

$$\varepsilon_s = C_\mu^{3/4} \dfrac{k_s^{3/2}}{0.07 \, \chi h}$$

where h - depth of the flow.

Initial and boundary conditions to the equation (6) For solution the problem about pollution spreading the distribution of the admixture concentration in the initial cross-section is needed to be known. The boundary conditions are conditions of the impenetrabity:

$$\begin{cases} \Phi(0,y,z,t) = \Phi_0(y,z,t) & (12) \\ q_{s,p} = 0 \end{cases}$$

388

where ϕ_o - function which must be known; q - admixture flow to the surface normal.

3   NUMERICAL SOLUTION

For creation of the discret analog of the differencial equations (1)-(6) and boundary and initial conditions (7) - (12) the control volume method is used. The control volume method was discribed in detail by Patankar in [11]. The curve-line ortogonal grid is automatically generating due to method of the conform reflection.
The iteration procedure based on the method of the establishment is used for the calculation of the 3-dimehsional velocity field. After the calculation of the velocity field the equation of the pollution transport is solved based on the march-method along longitudinal co-ordinate.

4   RESULTS

Due to methodic which was mentioned before the computer programm for solution the present problem was perfomed.

The computational results and experimental data [6] for the flow without admixture in the channel with two flood plains are presented in Tab.1 and in Fig.2 . The conditions of the flow were: flow depth of main channel was equal to 10 cm flow depth of flood plain was equal to S cm, width of channel was equal to 40 cm, width of main channel was equal to 20 cm, slope of channel was equal tc 0.00047, the channel was smooth.

TABLE 1 *Channel   with   two   flood   plains*

	Umax	Q	Tmax/Tav	Re=4*Uav*R/v
	cm/sec	l/sec		
measured	40.8	9.4	1.4	54 000
computed	41.3	9.8	1.5	57 000

Where index 'max' means 'maximum' and index 'av' means 'average', R - hydraulic radius. The calculation was carried out at the grid which includes 20*10 points. Computational results have a good agreement with experimental data. The maximum magnitude of the secondary currents is about 4% of Umax. Secondary currents influence essehtially on primary mean velocity distribution and boundary shear stress. The structure of the secondary currents is mainly composed of the flood plain vortex and the main-channel vortex. The strong inclined secondary currents are generated from the junction edge to the free surface.

computed ↑Z measured

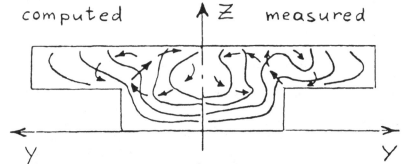

FIGURE 2 *Isovel lines of primary mean velocity U/Umax = 0.975; 0.9; 0.85; 0.75; 0.6 (solid lines) and two main vortexs of the secondary currents (pointers).*

The computational results and experimental data [12] for the salt water sprea-ding in the channel with flood-plain from the point-sourse are present in Fig.3. The conditions of the flow were: flow depth of main channel was equal to 5.1 cm, flow depth of flood plain was equal to 2.6 cm, width of channel was equal to 56 cm, width of main channel was equal to 37 cm, slope of channel was equal to 0.00047, the channel was smooth. Both computed and mesured results give 3-dimensional fields of the velocity and salt concentration. However, because of limited paper dimension, only depth-averaged salt spreading (plane spreading) are given.

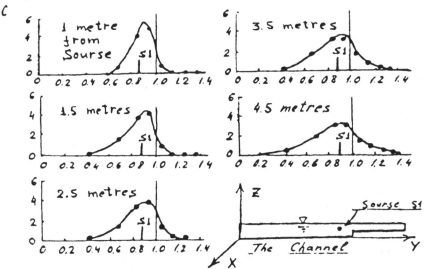

FIGURE 3 *A comparison of computed and mesured [12] depth averaged non-dimensional concentrations. The solid line is computed.*

390

The salt concentration measuring was carried out in 32 points in 5 cross-sections. The calculation was carried out on grid with 150*20*10 points (150 cross-sections from 3 cm each from other). Both experiment and calculation are shown that realising effluent in the deeper rather than the shallow channel gives a very much more rapid delution. Moreover, realising effluent in the deeper channel near the step of the cross-section (the point S1) leads to the maximum concentration moving towards the step, and realising effluent in the shallow channel near the step of the cross-section (the point S2 which is not shown in Fig.) leads to the maximum concentration moving away from the step. This conclusion is similar to that obtained by Wood and Tong Liang [12]. It is interesting to know that secondary currents in point S1 are rather strong ahd are directed towards the step, and secondary currents in point S2 are rather weak and are directed backwards the step. The virtue of these facts the author of present paper draws a conclusion that lateral pollution spreading (pollution transport across flow) is conditioned first of all by the secondary currents. Therefore, if rapid initial delution is important to reach then is a great advantage in realising the effluent in region with the maximum magnitude of the secondary currents. For examination of this hypothesis the calculation for the case when pollution sourse was disposed at the region of the inclined upflow from the junction edje was carried out. In this case the most rapid pollution delution was reached.

5   CONCLUSIONS

1) Presented mathematical model adequately corresponds the flow dynamic and mass transfer process in turbulent flow in the straight channels with complex cross-section with account of secondary currents.

2) For decrease of the pollution regions of the rivers and channels from the gutter it is necessary to creat the pollution influent in the region of the cross-section with maxsimum magnitude of secondary currents.

ACKNOWLEDGEMENTS

The author would like to thank Prof. Gipgidov A.D. for scientific leadership and to thank Dr. Prokofiev V.A. for help in the computer programm perfoming.

REFERENCES

[1] Kolman V.,(1984), Methods of the calculation of the turbulent flow, MIR, Moscow (in Russian).

[2] Launder B.E., Ying W.M.,(1972), Secondary flows in ducts of square cross-section, Journal Fluid Mechanics, Vol.54, pp.289-295.

[3] Reece G.E.,(1977), A generalized Reynolds-stress model of turbulence, Ph.D. Thesis, University of London.

[4] Tominaga I., Nezu I., Ezaki K.,(1989), Three-dimensional turbulent structure in straight open channel flows, Journal of Hydraulic Research, Vol.27,

[5] Kawahara Y., Tamai N.,(1989), Mechanism of lateral momentum transfer in compound channel flows, Proceeding of IAHR 23 Congress, Vol.B, pp.463-470.

[6] Tominaga I., Nezu I., Ezaki K.,(1989), Experimental study of secondary currents in compound open-channel flows, Proceeding of IAHR 23 Congress, Vol.A, pp.15-22.

[7] Launder B.E., Reece G.E., Rodi W.,(1975), Progress in development of Reynolds stress turbulence closure, Journal Fluid Mechanics, Vol.68, pp.537-566.

[8] Mironovsky A.L.,(1990), Modernizatuion of universal logarithmic velocity profile, Journal Energetika, Vol.6, pp.120-124, (in Russian).

[9] Roache P.,(1976), Computational fluid dynamics, Hermosa Publisher, Albuquerque.

[10] Anderson D., Tannehill J., Pletcher R.,(1986), Computational fluid mechanics and heat transfer, Hermisphere Publishing Corporation, New York.

[11] Patankar S.,(1984), Numerical methods of solution of mass transfer and liquid dynamics problems, Energoatomizdat, Moscow, (in Russian).

[12] Wood I.R., Tong Liang,(1989), Dispersion in open channel with a step in the cross-section, Journal of Hydraulic Research, Vol.27, No.5, pp.587-601.

Part 5
SEDIMENT TRANSPORT MODELLING

33 Modelling of the morphological processes in Venice lagoon

R. Warren, O. K. Jensen, J. Johnsen
and H. K. Johnson

ABSTRACT

A comprehensive dynamical 2 dimensional modelling system has been applied to understand and predict the morphological changes of Venice lagoon, Italy. The complex of models working together is:
    Hydrodynamic, advection-dispersion, waves, cohesive sediment transport and finally benthic biological models.

1. Introduction

The Lagoon of Venice is in a state of constant change with regard to its hydrodynamics, morphological and environmental state. The changes can be divided into natural trends and those due to human interventions. Since 1970 the natural changes have mainly consisted of an accelerated erosion of the barena and bassifondi of the lagoon, together with a sedimentation of the channels. In some areas the bassifondi (water depths 1-2 m) have been eroded by up to 50 cm and the narrow channels (ghebi) have been completely filled up with fine sediments.

The paper will present the investigations in the processes which have caused these severe changes in the lagoon, and the establishment of models suitable for describing the processes. The scientific background and the calibration of the models will be discussed in detail.

FIGURE 1   *Models and tasks considered for modelling the morphological changes.*

A sequence of models has been applied for the description of the physical and biological processes responsible for the displacement of the sediments. Hydrodynamics and advection-dispersion models describe the flow and spreading processes, in this case of the fine sediments. Wave models provide spatially and time varying wave fields in the lagoon, since waves are responsible for the stirring up of sediment during storm events. Sediment models calculate the erosion and sedimentation rate in a point as a function of suspended sediment concentration, vertical sediment transport, bed grain size distribution, vegetation coverage, shear strength of bed, current, wave climate, etc. Benthic models describe the biology and bed vegetation coverage varying both spatially and with time.

New developments have been necessary in the wave, sediment and benthic processes.

2. Registration of morphological evolution

A detailed bathymetric measurement campaign was carried out in 1970 covering all parts of the lagoon. Recently (1990) new soundings have been performed, not as detailed as in 1970, but covering major parts of the lagoon. The surveys indicate a drastic erosion of the tidal flats, which has taken place since the beginning of this century, but mostly the last 20 years' period.

The pollution of nitrogen and phosphorus has reached an alarming level. Dense algae growth prevents the sunlight reaching the sea bed, and without light the sea bed flora dies. Later, when the algae settles to the bed and decomposes, it uses all the available oxygen and the fauna also dies. With no flora or fauna to stabilize the bed, it becomes open for rapid erosion.

3. Hydrodynamic and spreading processes

The hydrodynamic model, solves the vertically integrated equations of continuity and conservation of momentum in two horizontal dimensions. These general equations are presented in Abbott et al (1981). The equations are solved by implicit finite difference techniques with the variables defined on a space-staggered rectangular grid. A 'fraction-step' technique combined with an Alternating Direction Implicit (ADI) algorithm is used in the solution to avoid the necessity of iteration. Second order accuracy is ensured through the centering in time and space of all derivatives and coefficients.

The advection-dispersion model solves the equation for a dissolved or suspended substance c in two dimensions (x,y) and time (t):

$$\frac{\partial c}{\partial t} + u\frac{\partial c}{\partial x} + v\frac{\partial c}{\partial y} = D_x\frac{\partial^2 c}{\partial x^2} + D_y\frac{\partial^2 c}{\partial y^2} + S$$

D_x and D_y are dispersion coefficients and discharge quantities and compound concentrations at source and sink points (S) are included on the right hand side of the equation. Current velocities and water levels are provided by the hydrodynamic part.

The equation is solved using the generally known QUICKEST scheme, see Leonard (1979) and Ekebjærg and Justesen (1991).

4. Simulation of lagoon waves and generation of wind drag coefficients

Islands, the littoral foreland, city of Venice, other land forms, plus the generally shallow and fetch limited situation in the Venice lagoon influences the atmospheric boundary layer, thus influencing both the wind speeds and the rate of transfer of momentum from the wind to the water body (i.e. the wind sheer stress).

From a number of wind stations placed inside the lagoon as well as in the Adriatic, the general wind speed variations have been established and expressed

397

as wind speed scaling factors in a 300 meter finite difference grid, where the factor 1.0 is taken for an offshore station. See figure 4.1.

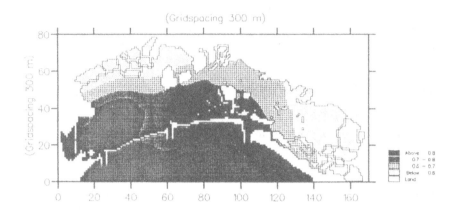

FIGURE 4.1 *Map of wind scaling factors in Venice lagoon.*

The scaling factor at every point is found as a weighted average of the scaling factors obtained from the stations. The applied method is similar to a procedure developed by the U.S. National Weather Service for estimating point rainfall at a given location using recorded values at surrounding sites. See Viessmann et al (1977).

The scaling map has been used for generation of spatially varying wind fields for the hydrodynamic and the wave modelling tasks.

Wind waves are handled in a model, which describes the propagation, growth and decay of short period waves in nearshore areas. The model takes into account the effects of refraction and shoaling due to varying depth, wind generation, energy dissipation due to bottom friction and wave breaking.

The wave model is a stationary, directionally decoupled parametric model. In this model, the wave action balance equation is parameterized in terms of a frequency-integrated zeroth and first moments of the wave action spectrum, m_0 and m_1, both as functions of the spectral wave direction. From the variables, wave height H_s, the mean wave period T_m and the mean wave direction θ, can be determined. Calibration and verification are carried out using data from points spread around the lagoon area. In addition one-dimensional fetch limited wave tests were set up, with typical fetch lengths and depths taken as representing the different zones of the lagoon. 'Shore Protection Manual' (1984) has at the same time been applied in the comparisons. The main purpose has been to determine the bed roughness, which can be used to characterize different

zones. In Figure 4.2 the simulated wave heights are compared with observed
values for a point in the Venice lagoon.

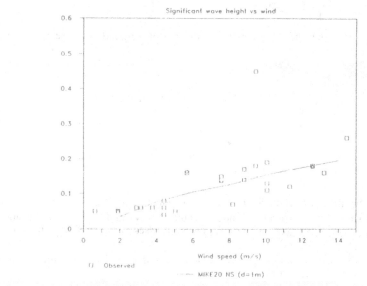

FIGURE 4.2   *Comparison between simulated and observed significant wave*
heights.

Traditionally the wind drag coefficient C_D (also called friction factor) has been
calculated using a wind drag formula, which is linear with wind speed (e.g.
Smith and Banke, (1975)). This formula does not directly take into account the
sea roughness, thus the influence of limited fetches and shallow water. In order
to include the influence of waves on the drag coefficient, a hybrid formulation of
the sea roughness models by Donelan (1990) and Kitaigorodskii & Volkov
(1965), following a method developed at Danish Hydraulic Institute, see Johnson
et al (1991). Using this method, time and spatially varying wind drag
coefficients are obtained, which are subsequently used in the hydrodynamic
simulations.

5. Cohesive sediment transport

The key requirements of the model were to describe the behaviour of sediment
during both tidal cycles and periods with high erosion rates due to wind waves.
   The cohesive sediment transport model has been built into the advection-
dispersion model described earlier. The model describes horizontal transport of
suspended cohesive sediment and detailed calculations of deposition/erosion rates
of the sediment represented as sink and source terms in the transport model.

Deposition : $S_D = W_s C_b P_D$

Erosion : $S_E = E P_E$

P_D P_E are probabilities for deposition or erosion respectively, see Krone, R.B. (1962).

$$P_D = 1 - \frac{\tau_b}{\tau_{cd}} \qquad \tau_b \leq \tau_{cd}$$

$$P_E = 1 - \frac{\tau_b}{\tau_{ce}} \qquad \tau_b > \tau_{ce}$$

The probability of pure horizontal transport equals unity for $\tau_{cd} < \tau_b \leq \tau_{ce}$. τ_b is the bed shear stress developed under combined wave and current motion and the threshold values τ_{cd}, τ_{ce} are critical shear stresses for deposition and erosion, respectively.

A 2-layer bed model was implemented into the model, i.e. fluid mud and under- consolidated bed layers.

The schematic representation of these two layers and properties of bed material similar to that in the lagoon are found in Parchure and Mehta (1985). A linear, increasing variation of the bed shear stress strength is assumed in the fluid mud and a constant in the consolidating bed.

Each of the processes, transport, deposition and erosion of cohesive sediment, are associated with different time scales.

During tidal high and low waters mainly pure horizontal transport of sediment occurs and this sediment deposits during slack tide on the bed as fluid mud, see Van Rijn, L.C. (1989), Mehta et al (1989), Mehta (1989), Tecter, A.M. (1986). Some of the fluid mud is re-entrained again during the following tidal high and low waters.

The under consolidated bed layer increases during the generally low energy tides and decreases during periods of high energy forcing functions, i.e. storm periods where wind waves are present. The period for this variation can take months.

The key processes, which influence the deposition, transport and erosion of cohesive sediment are: settling (due to gravity), flocculation, vertical transport, development of fluid mud and consolidating bed.

The model distinguishes between two settling states. For low concentrations, i.e. $C_b < 300$ g/m^3 practically only the settling of the individual floc particles will occur. For high concentrations, i.e. $300 \leq C_b < 10000$ g/m^3 the settling velocity will increase with the concentration due to flocculation. These processes are given in the following way, see Van Rijn (1989), Burt, T.N. (1986).

$$W_s = K \, C_b^m$$

K, m are coefficients.

The near bed concentration, C_b is the depth-averaged concentration through the vertical transport, i.e. a comparison of convective and diffusive transport represented by the Peclet number, see Teeter, A.M. (1986).

For sufficiently small Peclet numbers the presence of disruptive turbulent shear forces in the boundary layer implies the start of re-entrainment of fluid mud. The presence of combined current and wave motion implies larger turbulent shear stresses, which lead to additional break-up of the under consolidated bed.

For combined wave-current motion the eddy viscosity is strongly increased in the wave boundary layer close to the bed and the near bed current profile is retarded. The effect on the outer current velocity profile is described by introducing a "wave" roughness, K_w, which is larger than the actual bed roughness. It is assumed that wave motion is dominant close to the bed compared to the current, which means that the wave boundary layer thickness, ∂_w, and wave friction, f_w, have been determined by considering the wave parameters only, see Fredsøe, J. (1981), Jonsson, I.G. and Carlsen, N.A. (1976), Swart, D.H. (1974).

In figure 5 the model has been applied for a Bora storm event and the simulated suspended sediment concentrations are shown for a point as a solid line and observed values are marked with (+).

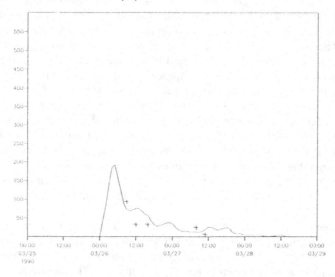

FIGURE 5   *Comparison between measured and simulated suspended sediment concentrations during a storm for a station in the central lagoon.*

401

6. Benthic vegetation coverage evolution

The erodability of the bed is an important parameter in the erosion rate calculation. It is not indifferent whether the bed is covered with healthy eelgrass or it is biologically dead with no flora or fauna to protect against unavoidable erosion. The vegetation coverage has changed a lot the last few decades, in the direction of a decreased distribution area.

The benthic model describes the interaction between phytoplankton, macroalgae (mainly Ulva) and rooted macrophytes (mainly eelgrass) and the influence on the growth of these species from suspended matter in the water through its effect on the water transparency.

The model includes sixteen state variables representing the three primary producers: phytoplankton, macroalgae, (carbon, nitrogen and phosphorus) and rooted macrophytes (shoot biomass and shoot density), organic matter in the water (detritus carbon), the inorganic nutrients and finally oxygen balance.

The model includes a description of the pelagic system since the interaction between the pelagic and benthic systems are of major importance. The total benthic model complex includes an eutrophication model emphasizing the benthic part. Seasonal variations are described in a range of biological and chemical determinants and they depend upon a number of forcing functions: water exchange (taken from the advection-dispersion model), influx of light, water temperature and nutrient loadings.

The carbon, nitrogen and phosphorus cyclus is shown in figure 6.1 exemplified by the carbon cyclus (nitrogen and phosphorous are in principle the same). An oxygen balance is included based on the processes of the carbon cyclus.

The pelagic system is described by the growth of phytoplankton, grazing of phytoplankton and the transformation of living phytoplankton and grazers to dead organic material (detritus) (process no. 1,3,4,5 and 6).. In addition phytoplankton and detritus are subject to sedimentation (process no. 2 and 9). The release of nutrients from the degradation of organic matter in the water and sediment and the corresponding oxygen demand are also included (process no. 10). The nutrient dynamics in phytoplankton are expressed by internal pools of nitrogen and phosphorus. This means, that algae growth can take place even though the nutrient concentrations in the water are very low. The most important factor concerning competition between phytoplankton and macroalgae is, from the phytoplankton point-of-view, the availability of nutrients.

The benthic vegetation in the model includes rooted macrophytes (eelgrass) and macroalgae (primarily Ulvae). Ulvae, as well as other algae, show an increasing growth rate with increasing temperature, but above a certain temperature level the population is known to potentially collapse. This temperature effect specific for Ulvae is included in the benthic model as the water temperature in the Venice lagoon is known to reach these high levels.

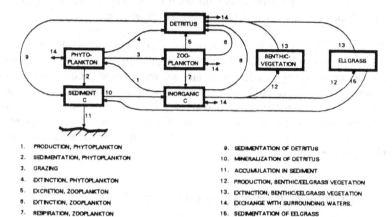

1. PRODUCTION, PHYTOPLANKTON
2. SEDIMENTATION, PHYTOPLANKTON
3. GRAZING
4. EXTINCTION, PHYTOPLANKTON
5. EXCRETION, ZOOPLANKTON
6. EXTINCTION, ZOOPLANKTON
7. RESPIRATION, ZOOPLANKTON
8. MINERALIZATION OF SUSPENDED DETRITUS

9. SEDIMENTATION OF DETRITUS
10. MINERALIZATION OF DETRITUS
11. ACCUMULATION IN SEDIMENT
12. PRODUCTION, BENTHIC/EELGRASS VEGETATION
13. EXTINCTION, BENTHIC/EELGRASS VEGETATION
14. EXCHANGE WITH SURROUNDING WATERS.
15. SEDIMENTATION OF EELGRASS

FIGURE 6.1   *Processes and state variables in the benthic model, exemplified by the carbon cyclus*

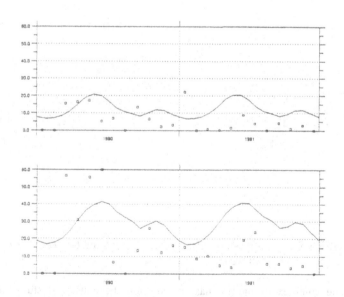

FIGURE 6.2   *Comparison between measured and simulated benthic vegetation (macroalgae), for two stations in the lagoon.*

Macroalgae will at significant biomasses have a severe shadowing effect on rooted vegetation. At higher biomasses a self-shadowing effect can be expected as well. The shadowing effect on eelgrass is one of the key points in the competition between macroalgae and rooted vegetation.

In general the expectations with the calibration of the benthic model was to reflect the variations over the year and also the conditions of competition between phytoplankton, macroalgae and macrophytes in the different parts of the lagoon. In figure 6.2 comparisons between measured and simulated concentrations of macroalgae (Ulva) in 1990 and 1991 are shown for two different areas of the lagoon. The unit for the concentrations is gCarbon/m^2. After calibration the model was applied for long term simulations to judge, in this case, the evolution of benthic flora.

7. Study of long-term morphological evolution

After all the models have been calibrated and are working together, the evolution of the morphology over the last two decades has been simulated with the result shown in figure 7.1. General human intervention has been included, such as the effect of dredging for vongole (a shell fish), shipping and dredging activities, to give an overall balance of the transport of cohesive material.

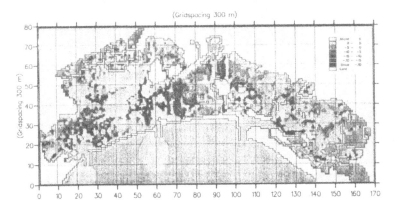

FIGURE 7.1   *Morphological evolution simulated between 1970 - 1990. (unit: mm/year)*

The complex of models has, in addition, been used to study the impact of various proposed human interventions and regulations. Each scheme is aimed at improving physical and biological conditions in the lagoon and thereby improving the state of the lagoon morphology. In figure 7.2 the effect of filling up a major navigation channel has been investigated in the advection-dispersion model. The changes in concentration of a conservative pollutant discharged at Fusina were extracted after a 28 day historical simulation.

8. Conclusion

Modelling has become an integral part of the decision process for the protection of aquatic environments. Although certain physical and biological processes are far from fully understood, our knowledge is sufficient to produce models of the ecosystem, which can distinguish between the effects of different environmental management policies. Technical and political decisions can, and are being made on the basis of modelling results. In this paper, one such modelling system and application has been described. It is an integrated system, which combines many of the processes which are important for environmental investigations and which is refined further as new knowledge becomes available.

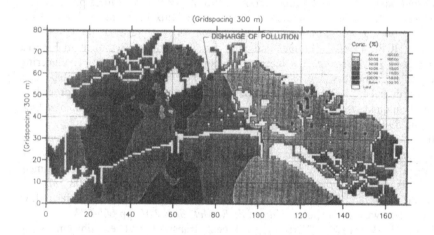

FIGURE 7.2   *Changes in pollution levels (in %), after having implemented dredging and filling up of major navigation channels*

ACKNOWLEDGEMENTS

The application for Venice lagoon has been carried out under contract to Consorzio Venezia Nuova, operating on behalf of Magistrato alle Acque di Venezia.

405

REFERENCES

Abbott, M.B., McCowan, A. and Warren, I.R., (1981), "Numerical modelling of free surface flows that are two dimensional in plan", *Transport models for inland and coastal waters*, Academic Press, pp 222-283.

Burt, T.N., (1986), "Field Settling Velocities of Estuary Muds", *Estuarine Cohesive Sediment Dynamics, Lecture Notes on Coastal and Estuarine Studies*, 14, Springer-Verlag, pp 126-149.

Donelau, M., (1990), "Air sea interaction", *The Sea, Ocean Engineering Science*, Vol. 9, Chapter 7, pp 239-292.

Ekebjærg, L. and Justesen, P., (1991), "An explicit scheme for advection-diffusion modelling om two dimensions", *Comput. Metha. Appl. Mech. and Eng.*, Vol. 88, pp 287-297.

Fredsøe, J, (1981), "Mean Current Velocity Distribution in Combined Waves and Current". *Progress Report No. 53*, ISVA, Technical University of Denmark.

Johnson, H.K. and Vested, H.J., (1991), "Effects of water waves on wind shear stress", Danish Hydraulic Institute.

Kitaigorodskii, S.A. and Volkov, Y.A., (1965), "On the roughness parameter of the sea surface and the calculation of momentum flux in the lower layer of the atmosphere", *Izvestiya Atmospheric and Ocean Physics*, Series 1, pp 973-978.

Krone, R.B., (1962), "Flume Studies of the Transport of Sediment in Esturial Processes", Hydraulic Engineering Laboratory and Sanitary Engineering Research Laboratory, Univ. of California, Berkeley, California, Final Report.

Leonard, B.P., (1979), "A stable and accurate convective modelling procedure based on quadratic upstream interpolation",*Comp. Meths. Appl. Mech. and Eng. 19*, pp 59-98.

Mehta, A.J., (1989), "On estuarine cohesive sediment suspension behaviour", *Journal of Geophysical Research*, Vol. 94, No. C10, pp 14303-14314.

Parchure, T.M. and Mehta, A.J. (1985), "Erosion of soft cohesive sediment deposits", *Journal of Hydraulic Engineering*, Vol III, No. 10, pp 1308-1326.

Smith, S.D. and Banke, E.G. (1975), "Variations of the sea surface drag coefficient with wind speed", *Quart. J. R. Met. Soc. 101*, pp 665-673.

Swart, D.H. (1974), "Offshore sediment transport and equilibrium beach profiles", *Delft Hydr. Lab. Publ. 131*, Delft Univ. Technology Diss., Delft.

Teeter, A.M. (1986), "Vertical transport in fine-grained suspension and nearly-deposited sediment", *Estuarine cohesive sediment dynamics*, lecture notes on coastal and estuarine studies, 14, Springer-Verlag, pp 126-149.

U.S. Army Corps of Engineers, (1984), "Shore Protection Manual", 4th edition, *Coastal engineering research centre*,East Belvoir, Virginia.

Van Rijn, L.C., (1989), "Handbook on sediment transport by current and waves", Delft Hydraulics, Report 461, pp 12.1-12.27

Viessmann, W., Knapp, J.W., Lewis, G.L. and Harbough, T.E., (1977), *Introduction to Hydrology*, Second edition, Harper & Row Publishers, New York.

34 Interaction between the river flow and bed forms in the estuarine reach of the Vistula River

L. Suszka

ABSTRACT

The interrelations between physical factors influencing the hydraulics of a non-tidal river estuary are described in this paper. The study is mainly based on the results of measurements carried out in the estuarine reach of the Vistula River. A preliminary model determining a coefficient of resistance to flow is proposed.

1   Introduction

Thanks to the now common use of computers mathematical modelling has developed rapidly over the last few years. The application of mathematical models in hydraulics enables, among other things, the prediction of water stages, flow velocity or even bed level variations. One of the indispensable parameters in the construction of these models is the coefficient of resistance to flow. It is essential to know this value if the model is to reflect the river flow precisely. Numerous papers concerning this problem have been written but the mathematical models contain the coefficients which have little in common with the physics of the phenomenon. In many models the constant value of the resistance coefficient is applied without taking into account the variations of the flow conditions. The estimation of this coefficient becomes much more difficult where the estuarine reach of a river is affected by the activity of the sea.

An attempt has here been made to present schematically the interrelations between such factors as: wind, waves, river discharge and fluctuations of the sea level, and their influence on the dimensions of bed forms in the estuarine reach of a river. The bed forms, in turn, influence the value of the coefficient of resistance to flow. These factors can be treated as secondary and ommitted in the case of a tidal estuary, but they may become dominant in the case of a non-tidal estuary.

The author has tried to describe roughly the interrelations between the above-mentioned factors. Any precise description of these interrelations would be very difficult to prepare and this is not the purpose of this paper, which should be treated merely as the preparation to further studies. They aim should be to supply the users of mathematical models with ready formulae with which to estimate the resistance coefficient which would take into account the lag time between any change in flow conditions and response in bed forms.

The results of systematic bed form measurements carried out in the estuarine reach of the Vistula River in the years 1987-1990 are used in this work. The measurements were carried out at two fixed longitudinal profiles, each one about 1 km long. The detailed description of the measurements and the analysis of their results are presented in other papers by the author (SUSZKA (1990) and (1991)).

2   Model Scheme

Let us imagine a sophisticated system of interrelations between physical factors which influence the resistance coefficient in a non-tidal estuary. It should be emphasized that the situation in which tidal waves occur is less complicated, as the effect of other factors is much smaller and can be neglected. Coming back to a non-tidal estuary, the following factors can be distinguished and treated as input parameters:

- the river discharge histogram, Q(t);
- the wind action i.e. its direction, velocity and duration which depend on the meteorological situation over the sea;
- the mean sea level;
- the river water temperature;

The wind action and the resulting sea level close to the estuary, together with the water discharge, are factors which are responsible for the local water depth and the friction slope in the estuarine reach of a river. This means that the application of an ordinary rating curve connecting the water discharge with water depth would be unsuitable. Besides, there is one more factor which can modify the predicted water depth and the friction slope in an estuary, namely wind waves moving upstream from the sea. This occurs, in particular, when the wind direction correlates with the river axis. This effect is not taken into consideration in the paper. When the water depth and the friction slope are already established, the bed shear stress responsible for sediment transport and bed form creation or destruction, can be calculated. In the meantime, the wind acting along the river may modify vertical distribution of shear stress and, what is particularly important, the value of the bed shear stress. This, in turn, will transform the existing bed form geometry until it is fully developed. This last process requires some time (lag time) to complete depending on dune dimensions in the earlier stage and the degree of unsteadiness of the whole system. The new dune dimensions and their 3-D pattern will influence the coefficient of resistance to flow which is additionally influenced by the water temperature. It will subsequently modify the water depth and/or the friction slope, which adjust the bed shear stress. The process described can be repeated many times until the state of equilibrium is attained. Let us imagine that in the meantime, one of the input parameters changed. In consequence, the whole process starts from the very beginning. The interrelation system described herewith is shown as the flow chart in Figure 1.

408

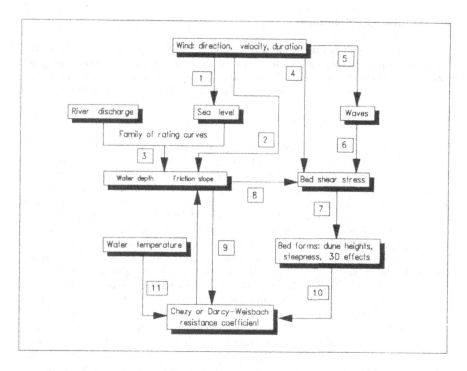

FIGURE 1 *Flow chart of the interrelation system in the estuary of a river.*

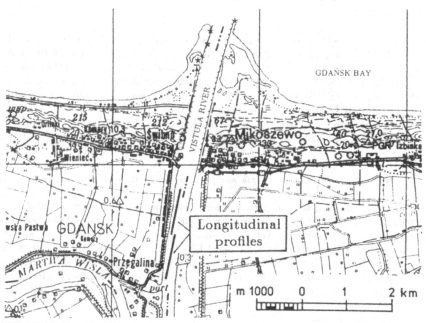

FIGURE 2 *Sketch of the estuarine reach of the Vistula River.*

409

3   Individual Relations

There are several partial functional interrelations between particular points mentioned in the preceding chapter. An attempt was made to establish these relations, using field data from the estuary of the Vistula River (see Figure 2). Particular relations are indicated in Figure 1 with numbers corresponding to those of subsections in this chapter.

3.1   Wind action vs. sea level

It would be very useful if information about wind over the sea were sufficient to specify the sea level in the downstream end of the estuary of a river. In reality, this relation is much more complex. Not only the baric situation covering a large region over the sea at the time, but also the preceding one will influence the sea level. An attempt was made to relate the wind velocity and its direction in the estuary of the Vistula River to the sea level. The correlation obtained was very poor. It should be noticed, however, that the Baltic Sea is specific in one respect. It is connected with the North Sea and, further, the Atlantic Ocean through the narrow Danish straits. If, for example, as a result of the baric situation, there is a large influx of water to the Baltic Sea or, just the opposite, some Baltic water flows out to the North Sea, the state of non-equilibrium between the Baltic and the North Seas will last for some time, even when the baric situaton returns to normal, hence it is difficult to obtain a good correlation between the wind and the sea level. The observations of wind direction and velocity and sea levels over a period of ten years show that the dependance between wind velocity and sea level is insignificant. In the case of northern winds the sea level increases with wind velocity. For southern winds the sea level decreases slightly when the wind velocity increases. In both cases standard deviations are very large so the relations obtained could only be used qualitatively.

3.2   Wind action vs. water depth

In the estuarine reach of a river, where the water surface slope is very small (in the case of the Vistula River it is of the order of 10^{-5}), the wind can modify the water depth and/or the friction slope considerably. This modification is due to the response of a complex process. It can be described as follows: the wind forces are transmitted to the water surface through the surface shear stress, which is usually calculated as $T_w = 1.5 \cdot 10^{-6} W^2$, where W denotes the wind speed 10 m above the water surface and T_w is kinematic shear stress (shear stress divided by the density of water). In the meantime, the distributions of the vertical shear stress and velocity are also modified. Finally, it leads to a variation of the water depth and/or friction slope.

3.3   River discharge vs. water depth and friction slope with sea level as a parameter

In the case of uniform flow in a river, there is a unique rating curve relating river discharge with the water depth. As far as an estuary is concerned, nonuniformity of flow appears. The variation of sea level will provoke a gradually varied flow in an estuary with backwater M1 or drawdown M2 (see FRENCH (1985) p.199). This will affect the ordinary rating curve. The influence of the sea level will decrease with the increase of river discharge . The example of the family of rating curves at the Świbno station (3 km upstream from the mouth of the Vistula River) shows, in Figure, 3 that it is impossible

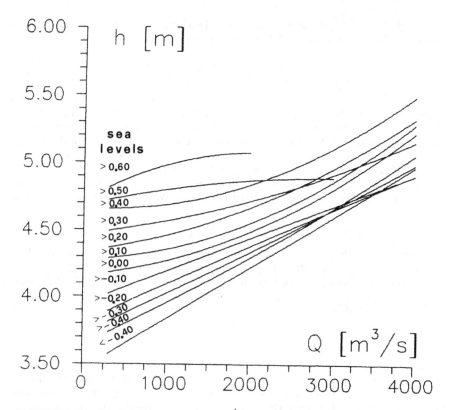

FIGURE 3 *Family of rating curves for the Świbno station (Vistula River) with sea level as a parameter.*

to approximate the relation between discharge and water depth in an estuary under a backwater effect by means of one rating curve. Detailed analysis of the water discharge and water stages at Świbno cross–section was made. The set of functional relationships between water depth and river discharge for different sea levels was obtained

$$Q = a + bH + cH^2 \tag{1}$$

where a, b and c are constants which depend on sea level.

3.4 Wind action vs. bed shear stress

The physics of this relation is described in the preceding subsection. An attempt has been made to propose a method of bed shear stress evaluation. There are different approximation methods of calculating the influence of wind shear stress on bed shear stress. For example NIHOUL (1977) proposes the "empirical bottom friction law".

$$T_b = DU^2 - mT_w cos\alpha \tag{2}$$

where T_b and T_w are the bottom and the surface kinematic stresses, $D = 2.1 \cdot 10^{-3}$ is the drag coefficient, m = 0.07 is the coefficient of the surface stresses transmission, α

is the angle between wind and flow directions. The improved bottom friction law has recently been proposed by KOŁODKO (1990) in the form of

$$T_b = (\frac{U}{C_k})^2 - \frac{1}{3}(R_T/C_k)(cos\alpha - n)T_w \qquad (3)$$

where C_k is the Chezy coefficient, R_T is the eddy Reynolds number, and n is the correction number (n = 0.5 is proposed by the author). Despite its simplicity, this formula reflects two ways of shear stress transmission: through changes in the velocity profile and through changes in turbulent viscosity. KOŁODKO (1990) proposes numerical values of his coefficients for the estuary of the Vistula River as $C_k = 13$ and n = 0.5. Another way of determining of the bed shear stress in the presence of wind shear stress is recommended by BUCHHOLZ (1988). His equation consists of two parts. The first one concerns the ordinary bed shear stress and the second evaluates the influence of wind shear stress:

$$\tau_b = \gamma R_h I_f - \tau_w \frac{B}{\chi} \qquad (4)$$

where τ_b is the bed shear stress, τ_w is the surface stress, R_h is the hydraulic radius, B is the width of an estuary, χ is the wetted perimeter and γ is the specific weight of water. Which of these two equations approximates the influence of the wind on the bed shear stress better needs to be verified. Precision of bed shear stress evaluation is of great significance for the determination of bed form transformation.

3.5   Wind action vs. waves

It would be very useful to predict parameters of waves which appear in the estuary of a river. There are two possible reasons of wave occurrence. The first one is the wind parallel to an estuarine axis as a result of which waves can easily develope in an estuary. The second possibility occurs when waves enter a river directly from the sea. Each should be examined separately. Due to the complicated character of this process, a none explicit formula is proposed in this model. In future, a more general model should include the relation between the wind action and waves. The importance of information concerning wave parameters occurring in an estuary is explained in the next subsection.

3.6   Waves vs. bed shear stress

It seems that wave action in the estuary of a river can modify the bed shear stress caused by river flow. The magnitude of oscillatory shear stress from waves depends on wave height and period. The question is whether the contribution of shear stress from waves to total shear stress is significant. If so, this would explain the reason why dunes in the estuarine reach of the Vistula River are smaller and shorter than results from calculations by means of formulae, thus the influence of waves on bed shear stress cannot be neglected. For example KACZMAREK et al. (1991) proposed a model of wave-current boundary layer which enables, on the basis of wave parameters, evaluation of the variation of friction velocity during the wave period. For example. the friction velocity for a wave which is 0.5 m high and has a period of 3 s can vary from 0.014 to 0.023 m/s depending on an equivalent sand roughness k_s. Such high values of friction velocity from waves, if added to friction velocity due to river current, can considerably modify the total shear stress. However, further studies of wave action on bed shear stress, particularly over a dune covered bed in the estuary are necessary.

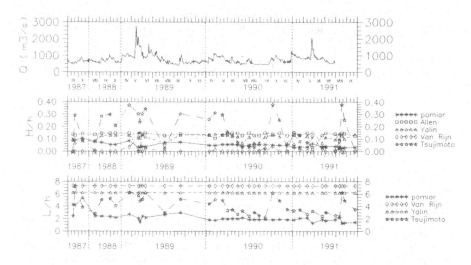

FIGURE 4 *Dune heights and lengths obtained from calculations are compared with the results of measurements.*

3.7   Bed shear stress vs. bed form geometry

This is a very interesting problem and so far unsolved. There exist different formulae connecting bed shear stress with the height and/or length of dunes. For example the theoretical one derived by YALIN (1964) or the empirical ones of ALLEN (1978), VAN RIJN (1984). TSUJIMOTO (1983) proposed the iterative procedure consisting of eight equations. As the final result dimensions of dunes and Chezy coefficient are obtained. The results of calculations by means of these and another formulae with those of field measurements in the estuary of the Vistula River are shown in Figure 4. Unfortunately, there is large scatter between results obtained with particular formulae. More important is that the results of measurements differ considerably from those calculated. In almost all cases dunes measured were about twice as short and twice as small as the average dimensions obtained from calculations with different formulae. Only in exceptional cases did measurements and calculations gave similar dimensions of dunes. Assuming that the formulae were verified with laboratory and field data, the discrepancy is due to the nature of flow in the estuarine reach of the river. The nonuniformity of flow and accompanying turbulent scales may have decisive significance in the creating of smaller and shorter dunes than would take place in uniform flow. It would therefore be very interesting to study this problem in a laboratory flume. Of course, the wave action on the bed forms through the bed shear stress, mentioned in preceding subsection, should be taken into account. One more factor which should not be ommited is lag time between the present flow characteristics, particularly bed shear stress, and response of dunes. It is well known that if flow variation is characterized with a certain time scale, the fully developed dunes corresponding to new flow conditions need much more time to adapt. This phenomenon was analysed and presented in papers by ALLEN (1978), FREDSOE (1979), SAWAI (1988), WIJBENGA et al. (1983) and TSUJIMOTO et al.(1990). It seems that the equations proposed by Tsujimoto et al. are the most convenient for practical application. They proposed two pairs of equations. The first describes the rate

413

of growth or decay of dunes when the water discharge variation is given. The second enables calculation of the new height and length of dunes after the period Δt.

$$\frac{dL}{dt} = \frac{\beta_L q_b}{(1-p)H}(1 - \frac{L - L_0}{L_e - L_0}) \tag{5}$$

$$\frac{dH}{dt} = \frac{\beta_H q_b}{(1-p)L}(1 - \frac{H}{H_e}) \tag{6}$$

$$L(t + \Delta t) = L(t) + (\frac{dL}{dt})_t \Delta t \tag{7}$$

$$H(t + \Delta t) = H(t) + (\frac{dH}{dt})_t \Delta t \tag{8}$$

where L and H are the dune lengths and heights respectively, β_L and β_H are the empirical constants ($\beta_H = 0.36$ and $\beta_L = 0.24$ for the growing stage of dunes, while $\beta_H = 0.72$ and $\beta_L = 0$ for the decaying stage of dunes), p is the porosity of sand $= 0.4$, L_0 the minimum dune length which appears first in dune development process from an initially flat bed, H_e and L_e are the fully developed dune height and length and q_b the sediment transport rate. It should be mentioned that the verification of the formulae was carried out in a laboratory flume and in case of natural rivers the values of β_L and β_H coefficients can differ.

It should be concluded that as yet, there is no reliable formulae for the calculation of fully developed bed form dimensions in the estuarine reach of a river. It was noticed that bed forms measured in the estuary of the Vistula River were always smaller and shorter than results from the formulae. Series of laboratory experiments have to be carried out to confirm author's findings from the field measurements. Additional analyses should be made to establish β constants to calculate the rate of growth or decay of dunes in natural estuaries.

3.8   Water depth and friction slope vs. bed shear stress

The total bed shear stress can easily be calculated for flow in an estuary when nonuniform flow is assumed. When other factors such as wind and wave action are neglected then the shear stress is

$$T_b = \rho g \cdot h \cdot I_f \tag{9}$$

where T_b is the bed shear stress, ρ the water density, h the water depth and I_f is the friction slope in an estuary.

3.9   Water depth and friction slope vs. resistance coefficient

This relation can be two-directional. The Chezy or Darcy-Weisbach resistance coefficients result from the water depth and friction slope given for a certain discharge. The first coefficient can be expressed as $C = U/\sqrt{(ghI_f)}$, the second being connected with the Chezy coefficient, C, through the relationship $f = 8/C^2$, where U is the mean flow velocity, g is the gravity and I_f is the friction slope. The inverse direction happens if the resistance coefficient is modified due to the growth or decay of bed forms. The reaction of the water depth or friction slope is immediate.

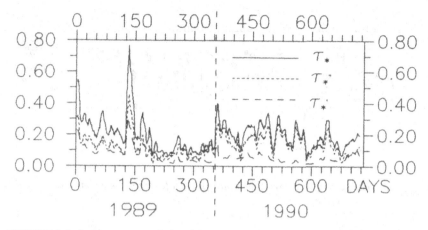

FIGURE 5 *Bed shear stress calculated for the estuarine reach of the Vistula River during the period of measurements. The contribution of the drag shear stress is indicated.*

3.10 Bed form geometry vs. resistance coefficient

One of the most important factors influencing the resistance to flow is the resistance from dunes covering river beds. A high contribution of shear stress from dunes to the total shear stress was observed in the estuary of the Vistula River, where it ranges from 60 to 80% of the total shear stress (see Figure 5). The Chezy coefficient for a river bed covered with dunes therefore differs from that determined for a flat bed. According to YALIN (1964a), the Chezy coefficient for a wavy bed can be evaluated by

$$C = \frac{C_F}{\sqrt{1 - \frac{H}{L}(ctg\varphi - \frac{H}{2h}C_F^2)}} \tag{10}$$

where C_F is the Chezy coefficient for a flat bed and

$$C_F = \frac{1}{\kappa}ln(11\frac{h}{k_s}) \tag{11}$$

where L and H are the length and height of dunes, h the water depth, φ the angle of repose of bed sand, k_s the equivalent sand roughness and κ the Von Karman constant. Yalin's equation and those proposed by VAN RIJN (1984), FREDSOE (1975), and TSUJIMOTO (1983) were used to calculate the Chezy resistance coefficient, C, in the estuarine reach of the Vistula River for the periods when the field measurements of bed forms were carried out. The results were transformed to the Darcy–Weisbach friction factor, $f = 8/C^2$ and compared with the results of calculations with the formulae published by HAQUE (1986) and VANONI (1975) (Figure 6). It can be seen that the deviation from f measured is not as important as in the case of bed form dimensions. It should be taken into account that in a natural estuary the resistance is due not only to bed forms, but also other factors, as can be seen in Figure 1.

415

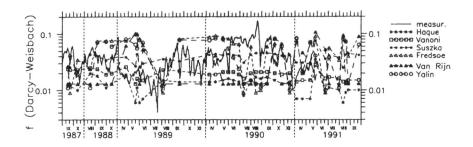

FIGURE 6 *Histogram of the Darcy-Weisbach friction factor measured and calculated by several authors.*

3.11   Water temperature vs. resistance coefficient

The water temperature effect on the resistance to flow is usually neglected if an open channel flow is considered. However, when long periods embracing different seasons are taken into account the water temperature can vary, from for instance, 0 to 22 C degrees, as in the case of the Vistula River. The direct effect of water temperature on the resistance coefficient, it means through the Reynolds Number, $Re = 4Uh/\nu$, can be ommited. Flow takes place in the fully rough zone (see FRENCH p.118) so the resistance coefficient does not depend on the Reynolds Number. However, bed form dimensions forming the main part of the resistance coefficient depend on the sediment transport rate which depends on the water temperature. This is seen explicitly in the Ackers–White formula, where the dimensionless particle size D_{gr} is a function of water viscosity, which in turn depends on water temperature. The effect of water temperature on bed form dimensions was already reported by VANONI (1975), but the explanation of this effect is still hypothethical.

4   Conclusions

The paper concerns the hydraulics of non-tidal river estuaries. As an example the estuarine reach of the Vistula River is assumed. The peculiarity of such estuaries is the interrelation of numerous factors, whose significance in tidal estuaries and especially in rivers beyond the influence of backwater, can be neglected. The author has tried to describe roughly the interrelations between particular factors such as river discharge, sea level, wind action, waves, bed shear stress and bed configuration. The bed forms, in turn, influence the value of the coefficient of resistance to flow. These factors can be treated as secondary and neglected in the case of a tidal estuary, but they may become dominant in case of a non-tidal estuary. Any precise description of these interrelations would be very difficult to prepare and was not the purpose of this paper, which should be treated merely as preparatory to further studies.

An attempt has been made to formulate the preliminary conclusions resulting from the analysis of measurements and observations carried out in the estuarine reach of the Vistula River. The most important are as follows:

   1. The hydraulics of non-tidal estuaries is governed by a sophisticated system of

416

interrelations of physical factors such as: river discharge, sea level, wind, waves, bed forms and water temperature.

2. In some of the relations the lag time has to be taken into account. A typical example is the lag time between the shear stress variation and the response of bed form dimensions, which need time to adjust to new flow conditions.

3. The resistance coefficient due to bed forms covering the bottom of the river estuary constitutes the most important part of the total resistance coefficient. The resistance depends on the relative height and steepness of dunes.

4. The bed form dimensions in the estuarine reach of the Vistula River are smaller than those calculated by means of formulae.

5. Waves and wind action in the estuary seem to be the main factors modifying the bed shear stress. This, in turn, creates variation in dune dimensions.

6. The problem considered in the paper needs further studies in field and laboratory conditions.

REFERENCES

ALLEN J.R.L.(1978), "Computational Models for Dune Time–Lag: Calculations Using Stein's Rule for Dune Height", *Sedimentary Geology*, 20, pp. 165–216.

BUCHHOLZ W. (1988), "Hydrodynamical Studies of River Estuary of Lower Odra", *Marine Institute Transactions*, No.699, Wydawnictwo Instytutu Morskiego, pp.7–27 (in polish).

FREDSOE J.,(1979),"Unsteady Flow in Straight Alluvial Streams: Modification of Individual Dunes", *Journal of Fluid Mechanics*, Vol.91, Part 3, pp. 497–512.

FRENCH R.H., (1985), *Open-Channel Hydraulics*, McGraw-Hill Company, New York.

HAQUE M.I.,(1986), "Form Resistance in Open-Channel Flow in the Presence of Ripples and Dunes", *Encyklopedia of Fluid Mechanics*, Gulf Publ. Co., Vol.2, Ch.4., pp.71-97.

KACZMAREK L., OSTROWSKI R. (1991),"Modelling of Wave-current Boundary Layers with Application to Surf Zone", *Hydrotechnical Archives*, Gdańsk, Vol. 39, No.1-2, pp 1-23.

KOŁODKO J. (1990),"On the Dynamics of Wind–affected River Flows", Proc. of the Polish–Bulgarian Seminar on Advanced Problems and Methods in Hydroengineering, Sofia, October 3–5, pp.53-66.

NIHOUL J.C.J., (1977), "Three-dimensional Model of Tides and Storm Surges in a Shallow Well-mixed continental Sea", *Dynamics of Atmospheres and Oceans*, 2, pp.29–47.

SAWAI K., (1988), "Transformation of Sand Waves due to Time Change of Flow Conditions", *Journal of Hydroscience and Hydraulic Engineering*, Vol.5, No.2 pp.1-14.

SUSZKA L., (1990),"Characteristics of Bed Forms in the Estuarine Reach of the Vistula River", Proc. of the Polish-Bulgarian Seminar on Advanced Problems and Methods in Hydroengineering, Sofia, October 3–5, pp.99-112.

SUSZKA L., (1991),"On the Influence of Dunes on the Resistance to Flow in the Estuarine Reach of the Vistula River", XXIV Congress IAHR, Madrid, Vol.A, pp.253-260.

TSUJIMOTO T., NAKAGAWA H., (1984), "Unsteady Behaviour of Dunes", International Conference on Hydraulic Design in Water Resources Engineering, 11-13 April, University of Southampton.

TSUJIMOTO T., MORI A., OKABE T., OHMOTO T., (1990), "Non-Equilibrium Sediment Transport: A Generalized Model", *Journal of Hydroscience and Hydraulic Engineering*, Vol.7, No.2, pp.1-25.

VANONI V.,(1975), *Sedimentation Engineering*, ASCE.

VAN RIJN L.C., (1984), "Sediment Transport. Part III: Bed Forms and Alluvial Roughness", *Journal of Hydraulic Engineering*, ASCE, Vol.110, No.12, pp.1733-1753.

WHITE W.R., PARIS E., BETTESS R., (1979), "A New General Method for Predicting Characteristics of Alluvial Streams", Report No. IT 187, Hydraulics Research Station, Wallingford.

WIJBENGA J.H.A, KLAASSEN G.L., (1983), "Changes in Bedform Dimensions under Unsteady Flow Conditions in a Stiaight Flume", Spec. Publs Int. Ass. Sediment., 6, pp.35-48.

YALIN M.S., (1964), "Geometrical Properties of Sand Waves", *Journal of Hydraulic Division*, ASCE, Vol.90, No.5, pp.105-119.

YALIN M.S., (1964a), "On the Average Velocity of Flow over a Movable Bed", *L'Houille Blanche*", No.1, pp.45-51.

35 Prediction of the siltation in the northern entrance of the Suez Canal for the second development stage

R. M. Hassan, E. El Sadek, L. C. van Rijn
and E. W. Bijker

ABSTRACT

The northern entrance of the Suez Canal consists of two intersecting approach channels; the deeper eastern channel and the western channel. A study is undertaken for the development of the Suez Canal (second development stage). There are four development alternatives to enable larger loaded tankers with draughts 56, 62, 68 and 72 feet for the four development schemes, respectively to use the Suez Canal. The development projects will require the deepening and widening of the eastern channel. An important part of the study deals with the prediction of the siltation rates in the eastern approach channel of the Suez Canal for the proposed development alternatives.

The present paper deals with the development and application of the SUTRENCH-model to predict the siltation rates in the eastern approach channel of the Suez Canal for the four proposed development alternatives. The SUTRENCH-model is a two dimensional detailed sediment transport model to be applied to pré-defined streamtubes covering the area of interest. As the model was originally developed for cohessionless sediment, the first stage of this research work is the development of the model to be applied to the mud sediment of the project area. The calibration of the model is performed for present conditions based on the recorded dredging quantities and the siltation heights calculated from soundings. The model is then used in the prediction of the siltation rates in the eastern channel for the four development alternatives. A senstivity analysis is performed for the SUTRENCH-model to test its senstivity to variations in wave heights and deviations in channel cross-section dimensions.

1. Introduction

The northern entrance of the Suez Canal (Figure 1) consists of two intersecting approach channels; the deeper eastern channel extending from Port Said bypass channel and extending offshore and the western channel passing by Port Said harbour. The western entrance channel of the Suez Canal is protected by two breakwaters. The western breakwater consists of a submerged part and an emerged part. The relatively deep Port Said bypass approach channel, eastern channel, was dredged in 1980 (first development stage) intersecting the western channel and extending about 12 km offshore without any sheltering effects.

A study is undertaken for the second development stage for the Suez Canal. There are four development alternatives that require to be evaluated. The development of the Suez Canal will require the deepening and widening of the eastern channel to enable the traffic of larger sized loaded tankers with draughts up to 56, 62, 68 and 72 feet for the four development alternative schemes, respectively to use the Suez Canal. The present paper deals with the prediction of the siltation rates in the eastern channel using existing data and numerical modelling techniques to determine the annual maintenance dredging for the development alternatives up till year 2015.

2. Environmental and sedimentological conditions

2.1 Flow pattern

Currents in the Port Said area are moderate and generally variable in direction and magnitude and have a large spatial variation, especially in the vertical direction. The only current measurements performed at the northern entrance of the Suez Canal were in 1936-1937 and 1938-1939. According to these measurements, the effect of the relatively recent dredged eastern approach channel is not included in the flow pattern.

When a current crosses an approach channel inclined to the current direction, it is refracted by the channel. Deviation in the flow pattern and change in streamtube width will occur resulting in the change of the average current velocity. Therefore, the ratio of the average current in the channel to the average current approaching the channel is a function of the current approaching angle relative to the channel direction and the difference in depth between the channel and the natural sea bed. A simplified current refraction approach (1) is developed to be applied to the northern entrance, using the previously mentioned field current measurements to determine the flow pattern for the present conditions, and for the four development alternatives. The year is divided into two seasons, summer (from May to October) and winter (from November to April). In order to minimize the error resulting from this simplified approach (1), smoothing of streamlines is performed to account for the sudden increase or decrease in the width of the streamlines. Figure 2 shows the flow pattern for the present summer condition. Flow patterns for the four development alternatives are also performed.

2.2 Wave height patterns

Deep water wave climate along the Egyptian Mediterranean are available from Summary of Synoptic Meteorological Observations. The analysis of data is illustrated in Figure 3 for the four seasons. Prevailing wave periods vary from 6 to 8 seconds.

Wave propagation runs using geo-optical approximation refraction models with a very coarse grid system were performed for the existing conditions for winter and summer seasons.

420

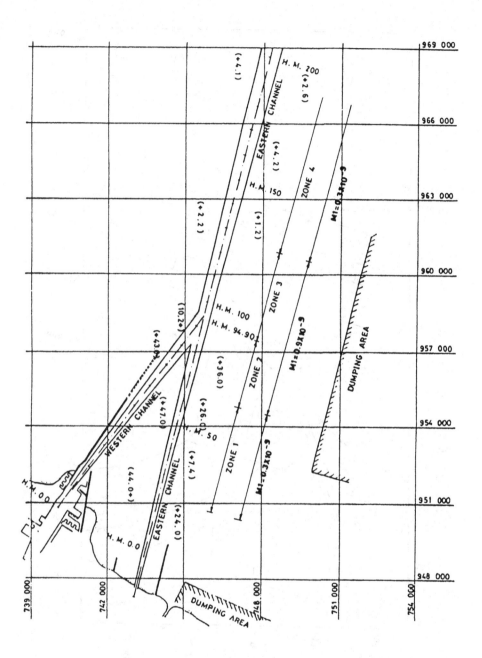

FIGURE 1   *Layout of the northern entrance of the SUEZ Canal showing median grain diameters of sea bed materials (µm), representative zones for calibration and M1 values*

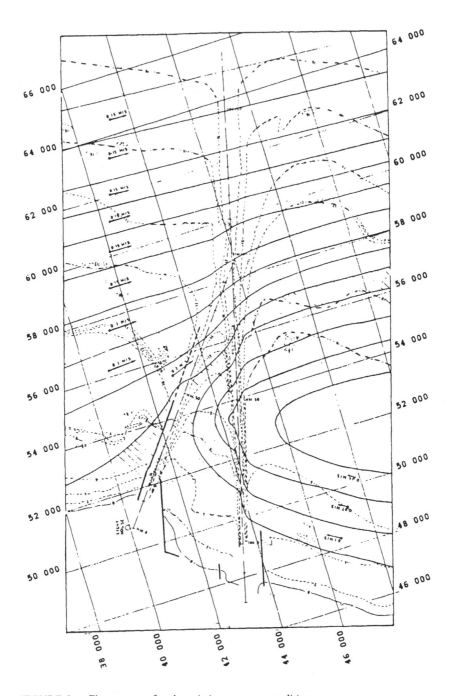

FIGURE 2   *Flow pattern for the existing summer condition*

For the winter season, the significant deep water wave height with 50 percent exceedance is taken as 1.2 meters and wave period 7 seconds for the north-west direction. For the summer season, the corresponding significant deep water wave height is taken 0.8 meter and wave period 6 seconds for the north-north-west direction. No wave height data are available to verify the predicted wave height patterns.

Due to the lack of field data, wave heights for the first and second development alternatives are taken as the present conditions. Due to the relatively deeper eastern channel for the third and fourth development alternatives, 10% increase in wave heights are considered.

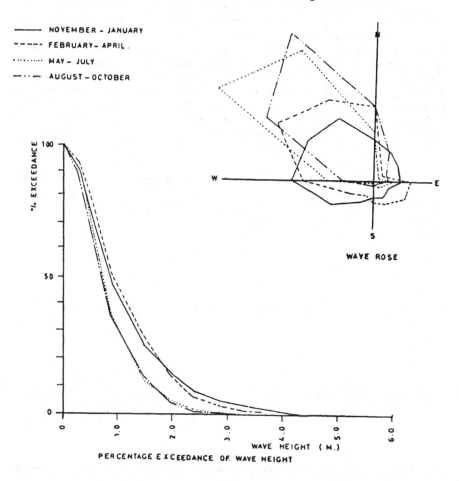

FIGURE 3 *Percentage of exceedance of wave heights and wave rose in deep water*

2.3 Sea bed materials

The sea bed material at the northern entrance of the Suez Canal is generally very fine and ranges from dark-grey very soft silty clay to sandy silt. The median grain sizes distribution at the northern entrance are shown in Figure 1.

2.4 Siltation rates, heights and dredging quantities

Siltation rates at the eastern channel are determined from periodic bathymetry surveys for different hectometers along the channel during periods not interrupted by maintenance dredging from 1980 to 1989. The periodic surveys are limited to the bottom width of the channel. Thus the siltation rates are only valid for the surveyed parts of the channel and are expected to be less than the maintenance dredging quantities. The average siltation height per year are thus determined along the channel based on the surveyed width. Based on the siltation heights and the nominal channel bottom widths, the average siltation rates along the eastern channel are plotted in Figure 4. On the same figure, the average annual dredging quantities are plotted assuming linear distribution of recorded dredging quantities in a recorded stretch. The average yearly total recorded dredging quantities for the eastern channel is 7.2 million cubic meters. The average yearly total siltation rates based on the channel design bottom width is 4.8 million cubic meters, i.e. two third of the corresponding recorded dredging quantities. By considering an increase of 20 to 40 percent in the siltation rates due to side slope effect, the siltation rates is equal to 6.2 million cubic meter per year. This discrepancy can be due to the underestimation of the siltation rates limited to the bottom width of the channel only or probably the overestimation of the dredging records.

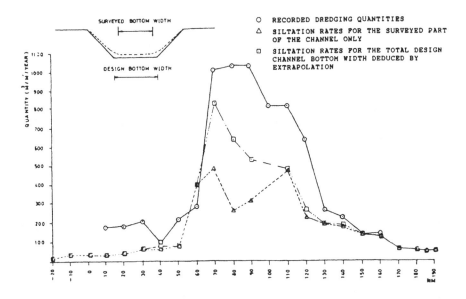

FIGURE 4   *Comparison between the calculated siltation rates and the recorded dredging quantities for the eastern channel*

2.5 Suspended sediment

The only data available of the suspended sediment concentration was from field survey performed in 1976 (2). Concentration of suspended solids measured during July were generally less than 20 ppm (weight by volume). Concentration up to 300 ppm were measured during stormier conditions in February.

3. Mathematical model for sedimentation in channels (SUTRENCH-model)

The SUTRENCH-model is a two-dimensional-vertical mathematical model for sediment transport by currents and waves that was developed at Delft Hydraulics (6),(8) to study morphological processes, specially the prediction of sedimentation in dredged channels.

The basic equation used in computing the concentration is the convection-diffusion equation assuming steady conditions and neglecting the transport by longitudinal mixing, expressed as:

$$\frac{\partial}{\partial x} (buc) + \frac{\partial}{\partial z} [b (w-w_s) c] - \frac{\partial}{\partial z} \left(b\epsilon_{s,cw} \frac{\partial c}{\partial z}\right) = 0 \tag{1}$$

where u = longitudinal velocity at height z above the bed; c = sediment concentration; w = vertical flow velocity; w_s = particle fall velocity of suspended sediment; $\epsilon_{s,cw}$ = sediment mixing coefficient by currents and waves; b = flow width; x,z = longitudinal and vertical coordinates, respectively. The following boundary conditions are to be specified:

Flow domain : initial bed levels, water depths, flow widths, wave characteristics, particle fall velocity, effective bed roughness, size, composition and porosity of bed material.

Inlet boundary : discharge, flow velocities, mixing coefficients, concentrations.

Water surface : net vertical transport is assumed to be zero.

Bed surface : bed concentration or upward sediment flux at bed as a function of local hydraulic and sediment parameters.

The bed boundary condition is specified at a small height ($z=a$) above the mean bed. Using this approach, the bed concentration or the upward sediment flux can be represented by its equilibruim value assuming that there is an almost instantaneous adjustment to equilibruim conditions close to the bed.

The bed concentration for silt is calculated according to the approach of Partheniades (3) as follows:

$$C_a = M1 \left(\frac{\overline{\tau}_{b,cw}' - \overline{\tau}_{cr,silt}}{\tau_{cr,silt}}\right) \tag{2}$$

in which C_a = bed boundary concentration; a = reference level above bed; $\overline{\tau}_{b,cw}' = \mu_c\overline{\tau}_{b,c} + \mu_w\overline{\tau}_{b,w}$ = effective bed-shear stress; $\overline{\tau}_{b,c}$ = current-related bed shear stress; $\overline{\tau}_{b,w}$ = wave related bed shear stress according to linear wave theory (based on significant wave height, length and period); μ_c = current-related efficiency factor; μ_w = wave related efficiency factor; $\overline{\tau}_{cr,silt}$ = critical bed shear stress for silt; M1 = material constant.

The upward sediment flux at the bed for the case of silt is defined as (4), (5), (6), (7):

$$E_a = - \left(\epsilon_s \frac{\partial c}{\partial z}\right)_{z=a} = w_sC_a = M1.w_s \left(\frac{\overline{\tau}_{b,cw}' - \overline{\tau}_{cr,silt}}{\tau_{cr,silt}}\right) \tag{3}$$

Either equation (2) or (3) can be selected as input parameters. Bed level changes are computed from a cross-section integrated sediment continuity equation; in the form:

$$\frac{\partial}{\partial t} (b Z_b) + \frac{1}{\rho_s}(1-p) \frac{\partial}{\partial x} (S_s + S_b) = 0 \tag{4}$$

in which t = time; b = flow width; z = bed level above a horizontal datum; p = porosity factor; S_b = suspended load transport; and S = bed load transport. The suspended load transport is computed as:

$$S_s = b \int_a^h u\ c\ dz \tag{5}$$

where h = water depth. In case of silty conditions, there may be a near-bed transport of fluid mud and the near bed transport of fluid mud $S_{b,fm}$ (m³/s/m) is represented by (6):

$$S_{b,fm} = M2\ (\bar{\tau}'_{b,cw} - \bar{\tau}_{cr,silt})\ (U_c + U_w) \tag{6}$$

in which $\tau_{b,cw}$ = bed-shear stress related to currents and waves; U_c = near-bed current velocity (= $0.05\bar{U}$); $\bar{U}$ = depth-averaged velocity; U_w = near-bed wave induced drift velocity; M2 = calibration coefficient.

4. Calibration of the SUTRENCH-model

The SUTRENCH-model is applied to the eastern approach channel using pre-determined streamtubes for the current refraction diagrams for the present conditions. The year is divided into summer and winter seasons. In the calibration of the SUTRENCH-model, the eastern channel is divided into four representative zones (Figure 1) chosen according to the trend of the siltation rates. The cross-sections of the channel at the different zones are determined from bathymetry surveys and the sea bed material characteristics (sizes, density of sea bed material, wet bulk density, porosity) are determined from surveys. The wave heights at the different locations are determined from the runs of the wave propagation model.

Three parameters are used for the calibration of Sutrench-model; the fall velocity, w_s, critical shear stress for erosion, $\tau_{cr,silt}$, and the material constant, M1. The fluid mud transport was assumed to be zero (M2 = 0). Values of the fall velocity for similar cases as for Port Said approach area are in the range of 0.1 to 1 mm/sec (9). Accurate values of the fall velocity can be only be obtained from in-situ field measurements, which were not available. Several computer runs are performed to determine the fall velocity as a part of the calibration procedure, yielding reasonable siltation estimates for all zones for a fall velocity of 0.3 mm/sec. Based on these trials, it was decided to apply a constant fall velocity of 0.3 mm/sec. In reality, there may be variations in the fall velocity along the channel because of differences in sizes of the bed materials. The differences are, however, small and do not justify a change of the fall velocity because other processes such as flocculation and presence of organic materials are important.

The critical bed shear stress for erosion should be in the range of 0.1 to 0.3 N/m² for consolidated silt deposits (9). In this calibration procedure, a value of 0.2 N/m² was chosen. More accurate values can be obtained from flume tests using natural mud-samples taken from the channel bed.

The material constant, M1, should be in the range of 10^{-8} to 10^{-10} m/sec (9) depending on the type of bed material. Since the fall velocity (w_s) and the critical bed shear stress ($\tau_{cr,silt}$) are taken constant, the M1 constant can vary along the channel to cope with variations in fall velocity, critical shear stress, etc., i.e. errors made in the estimation of the fall velocity, critical bed shear stress and the hydraulic parameters will be represented by variation of the M1 constant, which is a basic feature of any model calibration.

426

Calibration is performed for the different zones of the eastern channel based on the calculated yearly average siltation height and the concentration of the suspended sediment. The values of the M1 constant for different zones are presented in Figure 1.

Figure 5 shows the siltation heights calculated from the model together with the siltation heights calculated from periodic surveys.

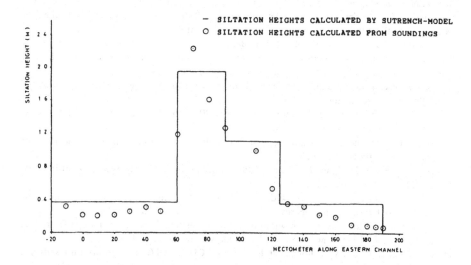

FIGURE 5   *Comparison between siltation heights calculated by the SUTRENCH-model and siltation heights from soundings*

5. Prediction of siltation rates for the development alternatives

After calibration, the SUTRENCH-model is applied for the prediction of the siltation rates in the eastern channel for the four development alternatives using pre-determined streamtubes covering the study area. The year is divided into two seasons each of six months. In order to predict the siltation rates in the eastern channel, the channel is divided into six representative zones. The choice of the representative zones is based on the expected trend of the siltation in the eastern channel. One cross-section is selected to represent each zone, and the siltation rate and height are assumed to be constant throughout the zone.

For each development alternative, there are six representative zones for summer and winter seasons. Thus forty eight different input data are prepared for the development alternatives.

The siltation rates in the eastern channel for the four development alternative schemes are predicted from the SUTRENCH-model as follows:
Alternative 1:   8.5   millions in-situ cubic meters per year;
Alternative 2:   9.0   millions in-situ cubic meters per year;
Alternative 3:   11.3   millions in-situ cubic meters per year;
Alternative 4:   13.2   millions in-situ cubic meters per year,
and may be compared to the 7 millions in-situ cubic meters per year for the present existing condition.

427

6. Senstivity analysis

A senstivity analysis was performed to determine the effect of the accuracy of the input data to the SUTRENCH-model (mud version). The senstivity of the model is first tested for the percentage increase in wave height. For the fourth development alternative, the wave heights are assumed to be 10% more than the present existing condition. To test the sensitivity of the SUTRENCH-model, a 20% increase in wave heights are considered. For a 10% increase in wave height for the fourth development alternative (20% increase in wave height than the existing condition) there is an increase in the siltation rate by 15%.

Secondly, the senstivity of Sutrench-model to deviation in channel dimensions due to dredging and/or survey tolerances, channel evolution and siltation is tested. The following results were obtained by considering a hypothetical channel cross-section:

- A decrease in the channel bottom depth from 20 m to 19 m resulted in a decrease of 5.8% in the sedimentation volume.
- An increase in the channel bottom depth from 20 m to 21 m resulted in an increase of 2.9% in the sedimentation volume.
- A decrease in the channel bottom width from 200 m to 150 m resulted in a decrease of 4% in the sedimentation volume.
- An increase in the channel bottom width from 200 m to 250 m resulted in an increase of 3.1% in the sedimentation volume.
- A decrease in the channel top width by 50 m by changing the channel side slope from 25:1 to 20:1 resulted in a decrease of 4.2% in the sedimentation volume.
- An increase in the channel top width by 50 m by changing the channel slope from 25:1 to 30:1 resulted in an increase of 4.8% in the sedimentation volume.

The above senstivity analysis shows that the SUTRENCH-model (mud version) is not sensitive to small and moderate deviation in the channel cross-section dimensions.

7. Conclusions

The SUTRENCH-model has been developed to simulate changes in the level of the sea bed (mud) of the Port Said area. The calibration of the SUTRENCH-model (mud version) is performed successfully for the present existing conditions (summer and winter) based on siltation heights calculated from periodic surveys. The calibrated model is applied for four development alternative schemes for the eastern approach channel after including the effect of the deepening and widening of the channel on the current and wave patterns, yielding siltation rates in the range of 8.5 to 13.2 millions in-situ cubic meters per year.

A senstivity analysis is performed for the SUTRENCH-model (mud version). The model is sensitive to variations in wave heights but not sensitive to small or moderate deviation in the channel cross-section dimensions.

REFERENCES

[1]   Hassan, R.M., (1992), Flow Across Oblique Channels, Journal of Engineering and Applied Science, Cairo University, (under publication).

[2]   Hydraulics Research Station, Wallingford, (1976), Siltation in Port Said Approach Channel, Ex 751.

[3]   Partheniades, E., (1965), Erosion and Deposition of Cohesive Soils, Journal of the Hydraulic Division, ASCE, Vol. 91, No. HY1.

[4]   Rijn, L.C., (1984), Sediment Transport, Part I: Bed Load Transport, Journal of Hydraulic Engineering, Vol. 110, No.10, pp.1434-1456.

[5]   Rijn, L.C., (1984), Sediment Transport, Part II: Suspended Load Transport, Journal of Hydraulic Engineering, Vol. 110, No. 11, pp. 1613-1641.

[6]   Rijn, L.C., (1985), Two Dimensional Vertical Mathematical Model for Suspended Sediment Transport by Currents and Waves, Report S 488-IV, Delft Hydraulics Laboratory, Delft, The Netherlands.

[7]   Rijn, L.C., (1985), Initiation of Motion, Bed Forms, Bed Roughness, Sediment Concentrations and Transport by Currents and Waves, Report S487-IV-Delft Hydraulics laboratory, Delft, The Netherlands.

[8]   Rijn, L.C., (1986), Sedimentation of Dredged Channels by Currents and Waves, Delft Hydraulics Laboratory, Communication No. 369.

[9]   Rijn, L.C., (1989), Handbook of Sediment Transport by Currents and Waves, Delft Hydraulics Laboratory, Delft, The Netherlands.

36 Simulation of near bank aggradation and degradation for width adjustment models

S. E. Darby and C. R. Thorne

ABSTRACT

The concept of basal endpoint control states that the long-term rate of retreat or advance of riverbanks, and hence adjustment of channel width, is determined by the sediment budget in the near bank zone. Previous attempts to simulate width adjustment by coupling one-dimensional aggradation-degradation models with bank stability algorithms are limited firstly by the assumption that bed scour and fill are distributed evenly over the cross-section and secondly because they neglect bank accretion mechanisms and lateral sediment transport fluxes. Improved models of width adjustment incorporating bank processes must be capable of simulating erosion and deposition specifically in the near bank zone.

In the model presented here, a method [30] to solve the shallow water flow equations, and so account for lateral shear stresses which significantly influence the flow in the near bank zone, is used to calculate the lateral distribution of discharge. The flow is then divided into segments, allowing the lateral variation of flow hydraulics and sediment transport over the cross-section to be simulated. The influence on the predictions of near bank scour and fill distributions of incorporating the lateral sediment transport fluxes in the model are also assessed, using sensitivity analyses. The more realistic estimates of near bank aggradation and degradation using the new model suggest that this approach provides a useful framework for incorporating bank process algorithms into the model to allow development of a predictive model of river channel width adjustment.

1. Introduction

Recently, attempts have been made to simulate river channel width adjustment by coupling one-dimensional aggradation-degradation models with bank stability

431

algorithms [1, 6, 22]. The theoretical basis for such models is the concept of basal endpoint control, which shows that long-term bank line retreat and advance is controlled by the sediment balance in the area adjacent to the bank, because bank retreat proceeds by combinations of both direct fluvial entrainment of bank material and mass failure of the river banks. Mass failure occurs when a critical threshold of stability is exceeded due to steepening of the bank angle by lateral erosion or by near bank degradation increasing the height of the bank. The near bank sediment balance determines the rate of degradation or aggradation and, thus, is strongly coupled to the mass stability of the river bank [28]. Bank accretion is also directly influenced by the near bank sediment balance, as accretion occurs when the quantity of sediment supplied to the near bank zone exceeds that removed. Since the sediment fluxes (Figure 1) are controlled primarily by the intensity of the flow in the near bank zone at the base of the bank, the rate of retreat or advance of any river bank is determined primarily by the near bank flow intensity. It is apparent then, that models using the basal endpoint control concept must be able to realistically simulate the lateral distribution of flow and sediment transport across the full width of a channel section, with particular emphasis on the near bank zone.

However, the previous approaches to width adjustment modelling based on sediment budgeting are limited in two main ways. Firstly, these approaches use one-dimensional sediment transport models which assume that aggradation and degradation are distributed uniformly across the bed of the channel cross-section, even though in natural river channels the magnitude of aggradation and/or degradation varies across the width of the channel in response to the lateral distribution of the flow and sediment transport. As the concept of basal endpoint control shows the rates of retreat and advance of the river banks, and thus width adjustment, to be controlled by the sediment budget in the *near bank* zone, it is of critical importance to predict the distribution of aggradation and degradation across the full width of the channel [2]. Secondly, previous approaches have neglected lateral exchanges of sediment occurring between the near bank and central flow regions (Figure 1).

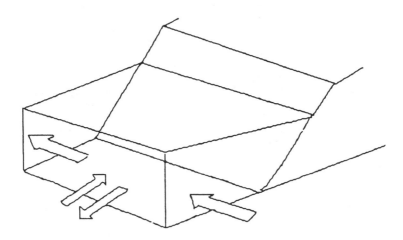

FIGURE 1 *Schematic representation of longitudinal and lateral sediment fluxes in the near bank zone*

432

As an initial step towards the development of a fully process based width adjustment model based on an aggradation-degradation model coupled with bank stability algorithms, a one-dimensional model applicable to straight river channels has been formulated, to predict more realistically the distribution of aggradation and degradation in the critical near bank zone. The model is based on the calculation of the lateral distribution of the flow at a cross-section and also includes lateral exchanges of sediment between the near bank and central flow regions. Sensitivity analyses have been performed to determine the influence of the lateral exchange mechanisms on the predicted distribution of aggradation and degradation.

2. Model development

2.1 Hydraulics

In order to predict the discharge in the near bank zone an approach based on the Lateral Distribution Method (LDM) developed by Wark et al. [30] was used to solve the depth integrated form of the shallow water equations (1) for the lateral distribution of unit discharge, q, at each model cross-section :

$$gDS - \frac{Bfq^2}{8D^2} + \frac{\delta}{\delta y}\left(Vt\,\frac{\delta q}{\delta y}\right) = 0 \tag{1}$$

where g = acceleration due to gravity, D = local flow depth, S = slope, f = Darcy-Weisbach friction factor, Vt = eddy viscosity and y = lateral co-ordinate. The use of equation (1) is an advance over methods which do not take into account the lateral shear stresses when calculating the lateral distribution of the flow such as stream tubes [19, 21] or Keulegan type splits [17]. This advance is of fundamental significance in this study, where the aim is to predict the hydraulics and sediment transport in the near bank zone, because it is in the near bank zone that the lateral shear stresses are greatest. The lateral shear stresses are modelled using the eddy viscosity concept. A simple but reliable eddy viscosity model appears to be one based on the bed roughness turbulence (J. B. Wark, personal communication, 1991), in which :

$$Vt = NEV\,(gDS)^{0.5}\,D \tag{2}$$

where, NEV = non-dimensional eddy viscosity coefficient, usually taken as about 0.16. The factor B relates the stress on an inclined surface to the stress in the horizontal plane [30], and is given by :

$$B = (1 + S_x^2 + S_y^2)^{0.5} \tag{3}$$

where, S_x = longitudinal bed slope and S_y = lateral slope.

At each cross-section, the slope and water surface elevation are required as input data for equation (1). These are obtained using the LDM in conjunction with a simple backwater routine based on the standard step method [8]. The total discharge obtained using the LDM is usually not identical to that used to generate the backwater curve, but these differences have been found to be minor. The LDM developed by Wark et al. assumes uniform flow and, therefore, uses the longitudinal bed slope as input to equation (1). However, the backwater routine solves the water surface profile for gradually varied flow. In these circumstances the longitudinal water surface slope determined from the backwater calculations is used as input to equation (1), though the

433

bed slope must still be used to calculate the B factor (J. B. Wark, personal communication, 1991).

Direct field observations have indicated that, for application of the concept of basal endpoint control, the near bank zone may be defined as the region extending a distance of about two bank heights from the channel margin towards the centre of the channel. Adopting this convention, the channel is divided into 3 flow segments at each cross-section. The unit discharge distribution is integrated laterally to give the mean unit discharge in each of the flow segments. The geometry of each segment then provides the remaining input data required for application of the sediment transport equations.

2.2 Sediment transport

The sediment continuity equation can be written for each individual flow segment :

$$\frac{\delta z}{\delta t} + \frac{1}{1-\lambda} \frac{\delta qs}{\delta x} = qlat \tag{4}$$

where, z = elevation of the bed, t = time, x = longitudinal distance along the channel, λ = porosity of the sediment and qs = total longitudinal volumetric sediment transport flux which may be determined from the Engelund-Hansen sediment transport equation [11] :

$$qs = \frac{q^2 S^{1.5}}{20 \ g^{0.5} \ D^{0.5} \ (SG-1)^2 \ d_{50}} \tag{5}$$

where, q = unit discharge, S = energy slope, g = acceleration due to gravity, D = flow depth, SG = specific gravity of the sediment, and d_{50} = median sediment grain diameter. In equation (4) the lateral exchange of sediments across flow segment interfaces is accounted for by the lateral outflow term which is numerically equal to the total volume of sediment transported across the segment interface per unit time. In this paper only the transport of uniform non-cohesive sand sized sediments is considered.

An aggradation-degradation model uses the solution of the sediment continuity equation to predict the change in bed elevation during a time step. In most aggradation-degradation models the procedure is usually to divide the longitudinal profile of the channel into finite reaches, and then compute the longitudinal sediment load at each cross-section. The bed elevation is then computed by solving the sediment continuity equation at the end of each time step. The approach adopted herein is similar, but the channel cross-section is also laterally divided into a number of flow segments.

A numerical solution scheme is required to solve the sediment continuity equation. An explicit finite difference scheme is used in this study. Equation (4) is written in finite difference form :

$$\Delta z = \frac{\Delta t}{1-\lambda} \left(\frac{qlat}{\Delta x} - \frac{\Delta q s}{\Delta x} \right) \tag{6}$$

where :

$$\frac{\Delta qs}{\Delta x} = \frac{qs^t_{j-1} - qs^t_{j+1}}{2\Delta x} \tag{7}$$

In equation (7), the subscript, j, represents the co-ordinate of an individual cross-section, while t represents the time step counter. By applying equation (6) in each model flow segment and at each cross-section, at the end of each model time step it is possible to use the calculated aggradation or degradation depths to update the cross-sectional geometry throughout the length and width of the model river. The updated geometry is then used as the starting point for the calculation of the water surface profile in the next time step, and the hydraulics and sediment continuity equations may again be solved. By repeating this sequence of computations for a series of time steps, it is possible to simulate the evolution of the channel geometry over time. In the following sections, expressions to calculate the net lateral outflow of sediment (net lateral sediment exchanges) from each flow segment are formulated.

2.2.1. Lateral Suspended Load Fluxes Einstein [9] noted that it is not commonly appreciated that suspended load is continuously being deposited against the banks of alluvial channels which transport suspended load. Parker [23] suggested that the suspended load is driven towards the near bank zone from the central region by a turbulent diffusive flux caused by a lateral concentration gradient of suspended sediment. That is, a flux of suspended sediment is supposed to be driven down the lateral concentration gradient by the mechanism of turbulent diffusion. Since the concentration of suspended sediment at any point in the channel is a function of turbulent intensity, which is itself dependent on flow depth, then a continuous lateral variation of suspended sediment concentration is expected to be the norm in river channels, due to the variation in flow depth across the channel section. Parker suggested the lateral diffusive flux of sediment is given by :

$$F_L = \varepsilon_y \frac{\delta \zeta}{\delta y}$$
(8)

where, F_L = lateral volumetric flux, ε_y = the lateral turbulent diffusivity coefficient and $\frac{\delta \zeta}{\delta y}$ = depth integrated lateral concentration gradient of suspended sediment. The lateral eddy diffusivity coefficient is approximated using [29] :

$$\varepsilon_y = 0.25 \, k \, U_* \, D$$
(9)

where U_* = shear velocity, D = flow depth and k = Von-Karman constant, here set equal to 0.40.

The depth integrated concentration of suspended sediment may be expressed as :

$$\zeta = \int_0^D c(z) \, dz$$
(10)

where the concentration of suspended sediment, c, at any vertical co-ordinate, z, may be calculated using [10] :

$$c(z) = E \exp \int_0^z \frac{V_s}{0.0077 \, U_* \, D} \, dz$$
(11)

435

where V_s = sediment fall velocity and the coefficient E represents an entrainment, or reference, threshold for the suspended sediment, given by the near bed concentration of suspended sediment. Numerous empirical formulations for the boundary concentration exist. Parker used the formulation [10] :

$$E = a \left(\frac{U*}{V_s}\right)^3 \qquad (12)$$

where a = an empirical coefficient determined from a suitable data set. A recent data set published by Garcia and Parker [14] enabled the coefficient, a, to be estimated. Best fit against this data set was achieved with a = 0.0003274.

The procedure to solve for the lateral suspended load flux is as follows. The depth integrated concentration of suspended sediment is found in each of the flow segments at each of the model cross-sections, by substituting the geometrical and hydraulic data previously computed into equations (12), (11) and (10). The average lateral concentration gradients across the boundaries of each flow segment at a cross-section are then determined using the difference between the depth-integrated suspended sediment concentrations in each flow segment divided by the width of the near bank flow segments. The lateral diffusive flux is then calculated using equation (8), with the value of the eddy diffusivity (equation 9) determined at the flow segment boundary itself. The flux is positive when directed towards the right bank.

Parker did not consider the influence of secondary currents on the transport of suspended sediments in his analysis. However, in the model presented herein, an attempt was made to assess the significance of convection of suspended sediments. The lateral convective flux of suspended sediments at any vertical co-ordinate, z, is given by

$$F_c = V_y (z) * c (z) \qquad (13)$$

where $V_y (z)$ is the lateral velocity at vertical co-ordinate, z. The lateral velocity, and hence the convective flux, is positive when directed towards the right bank. For simplicity, it was decided to model the secondary flow distribution using a traditional model of secondary velocity structure for straight channels [15, 25] where 2 secondary current cells occupy the channel cross-section, with the surface water converging at the centre of the flow and the the near bed water diverging from the centre of the flow (Figure 2). Secondary velocities were assumed to have a constant magnitude throughout each of the entire near bed and near surface regions, with the boundary between the near bed and near surface regions defined at half the flow depth.

The magnitudes of the secondary velocities were defined as a specified fraction of the downstream velocity at each of the flow segment boundaries. The secondary velocity magnitude was allowed to vary between 0 and 0.1 of the downstream velocity [7, 16, 24]. By defining the secondary velocity magnitudes as a free parameter, the influence of the secondary circulation on lateral exchanges of sediments can be assessed using sensitivity tests, even if the physics of the secondary flow phenomenon is clearly not adequately represented by the formulation adopted here.

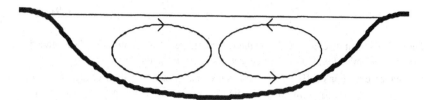

FIGURE 2 *Model representation of secondary cell structure*

The depth-integrated concentration is calculated for the near bed and near surface regions by applying equation (10), so that for the near bed region :

$$\zeta_{\text{near bed}} = \int_{0}^{D/2} c\,(z)\,dz \qquad (14)$$

and for the near surface region :

$$\zeta_{\text{near surface}} = \int_{D/2}^{D} c\,(z)\,dz \qquad (15)$$

In equations (14) and (15) c (z) is as given in equation (11). The net lateral convective flux is then given by :

$$F_c = F_{c\text{ near bed}} + F_{c\text{ near surface}} \qquad (16)$$

where the near bed and near surface region fluxes are calculated by substituting the specified values of the secondary velocities, together with the depth integrated concentrations calculated in (14) and (15), into equation (13).

2.2.2. Lateral Bedload Flux It has long been recognised that a lateral bedload component is generated by the gravitational drag on bedload moving in a longitudinal direction along a channel with a side slope [18, 26]. Parker [23] suggested that the gravitational bedload is related to the longitudinal bedload in a straight channel by :

$$q_{blg} = \frac{q_{bs}}{\mu}\frac{\delta D}{\delta y} = \frac{q_{bs}}{\mu}\tan\beta \qquad (17)$$

where, q_{blg} = lateral gravitational bedload flux, q_{bs} = longitudinal bedload flux, μ = a dynamic friction coefficient and β = side slope angle (positive if sloping down towards right bank). The longitudinal bedload flux is determined from the relation :

$$q_{bs} = q_s - q_{ss} \qquad (18)$$

where q_s is the total longitudinal transport flux, found using equation (5) and q_{ss} is the longitudinal suspended sediment flux, given by :

$$qss = q \zeta \qquad \qquad (19)$$

where q is the unit discharge and ζ is the depth integrated concentration of suspended sediment, determined previously from equations (10), (11) and (12).

The dynamic friction coefficient, μ, represents a parameter describing energy loss due to intermittent grain to grain contact which occurs during the transport of sediment over a sediment bed. Experiment and theory suggests that this may be assumed to be equal to the tangent of the static friction angle of the sediment [3, 4, 5, 13]. Typical sand grains have been found to have a friction angle with tangent of about 0.6 [4, 5, 13], the value used in this study. The side slope angle is found by dividing the difference in mean depths between adjacent flow segments by the width of the near bank flow segment when the flux is directed from the banks towards the bed and, when the flux is directed from the central flow segment towards the banks, by dividing the difference in flow depth between adjacent flow segments by half the width of the central flow segment.

Parker did consider the influence of secondary currents on the lateral transport of bedload. He assumed that the ratio of the lateral bed force, F_s, induced by lateral currents to the longitudinal force, F_x, obeys the relation [12] where :

$$\frac{F_s}{F_x} = \left(\frac{V_y}{U}\right)^2 \qquad \qquad (20)$$

where V_y = near bed lateral velocity (positive if directed towards the right bank) at the flow segment boundary and U = longitudinal flow velocity at the segment boundary, given by $U = \frac{q}{D}$. It follows that the lateral bedload flux due to the near bed secondary flow is given by :

$$qblc = qbs \left(\frac{V_y}{U}\right)^2 \qquad \qquad (21)$$

The net lateral bedload flux across each of the flow segment boundaries is, therefore, given by :

$$qbl = qbs \left\{ \frac{\tan\beta}{\mu} + \left(\frac{V_y}{U}\right)^2 \right\} \qquad \qquad (22)$$

2.2.3. *Net Lateral Sediment Outflow*   The net lateral exchange of sediment across a flow segment boundary, qlat, is found by summing each of the lateral sediment fluxes at the flow segment boundaries, so that :

$$qlat = qbl + F_L + F_c \qquad \qquad (23)$$

In the model presented here, only three flow segments were used to sub-divide the channel, so that equation (23) is solved at only two points. It should be noted that an assumption is made in the derivation of equation (23), the necessity for which is introduced when the flow is divided into segments. The longitudinal transport fluxes in each of the flow segments are assumed to be directed downstream. However, the position of the boundary of the flow segments runs parallel to the banks. When the width of the channel changes in a downstream direction, the longitudinal sediment fluxes in each of the segments, and the boundary of the flow segments are not parallel.

438

This leads to an effective lateral transfer of sediment between segments where the longitudinal fluxes intersect the flow segment boundaries. This is a reasonable approximation when the rate of change of width with distance along the channel is small (<5%).

3. Results

Two model runs were made in order to demonstrate the application of the model, and to make an initial assessment of the influence on the predictions of aggradation and degradation in the near bank zone of including lateral sediment exchange processes. A fictitious river reach was modelled, using 10 cross-sections. The initial geometry and roughness distributions (identical at each model cross-section) were used with the following input data : Q = 500 cumecs, S = 0.005, d_{50} = 1mm, λ = 0.2, SG = 2.65, Bank height = 4.0m, secondary current velocity coefficient = 0.1, time step duration = 180s. The response of the channel to reducing the longitudinal sediment inflow at the upstream boundary to zero was examined. The results are displayed in Figures 3 and 4. The first model run included the lateral sediment transport flux mechanisms, the other accounted only for the longitudinal sediment transport fluxes (qlat = 0).

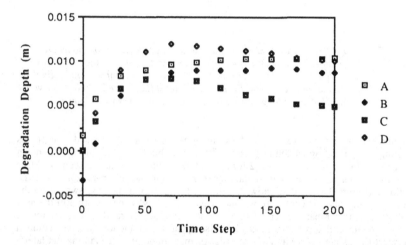

FIGURE 3 *Predicted degradation time sequences for near bank and central flow segments (curves A and B respectively) with lateral fluxes, and near bank and central flow segments (curves C and D respectively) without lateral fluxes*

Figure 3 shows the predicted degradation depths in the near bank and central flow segments at a representative model cross-section (No. 2). It is apparent that the model does indeed predict non-uniform degradation increments in the near bank and central flow zones. When lateral fluxes are neglected, the model predicts a more intense degradation rate in the central flow zone relative to the near bank zone (Figure 3, curves C and D). But the inclusion of lateral fluxes in the model analysis results in a relatively more intense degradation rate in the near bank zone, as compared to the central flow zone (curves A and B). It appears, therefore, that the predicted degradation sequences differ significantly when lateral sediment transport exchanges are considered.

439

This suggests that, at least for the model run conducted here, lateral sediment transport fluxes can significantly influence the predictions of the distribution of aggradation and degradation over the channel cross-section.

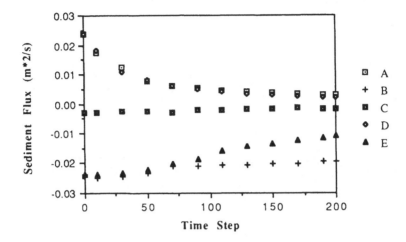

FIGURE 4   *Predicted near bank input and output sediment fluxes for model cross-section No. 2 with and without lateral sediment exchanges. Curves A, B and C denote, respectively, the longitudinal input, longitudinal output and net lateral fluxes when lateral fluxes are included. Curves D and E are the longitudinal input and output fluxes when lateral fluxes are excluded*

Figure 4 shows the relative magnitudes of the longitudinal and lateral sediment fluxes entering (+ve flux) and leaving (-ve flux) the near bank zone of the model reach characterised by cross-section No. 2 for both lateral sediment exchange scenarios. It is these fluxes that influence the degradation experienced in the near bank zone of cross-section 2 (Figure 3). The data show that the net lateral flux is directed away from the banks, flowing into the central flow zone. This explains why degradation in the near bank zone is greater than the degradation experienced in the central flow segment for the case when lateral fluxes are included. It is also apparent that the longitudinal sediment transport fluxes are approximately an order of magnitude greater than the net lateral sediment transport flux, which in this case is directed away from the near bank zone. The net lateral flux is in this case dominated by the lateral bedload flux because, despite the relatively large secondary current coefficient (0.1) used in this simulation, the suspension of sediment is limited by the presence of relatively coarse bed material.

Although the results show that the magnitudes of the lateral sediment trasport fluxes are relatively small, they also indicate that predictions of degradation are sensitive to the inclusion of the lateral sediment exchange mechanisms. Curves D and E on Figure 4 indicate that the predicted longitudinal fluxes significantly differ when lateral fluxes are considered. It is this divergence of the *longitudinal* fluxes (in this case the output flux) which appears to be primarily responsible for the differences in predicted distributions of near bank degradation. Thus, the sensitivity of the predicted degradation distribution to lateral sediment fluxes appears to be due to the indirect influence of the lateral fluxes on the longitudinal fluxes through the interaction of bed morphology, flow hydraulics and sediment transport.

4. Conclusion

It can be concluded that the model presented here is potentially capable of predicting realistic amounts of aggradation and degradation in the near bank zones. The results indicate that lateral sediment transport fluxes, although small when compared to the longitudinal transport fluxes, do have a significant influence on the predicted distributions of aggradation and degradation. Further work is required to analyse the interaction between the lateral and longitudinal sediment transport fluxes and predicted aggradation-degradation for a range of secondary current and sediment size scenarios. Also, it will be necessary to further investigate the influence of the lateral shear stresses on the near bank flow and sediment transport. The more realistic estimates of near bank aggradation and degradation produced by the new model suggest that this approach provides a useful framework for incorporating bank process algorithms into the model to allow development of a predictive model of river channel width adjustment, and this is to be the next step of the current research.

ACKNOWLEDGEMENTS

The first author is supported by a CASE studentship awarded by the Natural Environment Research Council (No. GT4/90/AAPS/43). Dr. Roger Bettess, of Hydraulics Research Limited, is thanked for the valuable and instructive comments that he has freely made throughout the course of this research.

REFERENCES

[1]   Alonso, C. V. and Combs, S. T. (1986) "Channel width adjustment in straight alluvial streams", In *Proceedings of the 4th Federal Interagency Sedimentation Conference*, Nevada, pp345-357.

[2]   Andrews, E. D. (1982) "Bank stability and channel width adjustment, East Fork river, Wyoming",*Water Resources Research*, **18**, pp1184-1192.

[3]   Bagnold, R. A. (1954) "Experiments on a gravity free dispersion of large solid spheres in a Newtonian fluid under shear", *Proceedings of the Royal society of London, Series A*, **225**, pp49-63.

[4]   Bagnold, R. A. (1966) "An approach to the sediment transport problem from general physics",*U. S. Geological Survey Professional Paper*, **422-I**, pp1-37.

[5]   Bagnold, R. A. (1973) "The nature of saltation and of bed load transport in water",*Proceedings of the Royal society of London, Series A*, **340**, pp141-171

[6]   Borah, D. K. and Bordoloi, P. K. (1989) "Stream bank erosion and bed evolution model", In *Sediment Transport Modelling*, S. S. Y. Wang (ed), Proceedings of the 1989 International Symposium of the ASCE, pp612-617

[7]   Brundett, E. and Baines, W. D. (1964) "The production and diffusion of vorticity in duct flow", *Journal of Fluid Mechanics*, **19**, pp375-394.

[8]   Chow, V. T. (1973) *Open-Channel Hydraulics*, McGraw-Hill, Singapore, pp680.

[9]   Einstein, H. A. (1972) "Sedimentation", In *River Ecology and Man*, R. Oglesby (ed), Academic Press, pp309-318.

[10]   Engelund, F. (1970) "Instability of erodible beds", *Journal of Fluid Mechanics*, **42**, pp225-244.

[11]   Engelund, F. and Hansen, E. (1967) *A Monograph on Sediment Transport in Alluvial Streams*, Teknisk Forlag, Copenhagen, pp62.

[12]   Engelund, F. and Skovgaard, O. (1973) "On the origin of meandering and braiding in alluvial streams", *Journal of Fluid Mechanics*, **57**, pp289-302.

[13]   Francis, J. R. D. (1973) *Proceedings of the Royal Society of London, Series A*, **332**, pp443-471.

[14]   Garcia, M and Parker, G. (1991) "Entrainment of bed sediment into suspension", *Journal of Hydraulic Engineering*, **117**, pp414-435.

[15]   Gibson, A. H. (1908) "On the depression of the filament of maximum velocity in a stream flowing through an open channel", *Proceedings of the Royal Society of London, Series A*, **82**, pp149-159.

[16]   Gulliver, J. S. and Halverson, M. J. (1987) "Measurements of large streamwise vortices in an open channel flow", *Water Resources Research*, **23**, pp115-123.

[17]   Hey, R. D. (1979) "Flow resistance in gravel-bed rivers", *Journal of the Hydraulics Division of the American Society of Civil Engineers*, **105**, pp365-379.

[18]   Hirano, M. (1973) "River bed variation with bank erosion", *Proceedings of the Japanese Society of Civil Engineers*, **210**, pp13-20.

[19]   Molinas, A., Denzel, C. W. and Yang, C. T. (1986) "Application of streamtube computer model", In *Proceedings of the 4th Federal Interagency Sedimentation Conference*, Nevada, pp655-664.

[20]   Olesen, K. W. (1987) "Bed topography in shallow river bends",*Communications on Hydraulic and Geotechnical Engineering (Delft University)*, **87-1**, pp265.

[21]   Orvis, C. J. and Randle, T. J. (1986) "Sediment transport and river simulation model", In *Proceedings of the 4th Federal Interagency Sedimentation Conference*, Nevada, pp665-674.

[22]   Osman, M. A. (1985) *Channel Width Response to Changes in Flow Hydraulics and Sediment Load*, Unpublished PhD thesis submitted to Colorado State University, pp170.

[23]   Parker, G. (1978) "Self-formed straight channels with equilibrium banks and mobile bed. Part 1. The sand-silt river" *Journal of Fluid Mechanics*, **89**, pp109-125.

[24]   Perkins, H. J. (1970) "The formation of streamwise vorticity in turbulent flow", *Journal of Fluid Mechanics*, **44**, pp721-740.

[25]   Prandtl, L. (1952) *Essentials of Fluid Dynamics*, Blackie, London, pp452.

[26]   Smith, T. R. (1974) "A derivation of the hydraulic geometry of steady state channels from conservation principles and sediment transport laws", *Journal of Geology*, **82**, pp98-104.

[27]   Thorne, C. R. (1978) *Processes of Bank Erosion in River Channels*, Unpublished PhD thesis submitted to the University of East Anglia. pp448.

[28]   Thorne, C. R. and Osman, M. A. (1988) "The influence of bank stability on regime geometry of natural channels", In *River Regime*, W. R. White (ed), Wiley, Chichester, pp135-147.

[29]   Van Rijn, L. C. (1984) "Sediment transport, Part 2: Suspended load transport", *Journal of Hydraulic Engineering*, **110**, pp1613-1641.

[30]   Wark, J. B., Samuels, P. G. and Ervine, D. A. (1990) "A practical method of estimating velocity and discharge in a compound channel", In *Flood Hydraulics*, W. R. White (ed), Wiley, Chichester, pp163-172.

37 Risk assessment for erosion of sediment deposits in river basins using GIS for data management

P. Ruland, U. Arnold and G. Rouve

ABSTRACT

Due to the industrial development in the close neighborhood of water courses, many small and medium size reservoirs are partially filled today with mud deposits made up of contaminated sediments. The work described in this paper deals with the development of a flexible, problem specific information system to support risk assessment and restoration planning of hazardous mud sites in river basins. First attention has to be given to risk of mud erosion and destabilization under extreme hydraulic boundary conditions. For this purpose, the dynamic information system under development combines a GIS, a family of linked simulation models (FE hydrodynamic flow model, boundary layer shear model, erosion risk model) and special tools for on-line interactive model manipulation and data analysis. In addition to a short description of the simulation models, special requirements to information management and spatial data processing are discussed. For the task given, an object-oriented GIS was selected from a group of commercially available products. The object-oriented features of this system facilitate the integration of multiple heterogeneous software tools. First applications of the developed models and of the information system prototype are illustrated in terms of case study results.

1   Introduction

Many river systems in old industrialized regions embody small reservoirs and lakes forming dead zones and sediment traps. During the time of growth a large number of mud deposits

443

developed into hazardous mud sites due to considerable amounts of adsorbed pollutants of any kind. Thus these mud sites can be seen as the rivers long term memory of its own pollution history.

Today these deposits are identified as one of the major risk to the water quality and the whole aquatic environment. In case of extreme hydraulic conditions such as careless weir operations or huge floods the deposits can be destabilized. In first consequence the destabilized mud load usually causes a strong BOD/COD increase leading to critically low Oxygen concentrations downstream which may be lethal for the fish population.

Moreover, spreading and sedimentation of heavily polluted mud from "old" layers may even cause a long term change of the biochemical conditions due to reinitialed chemical reactions between pollutants and the water constituents.

A simple solution to this problem could be dredging and depositing the mud on a safe site. However this is a rather costly method and appropriate sites are hardly available. If one seeks other possible solutions several aspects of the problem have to be investigated. In summary three areas of investigation can be identified.

1   The problem of sediment deposits in a river system has a physical-hydrodynamic aspect. The hydraulic condition under which sediment erosion occurs have to be determined. At least the risk of erosion during hydraulic stress situations has to be evaluated. A better understanding of the mechanisms involved can be achieved by simulating the actual transport of sediments. Feasible solutions like changing the operation of the reservoirs and weirs of a a river system can be defined.

2   The problem of sediment deposits contaminated by pollutants has of course a chemical dimension too. Questions like the following have to be answered. Provided that the sediments are mobilized, under what chemical conditions do pollutants dissolve into the water and to what extent. What are the impacts on the water quality and on human life.

3   While solving problems with hazardous mud sites in river system a large amount of information has to be taken into account making the problem very complex. First there can be lots of mud sites in the specific river system under consideration leading to a large amount of data and secondly the complexity of the relevant processes (hydrodynamic, hydrology, sediment, etc.) involved is rather high. Therefore efficient information management is a decisive key to problem solving in this area. As most of the information is spatially related the application of a Geographic Information System is most feasible.

At the Institute of Hydraulic Engineering and Water Resources Management (Aachen University of Technology) solutions in the area 1 and 3 are investigated. Correspondingly the paper describes simulation models and information management tools related to sediment transport in river systems.

The work on simulation models started with a model to simulate erosion risk. It was tested in a case study. The development of a 2-dimensional transport model is currently performed. Figure 1 shows the area under consideration which is situated at the river Agger one of the tributaries of the river Rhine.

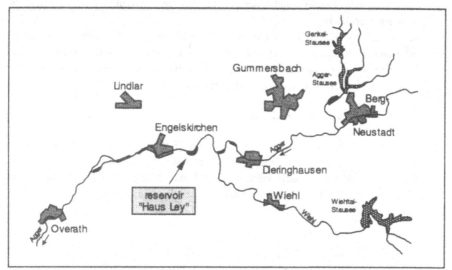

FIGURE 1   *Area map of the case study at the river Agger, one of the tributaries of the river Rhine*

Work on the combination of the simulation models with a GIS has been started. In addition to mapping the results of the modeling process the work focuses on the development of a dynamic information system, i. e. an interactive graphics oriented tool, which enables the user to operate the models on the map level and to integrate all necessary information for the decision making process.

Chapter two describes the simulation models being used throughout this study while chapter three deals with the incorporation of the model into the GIS. The paper describes the current stage of development and discusses related problems.

2   Simulation models for Sediment Transport in a River Environment

Simulating sediment transport in lakes, reservoirs and river basins involves quite a few different processes. A good overview is given for example in [7]. The following list summarizes the models required for a complete transport simulation:

- A hydrodynamic model computes the flow acting on the sediments. Usually a 2-dimensional model is used for this purpose. Large amounts of data for calibration and boundary conditions are required to build such a model.
- By means of a boundary layer model the shear stresses acting on the sediments can be determined. Input to this model are the flow velocities and frictions coefficients of the bottom.
- Sediment properties such as critical shear stress of erosion and deposition have to be determined by field and laboratory measurements. They provide the necessary data to

445

simulate the rate of erosion and deposition. In some cases flocculating occurs which can be modeled separately.

- The rate of erosion or deposition are the source-sink terms to the sediment convection-diffusion model. It simulates the concentration of sediment in the water column at given flow and sediment conditions. By keeping trace of erosion and deposition, it is possible to compute the river bottom changes with respect to time.

2.1   Models used in the current study

Utilizing all the above described models is a time and effort consuming process. An example can be found in [2]. In some cases the simulation of just the erosion risk may be sufficient. Figure 2 shows a chain of models and the required data serving this purpose. The authors utilized this scheme. It is described in more detail in the following sections.

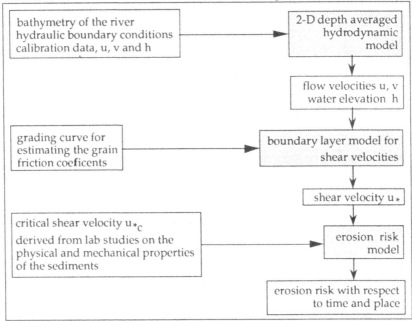

FIGURE 2:   *Model components for risk assessment of sediment erosion*

Hydrodynamic Model .   The hydrodynamic model used for this study is based on the well known RMA2V model [3]. By means of the FE method the two-dimensional vertically integrated shallow water equations are solved in their complete elliptic form. To account for the nonlinearity of the equations, a Newton-Raphson iteration scheme is used.

The time dependency is solved by using a finite difference algorithm (Crank-Nicolson) which is a composition of an explicit and implicit scheme and has proved to be quite stable.

The model yields the horizontal velocities u and v and the water elevation. During depletion or flooding of a reservoir or river basin the inundated area may change dramatically.

In the model this is, accounted for by eliminating or adding elements between iteration steps. This method requires large computing efforts as a fairly high spatial resolution is required in order to maintain flow continuity. On the other hand high spatial resolution increases computation stability. Therefore, fairly detailed meshes were used in the case study.

Boundary Layer Model   To be able to determine the erosion risk the acting force of the flow has to be computed. This can be done be using a flow formula for example the Darcy-Weißbach equation. The approach is based on the idea that each Finite Element can be regarded as a piece of the water column. Then the definition for shear velocity writes:

$$u_* = \sqrt{g\,h\,I}$$

(1)

I is the slope of the energy line. The depth averaged total velocity u_m in a water column can be computed by means of the results of the hydrodynamic model. Using the Darcy-Weißbach equation

$$\frac{u_m}{u_*} = \sqrt{\frac{8}{\lambda}}$$

(2)

the shear velocity u_* can be obtained. Equation 2 requires as input the flow resistance coefficient λ. It can be computed by equation 7 where the constants have been changed to reflect the difference between pipe and open channel flow [4].

$$\frac{1}{\sqrt{\lambda}} = -2,03 \cdot \log\left[\frac{4,4 \cdot v}{u_m \cdot 4h \cdot \sqrt{\lambda}} + \frac{k_s}{4h \cdot 3,71}\right]$$

(3)

Erosion Risk Model .   The motivation for developing an erosion risk model and its derivation is described in more detail in [5]. Here a short summary is given for convenience. The erosion risk model is based on the fact, that the onset of erosion does not start at a distinct critical value but is a smooth transition from rest to erosion. Two processes take part in the onset of erosion. They both can be described by statistical distributions.

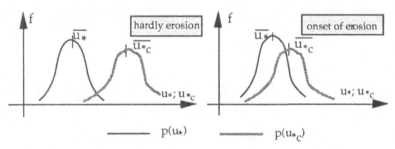

FIGURE 3   *Sketch of frequency distributions for different stages of erosion*

447

In Figure 3 the acting flow force is presented in terms of u* and the resisting force of the sediment in terms of u*c. Assuming the fact that both distributions were known, the following derivation could be made [8].

The probability or proportion of the sediment to be transported by a given flow velocity can be computed according to equation 4 (see Fig. 3):

$$P_u(u*) = \int_{u*_{min}}^{u*} p(u*_c)du*_c$$

(4)

If both frequency distributions (p(u*) and p(u*c)) are mutually independent one can compute the probability or risk of erosion R by solving the following equation:

$$R = \int_{u*_{min}}^{u*_{max}} p(u*)\left[\int_{u*_{min}}^{u*} p(u*_c)du*_c\right]du*$$

(5)

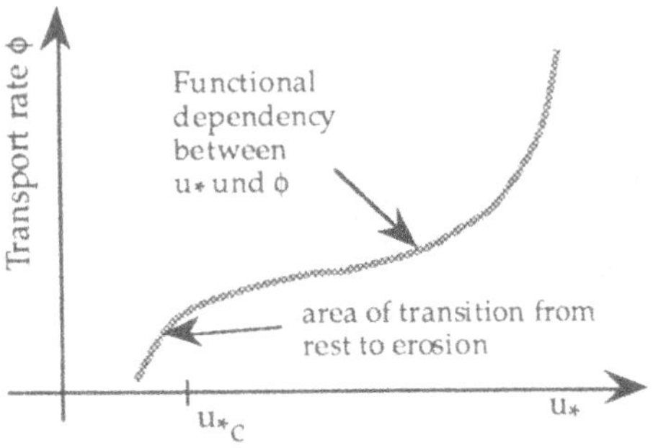

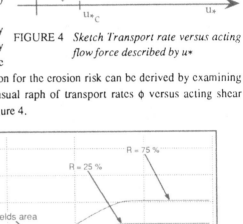

FIGURE 4   *Sketch Transport rate versus acting flow force described by u**

Unfortunately both of them are analytically unknown. The erosion risk R can vary between zero and one, one being the highest risk possible. An empirical function for the erosion risk can be derived by examining sediment transport measurements. The usual raph of transport rates φ versus acting shear velocity looks like the one sketched in Figure 4.

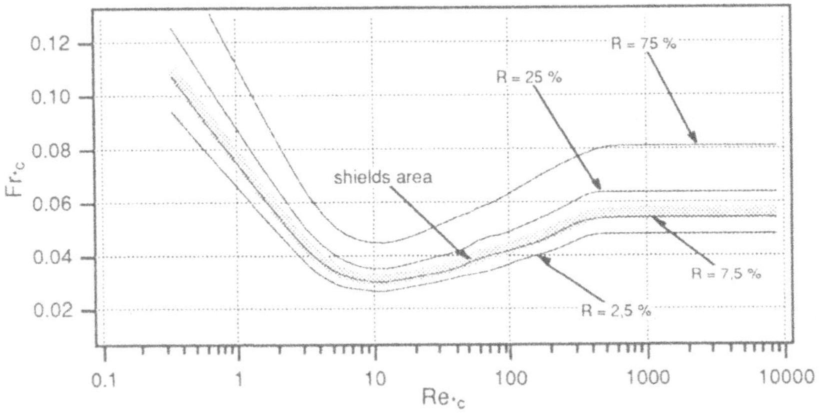

FIGURE 5   *Shields values compared with different risk classes*

By comparing the transport rates (lab and field studies) during transition from rest to erosion with mean critical values the following function can be derived [8]

448

$$R = \left(10\left(\frac{Fr_*}{Fr_{*c}}\right)^{-9} + 1\right)^{-1}$$

(6)

In this equation Fr_{*c} can be obtained according to Shields or by direct measurements. The graph of Shields values compared with those of different risk values is given in Figure 5.

Case Study Small Reservoir The model family as described in Fig. 2 was used in a case study, the reservoir "Haus Ley" (see Fig. 1). It is a small shallow basin with a length of about 800 m and width of about 90 m. The coefficients of the hydrodynamic model were calibrated by field measurements of flow velocities. The flow distributions of two test cases (1: depletion of the reservoir by weir operation, 2: a summer flood passing the depleted reservoir) were simulated.

Figure 6 shows a flow field after 60 % of the depletion has been performed. The flats being dryed are marked by a hatch style. The corresponding risk distributions are shown in Figure 7 (only the entrance area is drawn because erosion starts there). The critical values are chosen according to Shields. To demonstrate the sensitivity to changes of critical values the risk distribution is also plotted with critical values Fr_* assumed to be 50 % higher than Shields numbers.

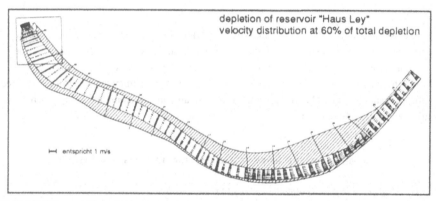

FIGUURE 6   *Flow field simulated by the hydrodynamic model*

2.2   Identifying Material Parameters

Figure 7 clearly indicates a high model sensitivity to varying critical values. This corresponds with the natural variability of the sediments. Their properties are highly influenced by cohesiveness and biological impacts changing the nature of the sediments dramatically. In literature an increase of the critical coefficient up to 7 times the shields value is reported. While decreases have been established sometimes too, commonly an increase of 2 to 3 times due cohesiveness and biological stabilization can be found.

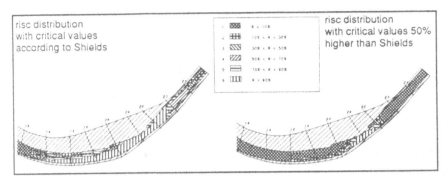

FIGURE 7   Riks distributions computed with the erosion risk model

In [5] the question of how to determine the critical parameter for erosion and deposition is discussed. Here it is concluded, that because of the numerous processes influencing the properties of sediments it is necessary to determine the critical value for erosion or deposition by field or lab measurements.

The authors plan to perform measurements in an annular flume to obtain these values for the case study.

2.3   Future Developments

An extension of the current model family is planned to simulate sediment transport. A Finite-Element model for the solution of the depth averaged convection, diffusion equation is available. Proper source and sink terms for erosion and deposition as reported for example by [1] will be incorporated in the model. The necessary material parameters are to be measured with an annular flume. The aim is to compute the amount of sediments being washed out during hydraulic stress events.

3   Data management using a Geographic Information System

3.1 Information related to sediment deposits in river systems

The need for information management in this area is commonly acknowledged. The list of related information types given below may illustrate that fact. It is obvious that information of different sources and from different disciplines have to be integrated and presented to the decision makers. For example information is needed about:

- environmental aspects   • water quality   • use of reservoir
- use of adjacent land   • governmental laws   • ownership questions
- operation dams, weirs   • aspects hydropower   • results simulation
- hydrologic impacts • sediment properties   • sediment pollution

Graphical tools to present this information can be:

Maps: • Landscape with area under consideration   • 3-D surface view
     • views on the sediments accumulated   • FE-Mesh
     • Contour graphic of the river and the reservoir
     • Maps showing flow velocities
     • color coded maps (thematic maps)

Related 2D graphs: • cross sectional views of the area
         • cross sectional view on any quantity of interest

The integration of this information into a GIS not only provides an easy to use tool for creating all these maps and graphs. Moreover, a GIS allows spatial queries to be made. For example a possible question could be: How many private properties are located along the river bank to determine the amount of people which could suffer from any impacts of mobilized sediment deposits. Another question could be to compute the size of the area in all reservoirs in a river system where the erosion risk is higher then 70 %.

A second aspect is that a GIS provides appropriate tools to create Finite-Element meshes resulting in less effort to set up the simulation models. Chap. 3.5 gives some more information on that subject.

3.2 What kind of GIS is suitable

Generally any GIS can be used for presenting spatially related results of simulation models. The major questions are how intense the desired link should be and whether the related data are stored inside or outside the GIS. In [6] criteria for a proper selection of a GIS system are evaluated. At the Institute for Hydraulic Research and Water Resources Management a detailed test and comparison of serveral differnt GIS-products lead to the selection of "Smallworld GIS". The system has been developed in England and is available on common workstations like "DEC", "SUN" or "IBM".

It would exceed the scope of this paper to give an overview about the main features of this GIS system. But three key-functions of the GIS are listed below. They are the reasons why Smallworld GIS has been elected by the authors:

• Integrated use of Raster and Vector Maps. Existing geographical maps can be scanned and transformed into geographical coordinate systems. It is not necessary to digitize complete topological maps by hand. Instead only the objects on that map which are of importance to the problem under consideration have to be digitized on screen.

• Definitions of the GIS in an Object Oriented Programming (OOP) Language. The complete GIS is programmed in an object oriented language called "Magik". This ultra-high level language significantly supports the user to customize any part of the GIS for his specific requirements. E. g. procedures and methods can be defined to perform simulation specific tasks.

• Object and Database: In Smallworld all geographic elements are viewed at as objects. They are organized in classes to share common properties. Objects of one class are stored in a

451

table of a relational database. The table itself is hidden from the users. Objects usually consist of non-spatial information, spatial information and some methods. These methods describe special behavior of objects belonging to one class. Objects can exist which do not have either non-spatial or spatial attributes.

The object structure of the GIS makes it fairly easy to establish objects such as FE-element of FE-node typically needed for simulation models. The objects serves as a vehicle to store all simulation related data in the database of the GIS having it available at any time. The next chapter illustrates the concepts and presents some results obtained so far.

3.3   Current state of development

Appropriate data structures for the hydrodynamic and erosion risk model have been defined. Figure 8 shows a part of the screen with the editors providing information for specific objects of the classes FE-Node, FE-Element and FE-Data.

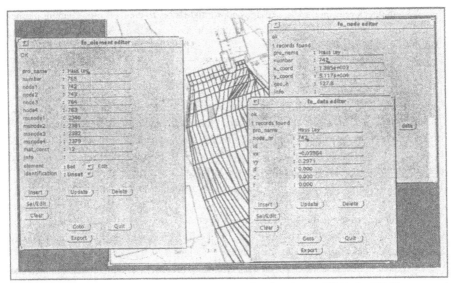

FIGURE 8:   *View on the screen with editors for the objects FE-Node, FE-Element and FE-Data.*

The FE-Element object contains slots to store the number of their corner and midside nodes. Each FE-Node object stores data being unique to itself like location and number. Besides, it points to an object FE-data were the data of simulation models is stored. The reason for storing this data separately from the Node or Element object is to be able to manage results of different simulation runs.

The appearance of an object can now be defined depending on the desired information. For example, the color of the elements can be set in dependency of the erosion risk computed for the element by the risk model. Or instead of drawing each node with a point style a method can

452

be added to the class of FE-Nodes which draws a vector with a length corresponding to the flow velocity computed by the hydrodynamic model for that point.

Finite Elements can be constructed with the system or meshes provided by grid generators can be edited to adopt them to the specific geometry of the simulation area. Figure 9 shows an example of a mesh produced by this procedure. Import and export procedures had to be defined to read old meshes and provide information of edited meshes in a suitable way for the simulation model.

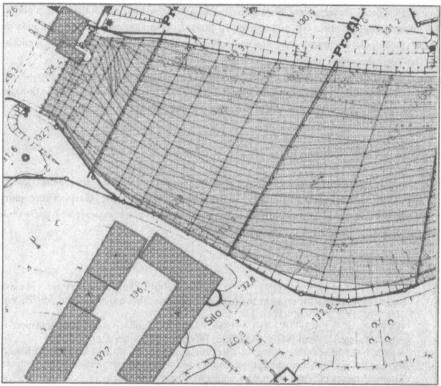

FIGURE 9   *View on the weir area of the reservoir "Haus Ley" togeter with the FE-mesch used for hydrodynamic modelling*

3.4   *Future developments*

The authors aim is to develop an information system with which not just only the results of simulation models can be viewed together with other information, but where the operation of models can be controlled too. Though the current available computer power does not allow a dynamic simulation for every model (for example the hydrodynamic model is very CPU intensive) some simulations like erosion risk can be computed realtime. While changing input parameters, the user can identify the best combination of coefficients by varying boundary conditions and simultaneously viewing the computation results.

The aim is to combine simulation models for sediment transport with a GIS system to obtain a dynamic, spatial information system for the management of hazardous mud sites in river systems. The development of a prototype is currently in progress.

4 Conclusions

The results of the developments performed so far at the Institute for Hydraulic Research and Water Resources Development have proved, that a combination of simulation models for sediment transport with a GIS is feasible. To limit the computational and data acquisition effort, the simulation models may not be to sophisticated. Instead measurements should be used to cover theoretical uncertainty.

The integration of GIS and simulation models can lead to a broader application of the models. A better information management can be provided. This might serve to obtain decision of higher quality not only in the field of sediment transport in river environments.

5 Acknowledgements

The study is funded by the Ministry for science and Research of the state North Rhine-Westphalia, Germany. Additionally it was sponsored by the German company "Moselkraftwerke Andernach GmbH" who operates the power plants at the river Agger. Their help in providing necessary data is acknowledged too.

REFERENCES

[1]   ARIATHURAI, R.; KRONE, R. B. (1976) "Finite element model for cohesive sediment transport"; Journal of the hydraulic division, Proceedings of the ASCE, Vol. 102, No. HY3, 1976, pp. 323 - 338

[2]   HAYTER, E. J.; MEHTA, A. J. (1986) "Modelling cohesive sediment transport in estuarial waters"; Appl. Math. Modelling 1986, Vol. 10, August, pp 294-303

[3]   KING, I. P.; NORTON, W.R.   "Recent applications of RMAs Finite Element models for two-dimensional hydrodynamics and water quality"; Proc. 2. Int. Conf. on Finite Elements in Water Resources, Imperial College, London 1978

[4]   ROUVÉ (Hrsg.) (1987) "Hydraulische Problem beim naturnahen Gewässerausbau"; Forschungsbericht DFG, VCH Verlagsgesellschaft, Weinheim

[5]   RULAND, P.; ROUVE, G. (1992) "Risk assessment of sediment erosion in river basins"; Fifth International Symposium on River Sedimentation

[6]   RULAND, U.; ARNOLD, U.; ROUVE, G. (1991) "Pollutant transport with fine cohesive sediments; modeling sediment erosion and transport processes in lakes and river basins"; International Conference on Computer Applications in Water Resources, July 3-6, 1991, Taipei, Taiwan, R.O.C.

[7]   SHENG, Y.,P. (1984) "Modelling bottom boundary layer and cohesive sediment dynamics in estuarine and coastal water"; Lecture notes on coastal and estuarine studies, Mehta, A. J. (ed); Estuarine cohesive sediment dynamics; Proceedings of a workshop on cohesive sediment dynamics, Tampa, Florida, USA, Nov.12-14, 1984

[8]   ZANKE, U: (1990) "Der Beginn der Sedimentbewegung als Wahrscheinlichkeitsproblem"; Wasser und x, Heft 1, 1990, pp. 40-43

38 Computations of erosion and deposition pattern in the Lower Karnafuli River, Bangladesh

L. A. Khan and M. K. Alam

Abstract

Chittagong Port, located on the right bank of the lower Karnafuli River, is the largest port of Bangladesh. Shoaling of the river is one of the major problems in the maintenance of navigational channel. To study the erosional and depositional features of the river, a simple numerical model has been developed. The sediment continuity equation is solved by finite difference method. Bed load is the dominant mode of sediment transport in the river. Therefore, the contribution of suspended sediment in the model formulation is neglected. The local erosion and deposition are computed by semi-empirical equations correlating sediment supply, local shear stress and the median sediment size. The hydraulic information necessary, to solve the sediment continuity equation, is obtained from a one-dimensional hydrodynamic model of the river system. Application of the model indicates that the computed yearly mean erosion and deposition are within 20 percent of the field data. The model is also used to study the effects of closing a secondary channel on the erosion and the deposition pattern in the lower Karnifuli river.

1 Introduction

Chittagong Port is situated on the right bank of the lower Karnafuli river, about 16 km. from the outfall. The port handles more than 80 percent of the export and import of Bangladesh. The present course of the navigational channel is maintained by extensive river training works[13]. The width of the navigational channel varies from 204 m to 335 m, while the mean depth ranges from 5.5 m to 9.0 m below the chart datum. The length of ship that can sail to the port is limited by a sharp S-bend, known as Gupta Bend. The draft of ship is restricted by siltation of the channel, and oscillations of the shoals under different hydraulic conditions[1, 11].

The Karnafuli river originates in the Lusha Hills in India, and follows a south-western course through the Chittagong Hill Tracts and the alluvial coastal plain in the south-eastern part of Bangladesh. The combined river system downstream of the Kaptai dam is shown in figure-1. Downstream of Chittagong the river falls into the Bay of Bengal at Patenga. The Halda is its main tributary. At Patenga the river is subjected to a strong tidal action. The fresh water inflow to the Karnafuli is controlled by the Kaptai dam, situated about 80 km from Patenga. The Halda is a flashy river with negligible fresh water inflow[10].

Figure-2 shows the navigational channel (channel deeper than 5.5 m) and the shoals in the lower Karnafuli in 1980. These shoals, requiring regular maintenance dredging, hinder the development of the port. From the early fifties the Chittagong Port Authority(CPA) has commissioned studies to identify alternate river training and remedial measures. The first systematic study was carried out by Hydraulic Research Station(HRS)[4]. The river training works in the lower Karnafuli are based on the recommendation of HRS. The Netherlands Economic Institute(NEI)[11] has studied the shoaling problem based on field data.

In 1985 the Department of Water Resources Engineering(WRE) and the Institute of Flood Control and Drainage Research(IFCDR), Bangladesh University of Engineering and Technology(BUET), Dhaka, Bangladesh, jointly undertook a mathematical model study of the Karnafuli river[1]. A simple one-dimensional erosion-deposition model was developed for a preliminary analysis of the problem. The hydraulic information, necessary for the solution of the sediment continuity equation, was obtained from a hydrodynamic model developed previously[6, 7]. This paper describes the erosion-deposition model and its application to the lower Karnafuli river. A discussion on the mode of sediment transport in the river is presented in section-2. The objective of the discussion is to identify the primary sediment transport mechanism. The governing equations and the solution algorithm are described in section-3. Calibration of the model is presented in section-4. An application of the model and discussion on results are in section-5. Finally, summary and conclusions of the study are listed in section-6.

2   Mode of Sediment Transport

The field data collected by NEI[11] and numerical model study[5] indicate that the lower Karnafuli is a well mixed estuary, with a maximum salinity intrusion length of about 20 km under lean flow conditions. The presence of saline water in an estuary causes flocculetion of suspended sediment particles[9] and a higher rate of shoaling. To simplify the analysis, the effect of saline water on shoaling is neglected.

Conventionally, sediment transport is divided into suspended sediment and bed load. In the lower Karnafuli the mean diameter of sediment particles is coarser than 0.15 mm and finer than 0.40 mm, with an average silt concentration limited to less than five precent[1, 11]. The longitudinal distribution of bed sediment (cross-sectional mean) in shown in figure-3. The suspension of sediment from river bed is determined by turbulent mixing characteristics of the flow. The ratio of the settling velocity of sediment particle and the entrainment velocity, α, determines the range over which suspended sediment transport is important[9]:

$$\alpha = \frac{\omega}{\kappa u_*} \qquad (1)$$

where ω is the fall velocity, κ is the von-Karman constant, taken as 0.4, and u_* is the shear velocity. In this study the fall velocity is computed by Mande and Whitemore equation[3], correlating sediment concentration with the fall velocity. For large values of α (≥ 5) negligible sediment entrainment occurs and suspended sediment does not play a significant role in the siltation of the river. In the lower Karnafuli river, the computed value of α varies from 7 to 13, suggesting that the bed load is the dominant mode of sediment transport[1]. This is consistent with the available field data[11].

3 Numerical Model

The rate of sediment transport in an alluvial river is related to bed shear stress generated by the flowing water. An increase in shear stress causes excess sediment outflow, thus causing erosion of the bed. If the shear stress in a reach is reduced then the sediment brought in will exceed the carrying capacity, and deposition will take place. For one-dimensional parameterization of the problem, the bed elevation η, is related to sediment transport rate q_s, by the sediment continuity equation[3, 9]:

$$\frac{\partial \eta}{\partial t} + \frac{1}{\gamma_s(1 - \lambda)} \frac{\partial q_s}{\partial x} = E - D \qquad (2)$$

where γ_s is the unit weight of dry sediment, λ is the porosity of sediment, E is the rate of sediment suspension from the bed, and D is the rate of deposited. Equation(2) can be used to compute the change in bed elevation if the sediment discharge, sediment suspension and deposition rates are known.

3.1 Bed Load

Near bed sediment movement occurs when the fall velocity is large compared to the bed shear velocity. In this study, the bed load is computed by Colby's equation for unmeasured sediment load[3]:

$$q_b = K u^\beta \qquad (3)$$

where K and β are empirical coefficents and u is the cross- sectional average velocity. This is an empirical equation derived from field data with grain size of sediment varying from 0.10 to 0.60 mm. The mean value of β is about 3.1. Though equation(3) was derived for uni-directional flows, it is assumed that

457

the equation can be applied to tidal rivers by approximating the flow as quasi-steady over each computational time step. As the contribution of suspended sediment is small, q_b essentially represents the total sediment discharge in the river.

3.2 Erosion Rate

The erosion of sediment particles from river bed by current is the main mechanism by which sediment is introduced back into the flowing water. Surface erosion takes place when the bed shear stress τ, exceeds the critical shear stress τ_c. In this study erosion is computed by the following equation[9]:

$$
\begin{aligned}
E &= M(\frac{\tau}{\tau_c} - 1) : \tau \geq \tau_c \\
&= 0 : \tau < \tau_c
\end{aligned}
\tag{4}
$$

where M is an erodebility constant. Equation(4) implies that the rate of erosion increases linearly with excess bed shear stress. Laboratory investigations indicate that M can vary from 10^{-8} to $2\text{x} \ 10^{-7}$ s/m[14]. The critical shear stress can be estimated by the following empirical relationship[9]:

$$
\tau_c = 166 d_{50}
\tag{5}
$$

where d_{50} is the median sediment size in mm, and τ_c is in g/m^2. The critical shear stress τ_c can vary from 1.8 to 3.0 N/m^2 for d_{50} varying from 0.2 to 1.0 mm.

3.3 Deposition Rate

An expression, similar to equation(4), that can be used for the computation of deposition is as follows[8]:

$$
\begin{aligned}
D &= \omega c(1 - \frac{\tau}{\tau_o}) : \tau \leq \tau_o \\
&= 0 : \tau > \tau_o
\end{aligned}
\tag{6}
$$

where c is the concentration of suspended sediment, and τ_o is the shear stress below which all initially suspended sediment will deposit eventually. In equation(6), the term within the parenthesis represents the probability that a particle will adhere to the bed. Experimental investigations indicate that τ_o depends on the properties of sediment and can vary from 0.04 to 0.15 N/m^2[8].

3.4 Solution Algorithm

To solve equation(2) it is necessary to determine the flow characteristics of the river as a function of space and time. This information is obtained from a one-dimensional numerical model of the Karnafuli river. The flow model[7] solves one-dimensional continuity and momentum equations for water by Preissmann four-point finite-difference scheme[2]. Equation(2) is solved by an explicit finite-difference method, forward difference in time and central difference in space.

458

The combined numerical model neglects the increase in the density of water due to the presence of the sediment particles and saline water. Under these conditions the water flow equations and the sediment continuity equation are decoupled, and the two system of equations can be solved sequencially. While solving the flow equations, the river bed is assumed fixed over a tidal period. Using the hydraulic information, the sediment continuity equation is solved for the tidal cycle, and the net erosion or deposition at the computational cells are determined. These results are used to update the geometry of the river, assuming erosion or deposition is proportional to the local depth of flow over a given cross-section. With the updated geometry, the process is repeated over the simulation period. The banks of the lower Karnafuli river are protected by extensive river training works[13]. Therefore, lateral migration of the banks are not allowed in the computations.

4   Model Calibration

The Karnfauli river system consists of three major branches, bounded by three external boundaries at Patenga, Kaptai and Pachpukuria (figure-1). Near Bakalia (figure-2) the river bifurcates and then rejoins, thus forming a looped river system. The lower Karnafuli, from Patenga to Halda Point, was discretized into smaller reaches of lenght 0.5 km each. A small distance steps, Δx, was used to incorporate the geometry of the important bars and to have a reasonable resolution of the channels forming the loop. The two upper branches, above the Halda Point, where the erosion and deposition rates were not of major interest in the study, were discretized with Δx of 2.5 km. The geometric properties at the computational grid-points were extracted from hydrographic charts supplied by CPA[1, 7]. The discretized river system consisted of six branches connected at three junctions, with three external boundaries.

Tides in the Karnafuli river is semi-diurnal, with tidal range varying from 5.2 m at spring tides to 1.5 m during neap tides. The hydrodynamic model was cailbrated by adjusting Manning's friction coefficient and comparing the computed water level with field data at Khal No.-10, Sadarghat and Kalurghat. Figure-4 shows the results of calibration and validation of the hydrodynamic model[7]. The maximum computed velocities were found to vary from 1.0 m/s to 1.5 m/s. The flood velocities were slightly lower than the ebb velocities. The computational time step, Δt, used in running the model was 15 minutes.

The erosion-deposition model was calibrated by computing erosion- deposition at grid-points, as functions of M and τ_o, and comparing the results with field data to obtain a reasonably good match. The field values of erosion-deposition were determined by superimposing hydrographic charts prepared in August 1983 and September 1984. The cross-sectional areas at the grid-point were determined and the difference was taken as either erosion or deposition, depending on whether the difference was positive or negative respectively. Out of 59 sections in the lower Karnafuli, hydrographic charts of the two periods were available at 17 sections. The erosion- deposition thus determined represents the net value over a period of one year. The direct use of these information for model calibration would require running the model over one year period, for different combinations of calibration parameters. This would be computationally very expensive. The problem was avoided by using a method similar

459

to that used by HRS[12]. The procedure consists of dividing tides in a given period (year) into classes (typically by tidal range) and multiplying the effect of each class by the number of times the tide occured in the period. In this study, tide at Patenga was classified into four categories. The September spring and neap tides (tidal ranges of 5.3 m and 2.6 m respectively) were considered to represent the average tidal conditions from May to October. The spring and neap tides in March (tidal ranges of 4.8 m and 3.4 m respectively) were taken as representative of the dry season average condition between November to April. Based on this analysis[1], the model was run for the four representative tidal cycles to compute the annual erosion-deposition rates. The distribution of d_{50} in the lower Karnafuli[11], as shown in figure-3, was assumed to remain unchanged over the simulation period. Figure-5 compares the computed erosion-deposition rate with the field data. The corresponding value of τ_o was 0.13 N/m^2. A mean value of M= 1.5×10^{-7}s/m was assumed. Figure-6 shows the corresponding computed erosion- deposition pattern in the lower Karnafuli river.

5   Results and Discussions

Figure-6 indicates that the lower Karnafuli river is undergoing significant morphological changes. The maximum annual average erosion rate is of the order of 0.46 mm per tidal cycle. The corresponding deposition rate is 1.67 mm. The mean deposition rate in the lower Karnafuli is about 0.78 mm per tidal cycle per year. During the study period, siltation was dominant in the Lower Karnafuli river, except at some short reaches. A comparison of the computed result with the field data (figure-6) indicates that the model has been able to reproduce the over all pattern of erosion- deposition rates quite well. At the sections where field data was available, the computed results are within 20 percent of the field values. The coefficient of correlation between the computed and field data is 0.936.

It should be pointed out that the yearly mean results are significantly different from short term rates of erosion and deposition. Within a tidal cycle each computational cell undergoes both erosion and deposition depending on velocity. However, the tidal average pattern does not change significantly. The present study is mainly concerned with the long term siltation of the lower Karnafuli river, so that tidal mean erosion and deposition are of primary interest. Figure-7 shows the results of computations for two cases i) spring tide (tidal range of 5.3 m) with yearly mean upstream inflow (250 cumecs), and ii) flood discharge (500 cumecs) with yearly mean tide (tidal range 2.4 m). The upstream inflow data were obtained from MOP[10]. The flood discharge has a much more pronounced effect on the erosion and deposition rates compared with higher tidal range. This is mainly due to higher residual (tidal mean) velocity associated with higher fresh water inflow into the river. The tides at Patenga, over a period of 12.42 hours, is almost cyclic[1, 11]. For no upstream inflow, the computed residual discharge and velocity are negligible.

As a practical application, the model was used to study the effects of closing the Bakalia right channel (looking north-east in figure-2). The Bakalia char(bar) lies about 7 km upstream of the port. The river is braided in this reach, causing unstable flow conditions at the port. In 1962 HRS[4] recommended the confinement of the flow to the deeper left channel to ensure stable channel conditions. In the numerical model, the closure of the right Bakalia channel was accomplished by neglecting the channel, resulting in a tree-type river structure consisting of three branches connected at a junction at the Halda Point. The flow is thus confined to the deeper left channel. The closure of the right channel reduces the flow area, in the effected region, by at most 17.5 percent of the combined flow area of the two channels. In applying the model, it was assumed that these changes in the river geometry are not significant enough to effect the boundary condition at Patenga. Figure-8 compares the erosional-depositional features before and after the closure. The closure of the Bakalia right channel reduces the tidal volume in the range of 2.0 to 3.0 million cubic meters, which correspond to about 3.25 percent of mean tidal volume. There is virtually no change in high and low water levels, but velocity increases considerably in the Bakalia channel and its vicinity. The maximum increase in velocity is of the order of 19.7 percent of the present maximum value. The figure indicates increased erosion at the upstream tip of the closure and a corresponding increase in siltation downstream. The closing of the channel may, therefore, result in considerable channel adjustments, such as increased erosion or deposition rates, and shifting of the navigational channel, and bars.

6 Summary and Conclusions

A one-dimensional numerical model for identifying the regions of erosion and deposition in a tidal river has been presented. As the bed load is the dominant mode of sediment transport in the Karnafuli river, the effects of suspended sediment in the siltation of the river was neglected. The bed load was computed by Colby's equation. The erosion and deposition rates were determined by semi-empirical equations. The computations of yearly average erosion and deposition were based on representative hydraulic conditions. The model results are in good agreement with the field data, indicating that siltation is significantly higher than erosion in the lower Karnafuli. As a practical application, the model was used to study the effects of closing a secondary channel at Bakalia. The flood flow condition causes increased erosional and depositional rates close to the Bakalia char. It should be noted that these results are based on simplified parameterization of the problem, and further studies in conjunction with physical models, will be necessary to arrive at an engineering decision.

7 Acknowledgements

This research was supported by the CPA. The field data for the model application were supplied by the CPA. The help and cooperations of WRE and IFCDR are greatfully acknowledged.

References

[1] Chittagong Port Authority, 1987, Mathematical Model Study of the Karnafuli River, Chittagong, Bangladesh.

[2] Cunge, A., Holly, F.M. Jr., and Verwey, A., 1980, *Practical Aspects of Computational River Hydraulics*, Pitman, London, U.K.

[3] Garde, R.J. and Ranga-Raju, K.G., 1985, *Mechanics of Sediment Transport and Alluvial Stream Problems*, Wiley Eastern Ltd., New Delhi, India.

[4] Hydraulic Research Station, 1962, The Karnafuli River, Report on Model Investigation, Wallingford, England.

[5] Khan, L.A., 1989, Calibration of Salinity Intrusion Model of the Karnafuli River With Sparse Field Data, In Falconer, R.A., et al., edited, *Hydraulic and Environmental Modelling of Coastal, Estuarine and River Waters*, Grower Technical Press, Bredford, U.K., pp.-301-09.

[6] Khan, L.A., 1987, A Numerical Model for the Simulation of Tides in Estuarial Networks, J. Indian Inst. Sci., 67, pp.93-108.

[7] Khan, L.A. and Ahsan, A.K.M.Q., 1986, Numerical Simulation of Tides in the Karnafuli River, In Schrefler, B.A. and Lewis, R.A., edited, *Microcomputer in Engineering: Development and Application of Software*, Pindridge Press, Swansea, U.K., pp.-497-506.

[8] Markofsky, M., Lang, and, Schubert, R., 1986, Suspended Sediment Transport in Rivers and Estuarues, In van de Kreeke, J., edited, *Physics of Shallow Estuaries and Bays*, Springer- Verlag, N.Y., pp.245-58.

[9] McDowell, D.M., and O'Conner, B.A, 1977, *Hydraulic Behaviour of Estuaries*, MacMillan Press Ltd., London, U.K.

[10] Master Plan Organization, 1984, Surface Water Availability, National Water Plan Project, Government of Bangladesh, Dhaka, Bangladesh.

[11] Netherlands Economic Institute, 1977, Chittagong Port Entrance Study, Rotterdam, the Netherlands.

[12] Rodger, J.G., 1980, Mathematical Models of Sediment Transport in Canalised Estuaries, In Sundermann, J. and Holz, K.P., edited, *Mathematical Modelling of Estuarine Physics*, Springer-Verlag, N.Y., pp.247-52.

[13] Shahidullah, S.M., 1981, The Training of the Karnafuli River Within the Port Reach, Chittagong Port Authority, Chittagong, Bangladesh.

[14] Sheng, Y.P., 1986, Modelling Bottom Boundary Layer and Cohesive Sediment Dynamics in Estuaries and Coastal Waters, In Mehta, A.J., edited, *Estuarine Cohesive Sediment Dynamics*, Springer-Verlag, N.Y., pp.-360-400.

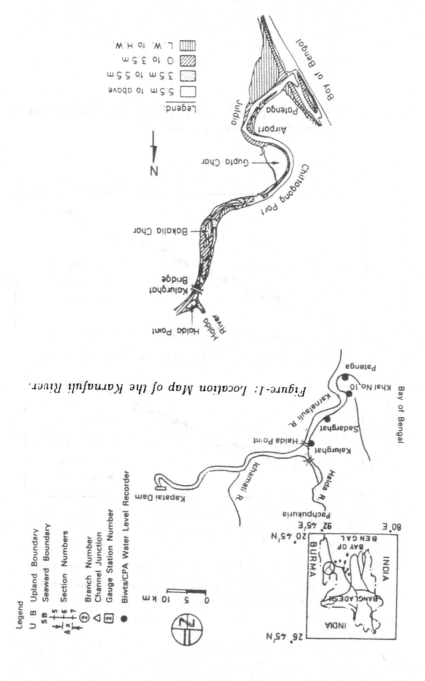

Figure-2: The Navigational Channels and Bars in the Lower Karnafuli River.

Figure-1: Location Map of the Karnafuli River.

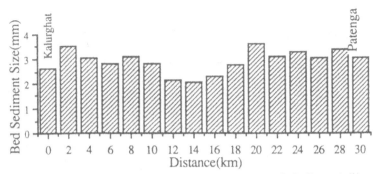

Figure-3: The Longitudinal Distribution of Bed Sediment Size in the Lower Karnafuli River.

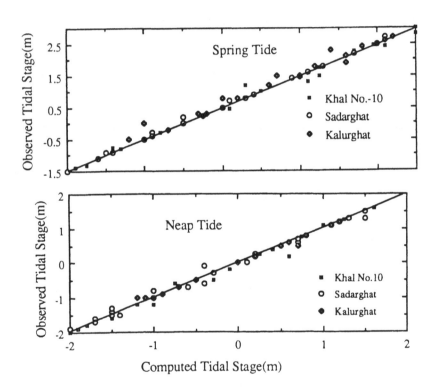

Figure-4: Calibration and Validation of the Tidal Flow Model, a) Calibration with Spring Tide and b) Validation with Neap Tide.

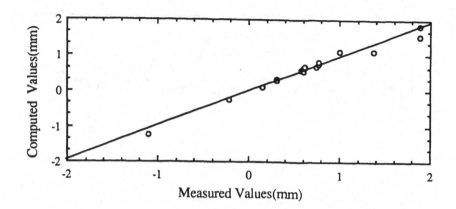

Figure-5: *Calibration of the Erosion-Deposition Model, and Comparison with the Field Data.*

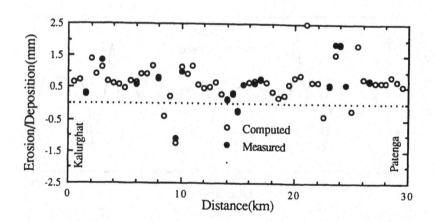

Figure-6: *Computed Erosion-Deposition Pattern in the Lower Karnafuli River.*

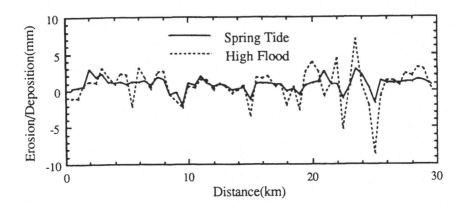

Figure-7: Comparisons of the Effects of Tidal Range and Flood
Discharge on the Erosion-Deposition Pattern.

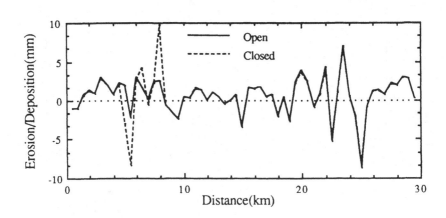

Figure-8: Effects of Closing a Secondary Channel on the
Erosion-Deposition Rates.

39 A study on problems in suspended sediment transportation

Z. Wang and A. Dittrich

ABSTRACT

A sophisticated theory of suspended load motion is presented in this paper based on numerical solutions of the diffusion equation with different boundary conditions. Wash load is distinguished from bed material load by the Rouse number Z according to the theory. Comparison with measured data from hyperconcentrated flows indicate that the new criterion even holds in the case of flows with high concentrations of cohesive sediment. Analysis and measured data indicate that serious erosion took place if $Z_b < 0.06$. A new parameter for the sediment-transporting capacity of bed-material load is suggested that allows to unify the measured data from rivers and pipe flows, low concentrated and hyperconcentrated flows. By taking the effect of the upper boundary in pipe flow on the momentum diffusion coefficient into account, a new solusion of the diffusion equation is presented which coincides well with various measured concentration distributions in pipe flows.

1. Introduction

Suspended sediment can transform into bed load when they fall on the bed, roll or move in saltation. On the other hand bed load can also be suspended if the particles are transported into regions where the turbulent velocity components are sufficiently high to hold them in suspension. The transition between bed load and suspended load is complex but, on statistical meaning, suspended load can be distinguished from bed load by the criterion $Z = \omega/\beta K U^* < 3$ or 5 [4]. Sediments transported in natural streams have a wide sieve curve. The larger sizes down to some minimum size are similar to those already composing the bed so that the transport rate can be adjusted to the capacity rate of the flow. The finer fractions which are more easily kept in suspension are not so likely to be found in the bed. Consequently, the portion of particles up to sizes smaller than those found in the bed can be transported under conditions less than the sediment-transporting capacity of the flow. As a result, the transport of this portion is independent of flow and depends only on conditions extraneous to the flow. This fine material has been given the name wash load to distinguish it from the bed-material load. By rule of thumb, the critical diameter of wash load is d_{b5} or d_{b10}, where d_{b5} and d_{b10} are the diameters 5% or 10% of sediment on the channel bed is finer. This criterion is convenient but it is not valid in all cases. For instance, in hyperconcentrated flood, sediment much coarser than d_{b10} can be transported over several hundreds of kilometers without deposition. It is essentially washed through the channel and should belong to the category of wash load.

It is well established that concentration distributions of suspended sediment in open channel flow follow the diffusion equation,

$$e_s \frac{dC_v}{dy} + \omega C_v = 0 \tag{1}$$

where C_v and ω are the volume concentration and fall velocity of suspended sediment, respectively and e_s is the sediment-diffusion coefficient. If e_s is assumed to be proportional to the momentum-diffusion coefficient, i.e. $e_s = \beta e_m$, the sediment concentration as a function of y is given in the form

$$\frac{C_v}{C_{vb}} = (\frac{H - y}{y} \cdot \frac{b}{H - b})^z \tag{2}$$

where H is the depth of the flow, C_{vb} is the concentration at the level y = b and the exponent

$$Z = \frac{\omega}{\beta K U^*} \tag{3}$$

is called Rouse number. With an appropriate selection of ß (usually between 1 and 1.5) the concentration distribution given by Eq.(2) coincides well with the distributions measured in open channel flows, although it is neither valid at the bed nor at the water surface.

Sediment concentration distributions in horizontal closed conduits with low average concentration (<1.5%) still follows Eq.(2), as found by Ismail [3]. However, laboratory studies of flow with suspended sediment at hyperconcentrations (C_v = 4% - 40%) in horizontal closed conduits by Charles [1], Stevens and Charles [5], Newitt (in Govier and Aziz [2]) and Wang and Qian [7] all indicate that the distributions are distinctly different than those suggested by Eq.(2). With sufficiently high concentration, the distribution curves were convex upward in the lower part of the conduit and concave downward in the upper part with the greatest gradient in the central zone. With further increase of average concentration nearly zero or even positive concentration gradients existed in the lower part of the flow.

2. A numerical solution of the diffusion equation

In the following, a two dimensional open channel flow with suspended sediment is considered. Suspended particles move together with water in the flow direction, fall towards the bed under the action of gravitation and diffuse in the opposite direction of the concentration gradient. The amount of suspended sediment transported through a unit area of a section in the flow field with a normal unit vector n is given by $q.n$, in which

$$q = (u\,i - w\,j)\, C_v - e_s \nabla C_v \tag{4}$$

where ∇Cv is the concentration gradient vector, i and j are the unit vectors in the flow direction (x-coordinate) and the vertical direction (y-coordinate), respectively. The mass conservation equation gives

$$\frac{\partial C_v}{\partial t} + \nabla \cdot q = 0 \tag{5}$$

Assuming u independent of x, ω independent of y and $\partial C_v / \partial t = 0$, Eq. (5) can be rewritten as

$$u \frac{\partial C_v}{\partial x} = \omega \frac{\partial C_v}{\partial y} + \frac{\partial e_s}{\partial y} \cdot \frac{\partial C_v}{\partial y} + e_s \frac{\partial^2 C_v}{\partial y^2} \tag{6}$$

Eq.(6) is the general diffusion equation. The boundary conditions can be formulated as follows:

(1)   On the surface $q \cdot j = 0$, or

$$- (\omega C_v + e_s \frac{\partial C_v}{\partial y}) = 0 \quad , at \quad y = H \tag{7}$$

where H is the flow depth;

(2)   The sediment concentration near the bed is determined by the bed load concentration and is constant in steady flow,

$$C_v(x,y_0) = C_{v0} \tag{8}$$

(3)   The sediment concentration at the entrance is assumed to be uniformly distributed

$$C_v(0,y) = C_{vi} \tag{9}$$

and the sediment-diffusion coefficient e_s is proportional to the turbulent momentum-diffusion coefficient e_m, defined as

$$e_m = U^* \frac{(1-\frac{y}{H})}{(\frac{du}{dy})} \tag{10}$$

Using the Prandtl-von Kármán velocity defect law, Eq.(10) can be written

469

$$e_m = K \, U^* \, y \, (1 - \frac{y}{H}) \qquad\qquad (11)$$

where U^* is the shear velocity, K is the von Kárman constant. Eq.(11) is not valid close to the boundary $y = H$ because diffusion of momentum and sediment at the surface still exist. Therefore, it is modified as follows

$$e_m = \left\{ \begin{array}{ll} K \, U^* \, y \, (\dfrac{1 - y}{H}) & , \; 0 < y \le aH \\[2mm] K \, U^* \, H \, a \, (1 - a) & , \; H \ge y > aH \end{array} \right. \qquad (12)$$

where a is a constant and found to be equal to 0.7 (see section 6 and Fig.7). The numerical solutions of Eq.(6) are computed with an explicit difference scheme.

Fig.1 gives the computed concentration profiles at different distances from the entrance under the boundary conditions $C_{v0} = 0.2$ and $C_{vi} = 0$, 0.2 and 0.4 with a Rouse number $Z = 2$. The results show that the concentration distributions are the same and stable for different incoming sediment concentrations C_{vi}, as long as $x > 60H$ which is the equilibrium distance for adjustment of sediment distribution. The results for $Z = 8, 4$, and 0.8 are similar, although the equilibrium distances are now larger but still less than 400H. In all these cases, the sediment concentration at distances > 400H is determined by the concentration of the bed load zone C_{v0} and the value of Z, and is independedent of the incoming sediment concentration.

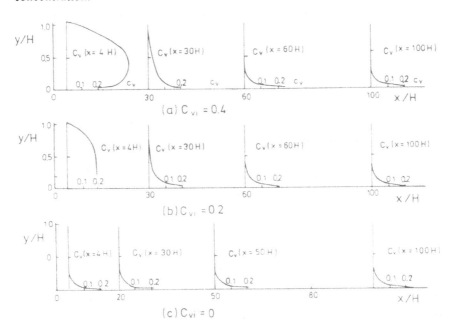

FIGURE 1: *Concentration distributions of suspended sediment (Z=2)*

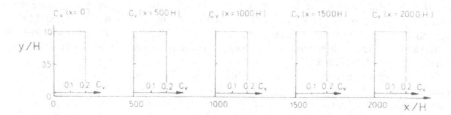

FIGURE 2: *Concentration distributions of suspended sediment (Z = 0.01)*

However, as Z is very small, the results are completely different. Fig.2 shows the concentration profiles for Z = 0.01. Very small Z values imply very weak settling and very strong turbulent diffusion. This is the case as the sediment is either fine or the liquid phase consists of clay suspension and the turbulence intensity is high. As can be seen from Fig.2, for small Rouse numbers the concentration profiles are uniform and only dependent on the incoming concentration C_{vi}.

3. Wash load and bed-material load

From the results presented above, it follows that a strong exchange between suspended sediment and bed material exists if Z is large. In contrary, the suspended sediment is essentially washed through the channel and no exchange with bed material will occur if Z is small. Thus, we can conclude that the sediment-laden flow with large Z values should be referred to bed-material load and that with small Z values to the category of wash load. According to the computed results, the transition from wash load to bed-material load is in the range of Z = 0.06 to Z = 0.1, i.e. the suspended sediment is wash load as Z < 0.06 and bed-material load as Z > 0.1. With the new criterion wash load and bed-material load are mutually transformable. During hyperconcentrated floods in the Yellow River and its tributaries in Northwest China, the high concentration of sediment and the existence of fine particles largely reduces the fall velocity of coarse particles. Consequently, the value of Z for coarse sediment can be smaller than 0.06. In this case, coarse particles with diameters larger than 0.1 to 0.2 mm can be transported as wash load.

Fig.3 shows the average weight concentrations S (kg/m³) measured in the Yellow River and its tributaries in the case Z < 0.06 versus the corresponding values measured at upstream gauging station S_{up}. The values of S_{up} were taken a time period T ahead, where T (=L/U) equals the distance between the two stations (more than 100 km) over the average velocity. The results prove that for sediment with Z < 0.06 the concentration does not change even after a travel over 100 km, and thus

$$S = S_{up} \quad as \quad Z < 0.06 \tag{13}$$

In the middle reach of the Yangtze River the average depth is about 4 m to 30 m, the average river slope is about 0.015% to 0.1% and the average sediment concentration is lower than 1 kg/m³. The critical diameter for wash load is 0.06 mm to 0.15 mm depending on the shear velocity U*. These values coincide well with the empirical values from investigations for a great project on the river (the Three Gorges Project) [8].

Thus, we can conclude from above that sediment carried by the flow can be referred to wash load if Z < 0.06, to bed-material load if 0.1 < Z < 3 which is supported by the diffusive force of the flow and has significant influence on channel bed deformation, and to bed load if Z > 5 which is little affected by turbulent diffusion and depends mainly on the tractive force of the flow.

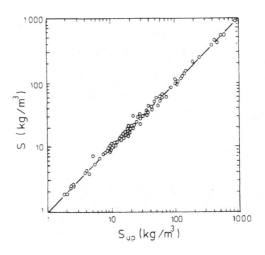

FIGURE 3: *Average concentration of wash load in the Yellow River and its tribu-taries versus upstream concentration S_{up}*

4. Sediment transporting capacity

The sediment-transporting capacity of a flow is the maximum rate of a particular sediment aggregation that can be transported by the fluid flow. Since wash load depends only on the incoming quantity and is independent of the local flow conditions, the following discussion about sediment-transporting capacity will only refer to bed-material load. If for any reason a local change in transport rate does occur, a corresponding local change in bed elevation has also to appear. An increase in the transport rate will result in local scour, and a decrease will result in local deposition. To represent the sediment-transporting capacity the average sediment concentration in equilibrium S_* (kg/m³) will be used. Fig.4 shows S_* related to C_{v0} and Z based on computed results, where S_* is calculated with $S_* = 2650 \cdot C_v$. The sediment-transporting capacity S_* increases with increasing C_{v0} and decreasing Z. If Z > 4, the Rouse number has little influence on the sediment-transporting capacity, because sediment with larger Z values move as bed load instead of suspended load and the sediment-transporting capacity depends only on the concentration C_{v0} of bed load. The bed load transport rate depends in turn on the dimensionless bed shear stress Θ.

Many formulas for sediment-transporting capacity have been proposed and practically used for open channel flow. However, these formulas, can not be used for hyperconcentrated and pipe flow. For instance, average sediment concentrations measured in horizontal pipe flows can deviate by a factor of more than 100 even in the case that the sediment-transporting capacity is in equilibrium. Fig.5 shows the relationship between the pressure gradient I, the average velocity U and the average concentration S in an experiment conducted in a 18 cm x 10 cm rectangular pipeline with 0.15 mm uniform quartz sand and water [6]. At U = 1.5 m/s, S_* could vary from less than 100 kg/m^3 to 900 kg/m^3 while there existed a thin layer of deposited sediment on the bed. Most of the available formulas for sediment-transporting capacity can not interpret this phenomenon.

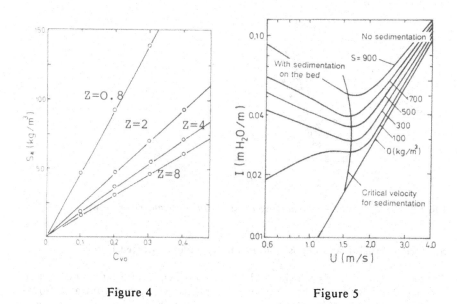

Figure 4   Figure 5

FIGURE 4: *Sediment carrying capacity S_* versus C_{v0} and Z*
FIGURE 5: *Pressure gradient as a function of average velocity and average sediment concentration in suspended sediment-laden pipe flow*

The fact that the sediment-transporting capacity S_* increases with increasing C_{v0} or the bed shear stress Θ and decreasing Z, it is found that, with $\Theta/\sqrt{Z}$ as a new parameter, the data of sediment-transporting capacity of rivers and pipelines can been unified. Fig.6 shows the relationship between S_* and $\Theta/\sqrt{Z}$, for data with $Z > 0.1$ and under equilibrium conditions. In hyperconcentrated flow the sediment-transporting capacity can increase largely owing to the reduction of fall velocity (consequently reduction of Z). In pipe flow with sediment near the bed moving as bed load, an increase in the pressure gradient or the shear stress Θ enhances bed load concentration and consequently increases sediment-transporting capacity S_*, even if the average velocity does not change.

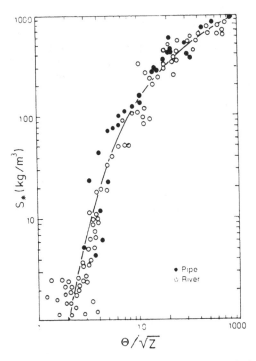

FIGURE 6: *Sediment-transporting capacity in pipe and river flow*

5. Serious erosion and deposition during hyperconcentrated floods

Because of significantly varying fall velocity with concentration and content of clay particles, the sediment-transporting capacity in hyperconcentrated flow is very unstable. For this reason, it is not unusual that serious erosion can be observed in the Yellow River and its tributaries during hyperconcentrated floods. Table 1 presents data of events that caused serious erosion in the Yellow River. The data illustrate that after the occurence of hyperconcentrated floods the river bed was dug 1 to 10 meters deeper in only 10 till 40 hours.

We have mentioned above that coarse sediment can be wash load during hyperconcentrated floods because the fall velocity of the particles is reduced considerably owing to high concentrations and the existence of fine particles. In this case, bed load can also transform into suspended load or even wash load. The river bed will be scoured contineously as far as the prevention of the bed load cover has been lost. At this stage, if all particles scoured from the bed are trans-

474

ported as wash load, no new bed load layer can form and thus serious erosion will occur.

At critical condition of serious erosion the Rouse number for bed material Z_b will be less than 0.06, with

$$Z_b = \frac{\omega_b}{\beta K U^*} \tag{14}$$

and ω_b is the group fall velocity of the river bed material.

TABLE 1 *Serious Erosion in the Yellow River*

Gauging Station	Year	Time	u [m/s]	S [kg/m³]	d_b [mm]	ω_b [cm/s]	Z_b	Note
Longmen Station	1964	July 06-07	0.210	695.0	0.200	2.30	0.047	Scoured 3.1m
Longmen Station	1966	July 18-19	0.220	933.0	0.160	1.70	0.018	Scoured 7.4m
Longmen Station	1969	July 27-28	0.200	752.0	0.170	1.90	0.036	Scoured 1.8m
Longmen Station	1970	Aug. 02-03	0.320	829.0	0.200	2.30	0.022	Scoured 8.8m
Longmen Station	1977	Aug. 05-08	0.240	821.0	0.250	3.20	0.042	Scoured 2.0m
Longmen Station	1979	July 24-25	0.085	424.0	0.220	2.40	0.236	No Scoure
Huayuankou Station	1977	Aug. 07-08	0.070	911.0	0.190	2.30	0.079	No Scoure

The data in table 1 confirm the suggestion and show that serious erosion took place as $Z_b < 0.06$ and did not occur as $Z_b > 0.06$. The smaller the value of Z_b, the more serious the erosion would be.

On the other hand, if floods at hyperconcentrations move through broad sections or over floodplains the average velocity and turbulence intensity will be reduced considerably, resulting in enhanced Z values. Parts of wash load particles transform into bed material load and, at the same time, a reduction of the sediment-transporting capacity of bed-material load will occur. Consequently, some coarser particles deposit and affect the suspension of finer particles because the fall velocity of the suspended sediment increases due to decrease in concentration. Consequently more sediment deposits on the bed. The chain reaction proceeds very fast if the size distribution of the suspended sediment is uniform. In this way, the river bed topography can change tremendously after only one or two days of hyperconcentrated flood.

6. Concentration distribution in closed conduit

The diffusion equation consists of two parts in closed conduit flow. For demonstration, let us consider a sediment-laden flow in a horizontal pipe line with rectangular cross section. For steady flow the diffusion equation for sediment in the lower part of the pipe follows the relationship given by Eq.(1). For the upper part of the pipe flow the diffusion equation can be written as (the y' axis is taken as positive from the upper boundary towards the pipe center)

475

$$e_s \frac{dC_v}{dy'} - \omega C_v = 0 \tag{15}$$

The distribution of e_m can be modified in the way

$$\epsilon_m = \begin{cases} K\, U^* \, y \, (1 - \dfrac{y}{y_m}), & y \le a y_m \\[2mm] K\, U^* \, y_m \, a \, (1 - a), & a y_m < y \le y_m \end{cases} \tag{16}$$

with y_m the vertical distance from the bed at which the velocity reaches its maximum value. In a first assumption, y_m has been taken equal the half pipe diameter ($y_m = D/2$). With $a = 2/3$, the distribution of the momentum diffusion coefficient measured by Ismail [3] is similar to the relationship given by Eq.(16). Fig.7 shows the distributions of ϵ_m in closed conduit according to Eq.(11) and Eq.(16), where a is taken as 0.7.

The von Kármán constant K' in the upper part of the pipe flow is different from K in the lower part and is assumed to be

$$K' = \alpha\, K \tag{17}$$

Usually α is slightly larger than one. Substituting Eq.(16) into Eq.(1) and (15), transforming the axis y' into y by $y' = D - y_m$, integrating the equations and determining the integration constants from continuity of concentration, the following concentration distribution will be obtained

$$\frac{C_v}{C_{vb}} = \begin{cases} \left(\dfrac{(y_m - y)}{y}\right)^Z & 0 < y \le a y_m \\[3mm] \left(\dfrac{(1-a)}{a}\right)^Z \exp\left(\dfrac{-Z}{(1-a)} \cdot \dfrac{y}{(ay_m - 1)}\right) & a y_m < y \le y_m \\[3mm] \left(\dfrac{1-a}{a}\right)^Z \exp\left(-\dfrac{Z}{a}\left(1 + \dfrac{\alpha}{(1-a)} \cdot \left(\dfrac{y}{ay_m} - 1\right)\right)\right) & y_m < y \le (2-a)y_m \\[3mm] \left(\dfrac{1-a}{a}\right)^{Z(1+\alpha)} \exp\left(-\dfrac{1+\alpha}{a}Z\right)\left(\dfrac{y - y_m}{2y_m - y}\right)^{-(\alpha Z)} & (2-a)y_m < y < D \end{cases} \tag{18}$$

where C_{vb} is the reference concentration measured at $y = 0.5 y_m$.

An experiment was conducted in a horizontal rectangular pipe line with cross section 18 cm wide and 10 cm high. The suspended load was quartz sand with diameter 0.15 mm and specific gravity 2.64. The group fall velocity of the suspended load followed the formula

$$\omega = \omega_0 \, (1 - C_v)^7 \tag{19}$$

where $\omega_0 = 1.89$ cm/s and the constant β equaled 1.3 for the chosen sediment. Fig.8 gives a comparison of the measured concentration distributions based on Eq.(18), with K and α determined from the measured velocity distributions [7].

476

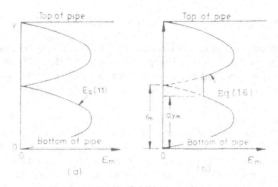

FIGURE 7: *The vertical distribution of the momentum diffusion coefficient*

The agreement is quite satisfactory. At low average velocities and concentrations the measured and calculated data result in distributions similar to those observed in open channel flow (Fig.8a). With increasing concentrations a turning point appears in the central zone and the distribution curve is convex upward in the lower part and concave downward in the upper part (Fig.8b). With an further increase in velocity the distribution curves tend to follow a linear relationship (Fig.8c). At very high concentrations the gradient of the concentration distribution curve approaches zero in the lower part of the flow (Fig.8d). If the effect of lift force due to velocity gradient (Magnus effect) is taken into account, a positive concentration gradient in the lower zone is possible.

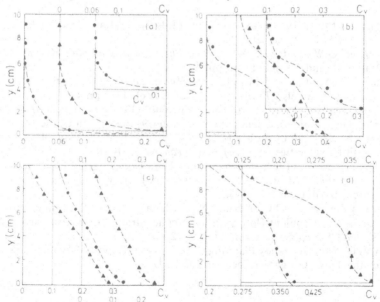

FIGURE 8: *A comparison of Eq. (18) with the measured concentration distribu-*
tions   - - -, Eq. (18)   ▲ & ●, measured data

477

7. Conclusions

Numerical solutions of the diffusion equation indicate that suspended sediment with Z values less than 0.06 distribute uniformly and depend only on the incoming concentration, and sediment with Z values larger than 0.1 depend on the local turbulence intensity and bed load concentration. It is shown that wash load can be distinguished from bed-material load if Z < 0.06. Coarse particles with diameters 0.1 to 0.2 mm in the Yellow River can be transported as wash load if the flow carries much fine sediment. If the Z_b values of the river bed material are less than 0.06 serious erosion occurs during hyperconcentrated floods. The data of sediment-transporting capacity in low concentrated and hyperconcentrated flow, river and pipe flow can be unified by the new parameter $\Theta/\sqrt{Z}$. The concentration distributions of hyperconcentrated suspended load motion in closed conduits differ strongly from those in open channel flow. Two diffusion equations, one is valid in the lower part and the other in the upper part of the pipe flow, have been used to solve the problem. By modifying the distribution of the momentum diffusion coefficient, the integration of the diffusion equations yield solutions that describe well the measured data.

ACKNOWLEDGEMENT

This research was partly supported by the Alexander von Humboldt Foundation

REFERENCES

[1] Charles, M.E., (1970), "Transport of solids by pipeline", *Proc. of Hydrotransport 1*, BHRA, Cranfield, Paper A3.

[2] Govier, G.W., and Aziz, K., (1972), *The Flow of Complex Mixture in Pipes*, Van Nostrand Reinhold Co., pp 792.

[3] Ismail, H.M., (1952), "Turbulent transfer mechanism and suspended sediment in closed channels", *Transactions*, ASCE, **Vol. 117**, pp 409-434.

[4] Qian Ning (Ning Chien), and Wan Zhaohui, (1983), *Dynamics of Sediment Movement*, Chinese Science Press, Chapter 9, Beijing (in Chinese).

[5] Stevens, G.S., and Charles, M.E., (1972), "The pipeline flow of slurries: transition velocities", *Proc. of Hydrotransport 2*, BHRA, Cranfield, Paper E3.

[6] Wang Zhaoyin, (1984), Mechanism of Hyperconcentrated Flows, *Doctorate Dissertation*, Institute of Water Conservancy and Hydroelectric Power Research, Beijing (in Chinese).

[7] Wang Zhaoyin, and Qian Ning (Ning Chien), (1984), "Experiment study of two-phase turbulent flow with hyperconcentration of coarse particles", *Scientia Sinica*, Ser.A, **Vol. 27**, No. 12, pp 1317-1327.

[8] YVPO (Yangtze Valley Planning Office), 1985, Report of Preliminary Design of the Three Gorges Project, (in Chinese).

40 Laboratory and field investigation of a new bed load sampler for rivers

M. T. K. Gaweesh and L. C. van Rijn

ABSTRACT

The problems of bed load measurements in sand bed streams are related to the sampler design and to the variability of the physical process of bed load transport. Efforts have been concentrated in developing a new (bag-type) bed load sampler. The basic criteria for the design of the sampler is that the lower edge of the sampler trap must fit closely to the stream bed. The sampler trap (mouth) is made movable in vertical direction over a distance of about 20 cm with respect to the sampler frame to adapt to the bed forms when the sampler is placed on the bed. Detailed experiments in a laboratory flume as well as field measurements have been conducted to determine the hydraulic and sampling efficiencies of the bed load sampler. The process of development and improvement of the sampler, its characteristics and the results of the laboratory flume experiments and the field measurements are explained.

1 Introduction

The measurement of bed load transport in sand bed streams is a significant problem. The basic problem is that the sampler must be placed on the stream bed where bed forms, ripples and dunes, are usually present. Therefore, the flow pattern and bed load movement in the vicinity of the sampler are altered to some extent. As a result, samplers do not catch material at the true rate and must be calibrated to determine their trapping efficiencies under different conditions of transport rates and bed material particle sizes.

Despite the various samplers that have been developed to measure the bed load, none has been universally accepted. Different techniques and types of bed load samplers are still not adequate for obtaining accurate measurements, particularly in sand bed streams. The problems of bed load measurements are mainly related to the sampler design and to the variability of the physical process of bed load transport. The sampling (trapping) efficiency of a bed load sampler depends on its hydraulic efficiency, the degree of contact between the sampler mouth and the bed, and on sampling disturbances generated at the beginning and at the end of the sampling period.

Typical problems of bed load sampling which are related to the instrumental design, can be classified as follows: a)The initial effect; sand particles of the bed may be stirred up and trapped in the sampler mouth when the sampler is placed on the bed (oversampling); b)The gap effect; a gap between the bed and the sampler mouth may be present initially or generated at a later stage under the mouth due to migrating ripples or erosion processes (undersampling); c)The blocking effect; blocking of the bag material by sand, silt, clay particles, and organic materials will reduce the hydraulic efficiency and thus the sampling efficiency (undersampling); d)The scooping effect; the sampler may drift downstream from the sampling location during lowering it to the bed, and may be pulled forward (scoop) over the bed when it is raised up again acting as grab sampler (oversmapling).

Typical sampling problems related to the variability of the physical processes of bed load transport are: a) the number of measuring locations along a bed form, b) the number of measurements at each location, and c) the sampling duration of each measurement.

In view of these mentioned problems of bed load sampling, a new bed load sampler (called the Delft-Nile sampler) has been designed and improved by Delft Hydraulics (DH) in cooperation with the Hydraulics and Sediment Research Institute (HSRI).

2 Sampler characteristics

The new bed load sampler (Figure 1) has been designed with unique characteristics which permit accurate measurements of actual bed load transport in sand bed streams. The unique feature is a movable sampler mouth (width = 0.096m, height = 0.055 m) which can move in vertical direction (upward and downward) over a distance of about 0.20 m with respect to the sampler supporting frame. This movable mouth construction was designed to allow a perfect contact between the lower edge of the sampler mouth and the forms of the stream bed, and to reduce the gap effect as much as possible. A nylon bag (0.18 x 0.32 m^2) with a mesh size of 250 μm is connected to the sampler mouth. At the upper side of the bag, a nylon patch (0.10 x 0.15 m^2) with a mesh size of 500 μm is present to reduce the clogging of the bag by fine particles as much as possible. The bottom side of the sampler mouth has a sharpened edge and a slope of 1:10 to facilitate its insertion into the sand bed surface. The bed load particles can easily enter the mouth and are then trapped in the rear side of the mouth. In this way, the erosion processes of the bed at the front of the sampler can be avoided, so that representative samples can be obtained.

2.1 *Hydraulic efficiency*

The hydraulic efficiency of the bed load sampler, defined as "the ratio of the sampler intake velocity to the natural stream velocity", was investigated by measurements in a laboratory flume. The mean velocity through the entrance of the sampling bag was measured using a small Ott-propeller meter. This later velocity was compared with the mean flow velocity measured simultaneously at the same height at a section 2m upstream of the sampler mouth. Ten successive measurements of 30 seconds each were carried out. Three types of bags were tested: 250 μm nylon bag, 150 μm nylon bag, and an impermeable plastic bag. At the upper side of each bag a nylon patch of 500 μm was present. The impermeable bag is supposed to simulate a nylon bag which is fully clogged by fine silt, clay, and organic material as may be present in natural conditions [1]. Four filling percentages (by volume) of the sampler bag were tested: 0%, 25%, 50%, and 75%. The results are given in (Table 1).

The hydraulic efficiency of the 250 μm-bag and the 150 μm-bag, both with a patch of 500 μm at the upper side, are about unity for filling percentages in the range of 0 % to 50 %. A filling percentage of 75 % reduces the hydraulic efficiency to about 0.75. Hydraulic efficiency larger than one is caused by the geometry of the sampler. The hydraulic efficiency of an impermeable plastic bag with a patch of 500 μm at the upper side, simulating a fully blocked bag, is about 0.8 for a filling percentage in the range of 0 % to 50 %. A filling percentage of 75 % reduces the hydraulic efficiency to about 0.70. Based on these results, a maximum filling percentage of 50 % is advised to be used. Blocking of the nylon material by fine sediments will result in a hydraulic efficiency of about 0.8 which still considered as a high efficiency.

TABLE 1 *Hydraulic efficiency of the new bed load sampler*

velocity in sampler mouth (m/s)	mesh size bag (μm)	mesh size patch (μm)	hydraulic efficiency			
			filling percentage			
			0 %	25 %	50 %	75 %
	250	500	1.04	1.01	0.98	0.91
0.5	150	500	1.07	1.04	0.98	0.94
	(impermeable)	500	0.82	0.82	0.82	0.82
	150	500	1.05	1.02	1.01	0.75
0.8	150	500	1.08	1.03	0.99	0.75
	(impermeable)	500	0.84	0.80	0.80	0.70

3 Investigation of sampling efficiency

The sampling efficiency is defined as "the ratio of the bed load transport measured by the sampler at a certain location during a certain period to the true bed load transport at the same location during the same period if the sampler had been absent" [5]. The sampling efficiency of many samplers is not constant but varies with the transport rate and sediment size [7]. An experimental program was prepared to investigate the sampling efficiency of the new bed load sampler for different transport rates and particle size distributions of bed material.

3.1 *Laboratory experiments*

The experimental work was performed in a laboratory flume at HSRI. The flume is 26 m long with a square cross section of (1 x 1 m²) and is provided with glass side walls to facilitate visual observations. A layer of 0.2 m thickness of sand (median particle size = 400 μm) was placed on the flume floor. The flume is operated with a sand feeding system. A sand trap is located at the downstream end of the flume to collect the eroded sand during the run time. Flow discharges up to about 0.25 m³/s can be delivered into the flume inlet. The bed load sampling was performed at a fixed section located at the middle of the flume length.

The sampling efficiency was studied by conducting a series of tests of different transport rates. The daily average rate of the sand feed and trap systems was used as an estimate for the true bed load transport rate at the measuring location to be compared to the average bed load transport rate measured by the sampler.

The sampled bed load transport rate (q_b) was determined as:

$$q_b = (G - G_o) / b\,t \tag{1}$$

where G = dry weight of sand catch for a sampling period (t); G_o = dry weight of sand catch due to initial and scooping effect; b = width of sampler mouth; and t = sampling period. Values of (G_o) were determined by measurements using instantaneous bed load sampling; the sampler was lowered into the bed and immediately raised up after the mouth had touched the bed.

Different sampling periods of 1, 3, 5 and 10 minutes were used in each test. The maximum sampling period was selected to provide samples that would not fill more than 50 % of the capacity of the sampler bag. For each sampling period, individual samples were consecutively collected. The total sampling time of the individual samples was kept constant. The collected samples exhibit large cyclic variations in the measured bed load transport rates. Laboratory and field observations of large variations in the measured bed load transport rates have also been reported in the literature [2] and [6]. The mean bed load transport rate was determined based on statistical analysis of the moving average which essentially smoothed out the fluctuations of the individual bed load transport rates. The sampling efficiency was determined as the ratio of the sampled to the true bed load transport rate. The results of test 3 and 4 with a mean flow velocity of 0.42 and 0.51 m/s, respectively, are shown in (Table 2).

TABLE 2 *Sampling efficiency of the new bed load sampler*

test number	sampling period (minutes)	sampling efficiency
	10	0.81
	5	1.29
3	3	1.42
	1	1.23
	10	1.07
	5	0.85
4	3	0.92
	1	1.32

The values of sampling efficiency, presented in (Table 2), varied between 81 % and 142 % indicating high trapping efficiency of the new sampler. Comparison of the median particle size of the trapped samples was identical to that of the bed load transported in the flume indicating representative bed load sampling.

A theoretical treatment of bed load transport variability under constant flow conditions was given by Hamamori [4]. He derived a probability distribution function of relative transport rates for the case of secondary ripples on top of primary dune bed forms. The ratio of each individual bed load transport rate divided by the long-term mean rate is the relative rate for each individual. Figure 2 shows the theoretical distribution of Hamamori and the distributions of relative individual transport rates measured by the sampler during test 2. The theoretical distribution shows that individual transport rates vary from zero to four times the long-term mean rate and that 60 % of all rates are less than the mean. The distributions of the relative transport rates measured by the sampler correspond closely with the theoretical distribution.

483

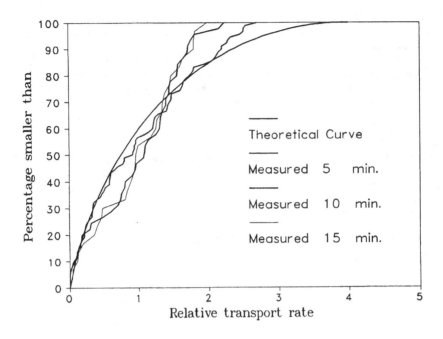

Y-axis: Percentage smaller than (0, 10, 20, 30, 40, 50, 60, 70, 80, 90, 100)

X-axis: Relative transport rate (0, 1, 2, 3, 4, 5)

Legend:
Theoretical Curve
Measured 5 min.
Measured 10 min.
Measured 15 min.

FIGURE 2 *Distributions of relative transport rates (laboratory)*

4 Field measurements

The inherent variability of bed load transport process was investigated in the field using the new sampler. Bed load measurements were performed to provide information about the number of measuring locations along a bed form, the number of measurements at each location, and the sampling duration of each measurement. The measurements were conducted in the Nile river, in Egypt, at Bani Mazar. At that site, the width of the river is 570 m and the river bed consists of medium to coarse sand with median particle size ranges between 390 µm and 480 µm. The measurements were performed at the deepest part of the main channel, and sampling was done from a survey boat. Sounding of the bed forms showed the presence of dunes with an average length of 20 m and height of 0.20 m. Five sampling locations with intervals of 5 m were selected over a longitudinal section with a length almost equal to the dune length. The water depth at the sampling locations was about 4 m and the mean water velocity ranged from 0.85 to 0.93 m/s.

During a four-day period of nearly constant water discharge, four sets of bed load samples were collected using the new sampler. The first and second set consisted of 30 and 50 individual samples with a sampling duration of 5 and 3 minutes, respectively. The number of bed load samples of both sets was equally distributed over the five measuring locations. The sum of the individual samples and the sampling duration of the third and fourth set were identical to that of the first and second set, respectively. However, all the bed load samples of the third and forth set were collected at only one measuring location. The bed load samples of each set were collected consecutively with an approximate four-minute interval between samples. A number of buoys were used to facilitate positioning of the survey boat at exactly the same measuring locations during the mentioned period.

4.1 *Analysis of the field data*

Individual bed load transport rates measured at the five sampling locations exhibit larger cyclic variations compared to that measured at only one location. Maximum transport rates occurred near the bed form crest (high velocities), while minimum transport rates occurred near the bed form trough (low velocities). The mean bed load transport rate was determined based on statistical analysis of the moving average. However, accurate determination of the mean bed load transport rate at a fixed sampling location requires sequential sampling over a period long enough for a migrating bed form to pass the sampling location. This means that, in rivers with low migration velocity of bed forms, a very long period of measurements and an enormous number of samples would be required to obtain a proper mean of the bed load transport rate. Therefore, representative sampling should be carried out at different locations along the bed form.

When the relative bed load transport rates of the four sets were plotted as cumulative frequency distributions, they showed good agreement with Hamamori's theoretical distribution. Figure 3 shows the theoretical distribution of Hamamori and the distribution of the relative individual transport rates measured along the bed form of the second set.

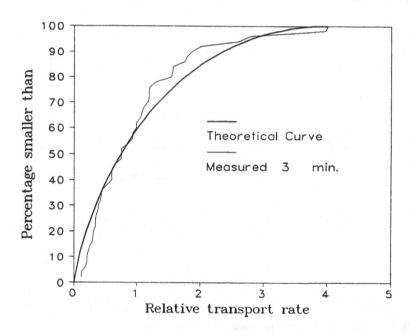

FIGURE 3 *Distributions of relative transport rates (field)*

The measurements were used to determine the number of bed load samples required to obtain a proper mean when samples are collected at different locations along the bed form length. Figure 4 shows the variation coefficient of the mean bed load transport rate as a function of the number of samples for the measurements carried out along the bed form length of the second set. The variation coefficient (V), also defined as the dimensionless standard deviation, is determined as follows:

$$V = \sigma / (\mu \, N^{0.5})\tag{2}$$

where σ = standard deviation of measured bed load transport rates, μ = arithmetic mean of measured bed load transport rates, and N = number of samples. According to the results depicted in Figure 4, the value of the variation coefficient of 50 samples is 0.12. This means that, if 50 samples were collected, then the mean bed load transport rate will be within ± 12 percent of the true mean. Similarly, the figure shows that if only 20 samples were collected, then the mean bed load transport rate will be within ± 22 percent of the true mean. Similar results were obtained by statistical analysis for samples collected at random positions from several bed forms [3].

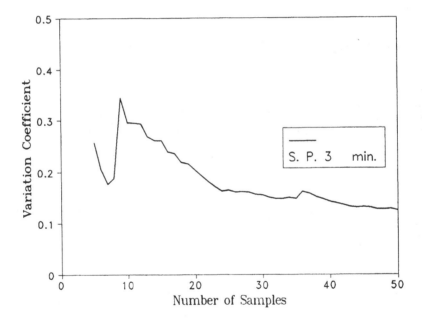

FIGURE 4 *Variation coefficient determined from the field data*

However, the number of samples necessary for a substantial mean of bed load transport rate depends upon the accuracy required, the time scale of cyclic rate variations, and whether the sampler is static (fixed at one location) or can be moved relative to the transport process [2].

Some of the measurements showed a large sand catches (the sampler bag was half full) in 5 minutes duration with depth-averaged velocities in the range of 0.85 to 0.93 m/s. Accepting a maximum filling percentage of 50 % of the sampler bag, the weight of the corresponding sand catch is about 2000 grams. The bed load transport formula of Meyer-Peter-Muller has been applied to determine the maximum sampling period for bed material of a median particle size of 500 μm, and depth-averaged velocities in the range of 0.75 to 1.5 m/s. The maximum bed load transport rate was assumed to be equal to three times the mean transport rate. The results are given in (Table 3).

Based on the results shown in Table 3, it is advised to use a maximum sampling period of 5 minutes for depth-averaged velocities up to 0.8 m/s. At higher velocities up to 1.5 m/s (higher transport rates) the maximum sampling period must be reduced (between 30 and 180 second) or a larger bag should be used. The bed load samples of each set were analyzed to determine the grain size distribution. Similarly, bed material samples collected by a grab sampler at the same locations were analyzed. Comparison of the D_{10}, D_{50}, and D_{90} - values of the bed material and the bed load samples shows close agreement indicating representative sampling of the bed load sampler.

TABLE 3 *Maximum sampling period*

depth-averaged velocity (m/s)	maximum sampling period (seconds)
0.75	450
1.00	120
1.25	50
1.50	30

5 Conclusions

A new bag-type bed load sampler was designed with unique characteristics which consists of a movable sampler mouth to allow a perfect contact between the lower edge of the sampler mouth and the forms of the stream bed. The sampler was tested under laboratory and field conditions and has proven to be highly efficient under both conditions. The hydraulic efficiency was found to be close to unity for filling percentages up to 50 % of the capacity of the sampler bag. The sampling (trapping) efficiency was studied in a laboratory flume with sand bed (D_{50} = 400 μm) by conducting a series of tests of different transport rates. Comparison between the sampled rates with the corresponding true rates indicates high trapping efficiency. Comparison of particle size distributions of the trapped samples showed close agreement with that of the bed material, indicating representative sampling. Based on analysis of the field investigation, it is advised to use a maximum sampling period of 5 minutes for depth-averaged velocities up to 0.8 m/s. At higher velocities (up to 1.5 m/s), the maximum sampling period must be reduced (between 30 and 180 seconds)or a larger sampler bag should be used. Laboratory experiments are currently being executed at HSRI to expand the present investigations to a wider range of flow conditions and bed material particle sizes.

ACKNOWLEDGEMENTS

Dr. M. M. Gasser the director of the Hydraulics and Sediment Research Institute at Delta Barrage in Egypt is gratefully acknowledged for his support and providing the laboratory and field facilities to perform this research program in close cooperation with the Dutch technical assistance project. Eng. A.M.Amin of HSRI is acknowledged for his assistance in executing the laboratory and field investigations. The survey team is also acknowledged for their cooperation during the field work.

REFERENCES

[1]   Beschta, R.L., (1981), "Increased Bag Size Improves Helley-Smith Bed Load Sampler for Use in Streams with High Sand and Organic Matter Transport", Symposium Erosion and Sediment Transport Measurement, Florence, Italy.

[2]   Carey, W.P., (1985), "Variability in Measured Bed Load Transport Rates", Water Resources Bulletin, American Water Resources Association, Vol.21, Paper No. 84156, U.S.A.

[3]   De Vries, M., (1973), "On Measuring Discharge and Sediment Transport in Rivers", Publication 106, Delft Hydraulics, Delft, The Netherlands.

[4]   Hamamori, A., (1962), "A theoretical Investigation on the Fluctuations of Bed Load Transport", Report R4, Delft Hydraulics, Delft, The Netherlands.

[5]   Hubbell, D. W., (1964) "Apparatus and Techniques for Measuring Bed Load", Water Supply Paper 1748, United States Geological Survey, Washington D.C.

[6]   Hubbell, D. W., et al., (1985), "New Approach to Calibrating Bed Load Samplers", *Journal of Hydraulic Engineering Division*, ASCE, Vol.111, No. 4, Proc. Paper 19655.

[7]   Novak, P., (1957) "Bed Load Meters-Development of a New Type and Determination of their Efficiency with the Aid of Scale Models", *Transactions, International Association of Hydraulic Research*, 7 th General Meeting, Vol. 1, Lisbon, Portugal.

Part 6
EXPERT SYSTEMS

41 A microcomputer-based tool for the investigation of coastal hydrodynamic systems

D. G. Kleinschmidt and B. R. Pearce

Introduction:

Those of the modeling community who are involved in the investigation and simulation of processes in estuarine regions have found a powerful and flexible tool in the mainframe computer. Since the development of effective microcomputers, modelers have often utilized them to assist in the preparation of input data sets and in the presentation of output data in plotted or tabular form. As processor speeds and memory capacities increase, microcomputers are being used for numerical modeling as well.

In an effort to develop programs that are not only easier to use and control but whose output is more meaningful and understandable, modelers are beginning to write models with graphical user interfaces (GUIs). The term GUI describes an interface in which the operator's interaction with the program (model) running on it is via the display screen and a pointing device. The experiment described herein takes the process one step farther in that the model runs fast enough that one views the model's output as a picture or movie instead of "numbers" and can interact with the model in real time (or "on the fly") via the GUI.

This has some very real benefits. The human brain can process some 10 - 100 gigabits of visual data per second as color and motion. By comparison let the reader judge his or her ability to read and interpret numerical model output as text or ASCII files. The difference is at least 6 orders of magnitude! This means that visually the user can much more quickly interpret model results, readjust necessary model parameters and finally can much more quickly reach a usable solution. This article describes a model that was developed with the intent of taking a step toward these goals.

The following includes a brief description of the project for which a model was required; a brief description of the physical basis for the model; and, finally, a description of and results from the GUI model PLUME. Before continuing, the authors wish to state that the goals of the project did not include calibration and verification of the particle tracking model, rather the goals were to test the viability of combining the display of hydrodynamic data and a running model via a GUI. We should note that this application was developed using "C". Upon reflection, the authors are confident that an object oriented programming (OOP) language would be helpful in facilitating the GUI development.

Project description:

The goal of the project was to study the applicability of a microcomputer based numerical model for the analysis of a hydrodynamic system. Three major criteria should be met by such a model: 1. The model should run fast enough to enable a viewer to observe the hydrodynamic processes of interest during the model run. 2. The model should present results in a variety of graphical formats. 3. Appropriate aspects of the model's operation should be controllable by the viewer while the model run is in progress, using an interactive graphical format.

An ongoing project provided a modeled area of suitable size and an input data set consisting of a time series of velocity vectors and water depths for each wetted cell in the grid. The project involved determining the location and extent of a waste heat plume produced by a fossil fueled power plant located adjacent to the Delaware River. Figure 1 shows the section of the Delaware River Estuary comprising the study area. The models used in the base study are described by Foster [3], and Pearce et al. [7].

The particle tracking microcomputer model (Plume) is written in the C programming language [5] and currently runs on a Macintosh IIci microcomputer. The Macintosh was selected primarily because of its availability, its graphic user interface functions, and its ability to handle large data sets in RAM.

To some extent, all three of the above criteria have been realized. Results can be presented as temperature rise isotherms or particle positions. Interactive control of the model's operation, via the GUI, has proven to be essential, allowing the operator to observe particle positions and to display, at the same time, other data of interest such as velocities. For example, the GUI allows the operator to select locations within the modeled area where velocity vectors are to be displayed and also allows them to be turned on or off as the model run proceeds.

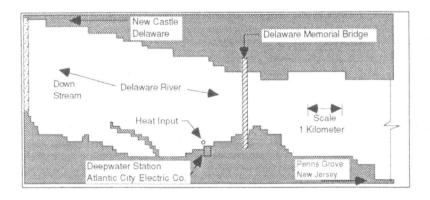

FIGURE 1 *Modeled area*

Models Involved:

The models which were used in the heat transport study consisted of one to establish the hydrodynamic properties of the study region, and one to simulate the transport of a substance contained within the water column. The hydrodynamic model which generated the Input data for the Plume model was based on the conservation of mass and momentum in two dimensions, and was solved by a vertically averaged, explicit, finite difference technique. The two dimensional equations for incompressible flow which for the basis for this solution are described by Reid and Bodine [8], and others.

The second half of the modeling problem is the transport of the substance in question. The mass transport equation which forms the basis for this analysis is:

$$\frac{\partial C}{\partial t} + U\frac{\partial C}{\partial x} + V\frac{\partial C}{\partial y} = \frac{1}{H}\left[\frac{\partial}{\partial x}\left(HD_x\frac{\partial C}{\partial x}\right) + \frac{\partial}{\partial y}\left(HD_y\frac{\partial C}{\partial y}\right)\right] + \lambda\left(C - C_e\right) + Q_s \qquad (1)$$

where:

C   is the concentration
U   is the x velocity
V   is the y velocity
H   is the total water depth
D_x is the diffusivity in x
D_y is the diffusivity in y
λ is the decay coefficient
Q_s is the source or sink rate

For the Plume project, C represents a concentration of heat (joules per cubic meter), output as a temperature rise above ambient in degrees centigrade, lengths are in meters, and time is in seconds. The decay term was set to zero to reflect the worst case condition of no heat loss to the atmosphere.

The model (Lagrangian) is based on the idea that a substance of interest can be discretized into particles thereby allowing a plume to be modeled by following them as they move freely through the modeled area. In effect the observer is riding with each particle through the model domain. It has been shown that the Lagrangian model will provide a complete solution of the mass transport equation approaching the analytical solution as the number of particles approaches infinity [2]. In practical terms this means that the accuracy of the solution is a function of the number of particles within the model. Because the computational effort is directly proportional to the number of particles within the system at any given time, preliminary investigations can be conducted quickly. Final analysis may then be performed by increasing the number of particles.

Another feature of the Lagrangian model is that the computations required are relatively simple. A given particle's motion and even its properties are a function of it's current location within the model. Particles are acted on by two processes, one being advection and the other diffusion. (See Figure 2) Advection is the change in position due to a velocity applied over a time period. Diffusion is a displacement due to the turbulence of the fluid body and is represented in the model by random displacements in X & Y within predefined ranges. Referred to as the step length (Lx & Ly), the range of the displacement is the mechanism whereby diffusion is quantified [1].

The input data set required by this model consists of a description of the hydrodynamic

493

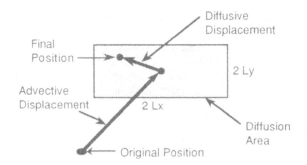

FIGURE 2   *Advection & diffusion displacement schematic*

model area and the time series of velocities & water depths. Also needed are values for various parameters used within the model during the run.

The boundary conditions used are as follows: Particles crossing a wet end of the model area are allowed to leave and are then removed from the model permanently. Particles attempting to cross a solid boundary are returned to their original position and remain there until the next time step. This represents an "approximation" of the actual physical process but is necessary to conserve mass (or heat).

Plume model overview:

The package consists of two programs, the model and an input data preprocessor. The preprocessor is used to take the data output by the hydrodynamic model and condense and organize it into the format required by Plume. By using a preprocessor, many different mainframe models, each with different output data formats, can be used to provide the required input data. The following paragraphs describe the Macintosh Event Manager, the model's Input/Output routines, and the Time Step Processing routine.

The heart of the Macintosh operating system is the "Event Manager". The Event Manager is responsible for detecting the occurrence of events (mouse clicks, keys pressed etc.) and sending the appropriate data to the program. The program is then responsible for deciding what type of event has occurred and what action, if any, is required. This enables a program to have full access to menus and windows and the ability to manipulate their contents. In Plume the main processing loop waits for events. If, in a pass through the loop, an event is present the program sends the event record to the appropriate routine. If no event is detected the next model time step is processed. This allows full control of the model run, enabling the operator to stop the execution, change parameters or request an output display, and either resume execution or restart the model as appropriate.

Within Plume, the implementation of the solution is carried out within the time step processing function. The "Source" is simulated by adding a number of particles to the model equal to the heat output for that time step. A particle is a dimensionless object containing a quantity of energy . The number of particles is equal to the heat output of the station divided by the amount of heat carried by each particle. The Advective Term is handled by the

494

following equations:

$$\Delta X = U(x,y) \, \Delta t \qquad\qquad (2a)$$
$$\Delta Y = V(x,y) \, \Delta t \qquad\qquad (2b)$$

Where:

U(x,y)   is the velocity in the x direction at point (x,y) and time = $n\Delta t$
V(x,y)   is the velocity in the y direction at point (x,y) and time = $n\Delta t$
n   is the number of the current model time step

As stated in a previous section, the diffusion of suspended particles can be simulated by the use of a random displacement in addition to the advective displacement. Koutitas [6], for example, shows that the step length, L, or range of the random motion is related to the diffusion coefficient, D, by:

$$L_x = \sqrt{6 \, \Delta t \, D_x} \qquad\qquad (3a)$$
$$L_y = \sqrt{6 \, \Delta t \, D_x} \qquad\qquad (3b)$$

The long and relatively narrow nature of the Delaware River Estuary in the modeled area produces a predominantly longitudinal current which is strong enough to produce additional turbulent diffusion in the longitudinal direction. This turbulent diffusion has been described by Jozsa [4], and others, as a linear function of the local shear velocity and water depth and allows better definition of the diffusion coefficient for conditions of low flow or shallow depth. The coefficients as used in Plume are given by:

$$D_x = 6 \, H \, U \, \sqrt{\frac{f}{8}} \qquad\qquad (4a)$$
$$D_y = 1.5 \, H \, U \, \sqrt{\frac{f}{8}} \qquad\qquad (4b)$$

Where:

H   is the depth of the water
U   is the longitudinal velocity
f   is the Darcy - Wiesbach friction factor

Display functions which operate on the plume, like the temperature plotting display, are not activated until the end of a time step due to the fact that the next event is not requested from the Event Manager until the time step processing routine has returned control to the main function.

Program elements:

Data input module:

It was decided that in order to maintain flexibility and minimize setup time, the major data set would be preprocessed in a separate program. It was decided to utilize a straight binary format for the input data set. The advantages of doing so are twofold. First, the data set size is reduced by a factor of 2, and second, the time required for loading the data into the program is reduced by a factor of 9. The bulk of the time reduction is due to the fact that the values no longer had to be translated from ASCII to integer format as they were read in.

The data output from the tidal model used in the development of Plume consisted of an X & Y coordinate pair, U & V velocities, and the water depth for each cell and each time step. The origin for the coordinates output from the flow model was in the lower left of the model grid and the origin for the Macintosh's display is in the upper left. The input program condenses the data, and flips the columns to align the data with the Macintosh's display. It also sets up the data structure in memory that is used within the main program. Once this is accomplished, the data is written to disk in a binary stream format. Assembling the data structure is not strictly necessary for the operation of the model, but does serve to insure that there will be fewer problems when the data is loaded into the main program for a model run.

As Plume is improved and used in different projects, the model itself should not require any new programming to adapt it for use. The input data preprocessor will likely have to be changed and recompiled for each new data format. In future versions the authors would like to have "on board" velocity calculations, allowing for the user to change day, and/or tidal conditions, and/or winds, etc..

Data structure:

In Plume the basic data structure used is the cell. It consists of 3 integers each 2 bytes long. The main data set includes a coordinate page followed by 50 velocity vector pages as shown in Figure 3. The grid used in the hydrodynamic model consisted of 122 columns by 44 rows or 5368 cells for each page of data in Plume. The coordinate page cells contain an X & Y coordinate and a cell type code which is used to differentiate between model grid element types. The velocity vector pages contain the U & V velocities and the water depth. Each time step of the tidal data requires one page of cells.

The next largest data structure is the particle list. Each particle structure contains an X & Y coordinate, and a pointer to the cell on the coordinate page which contains that coordinate. This allows access to the velocity data for that particle by simply adding an offset to it's pointer to get to the appropriate tidal data page. The offset can be calculated by multiplying the number of the current tidal cycle time step and the total number of cells in one page of the grid. In fact the offset is simply incremented by the number of cells per page each time the tidal cycle time step is incremented, and stored in a global variable. Other data structures are used for the velocity vector display list, and a list used during the temperature display process.

Operator interface:

The normal display setup for Plume consists of the model area which occupies the top portion of the screen. Beneath that is the graphical time display which consists of a sine curve upon which a small square marker is placed to indicate the progress of the model within the current tidal cycle. Beneath that is a text box which indicates the number of the current tidal cycle, and finally below that is a text box which contains the total number of heat particles within the model. As menu items are selected, appropriate prompts are displayed in text boxes which are removed when the action is completed.

496

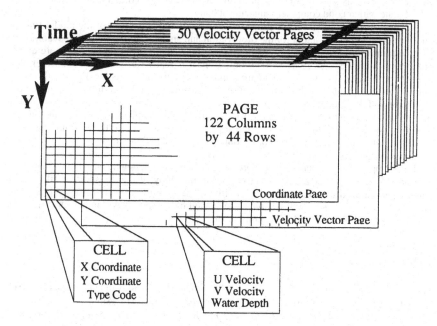

FIGURE 3 *Coordinate & velocity data structure*

Plume starts execution by loading the input data set during which time the model display is blank. It then draws the model area on the screen along with the graphic time display. Execution proceeds by inputting heat particles at the existing outfall location, incrementing the time display, and updating the text boxes as required. At any time during the model run the operator can select a menu from the top menu bar on the screen. When the current time step has been completed the menu will be displayed and the item selection can be made. If printed output is desired the model can be halted and a screen capture desk accessory used to copy the desired portion of the screen either to a file or a printer. The figures used in this paper which show the model were made in this manner.

Model operation, results & conclusions:

The performance of the model has proven to be better than first expected. Two relatively separate applications have emerged. One is the model's use as a data visualization tool. This is shown well when the number of particles is reduced to a minimum and the hydrodynamic processes are seen through the placement of velocity vectors at selected locations, see Figure 4. The behavior of the currents is easily highlighted and the output of the hydrodynamic model is seen in a way that is not evident with conventional output. The other application is the performance of the model in a conventional sense. When the model is run with a few particles input per time step the heat plume's behavior can be observed, although somewhat roughly, see Figure 5. This is an ideal condition for the rapid evaluation of potential outfall locations or other factors which may affect plume behavior. When the model is run with a large number of heat particles, and the model time step is shortened, the most accurate results

497

can be observed, as shown in Figure 6, though it should be noted that effective observation is best done with a good computer magazine at hand.

In conclusion, the goal of this project was to study the feasibility of the model running on the microcomputer as a tool for the observation and analysis of hydrodynamic features and pollutant transport. Plume has proved to do that quite satisfactorily though there are quite a number of "rough edges" to be worked on before it is ready for general use. Further work needs to be done on the operational controls and significant study is required before the model output can be accepted without verification by a similar mainframe model. The relative roughness and the magnitude of the temperature rise may be due to any one of, or combination of, several factors. The principal factor is the relatively few number of particles, and another is the method by which the heat of each heat particle is distributed to the surrounding water. It is anticipated that a faster CPU would contribute significantly to the quality of the results. These factors are prime candidates for future work.

It is the opinion of the authors that cooperation between supercomputers and microcomputers will prove to be a powerful and effective tool in the study of the transport of all types of substances within the nearshore marine environment. And it is expected that as technology advances, more of the load will be able to be carried by computers within design offices and on desks, instead of those in the computer centers.

References:

[1] Bugliarello, G., and Jackson, E.D., (1964), "Random walk study of convective diffusion", *Journal of the Engineering Mechanics Division*, ASCE, EM4, August

[2] Fisher, H.B. etal., (1979), *Mixing in Inland and Coastal Waters*, Academic Press, N.Y.

[3] Foster, D.L., (1991), "Numerical thermal plume simulation using a Lagrangian tracer technique", *Ms thesis*, Department of Mechanical Engineering, University of Maine, January

[4] Jozsa, J., (1989), "2-D Particle tracking model for predicting depth-integrated pollutant and surface oil slick transport in rivers", *Proceedings: Hydraulic and Environmental Modelling of Coastal, Estuarine and River Waters*, Newport, Rhode Island

[5] Kernighan, B.W., D.M. Ritchie, (1988), *The C Programming Language, Second Edition*, Prentice Hall,

[6] Koutitas, C.G., (1988), *Mathematical Models in Coastal Engineering*, Pentech Press, London.

[7] Pearce, B.R., D.L. Foster, V.J. Schuler, P.V. Suesy, V.G. Panchang, (1991), "Numerical thermal plume simulation using a Lagrangian tracer technique", *Technical Report*, Department of Civil Engineering, University of Maine, April

[8] Reid, R.O., and Bodine, B.R., (1968), "Numerical model for storm surges in Galveston Bay", *Journal of the Waterways and Harbors Division*, ASCE, WW1, February.

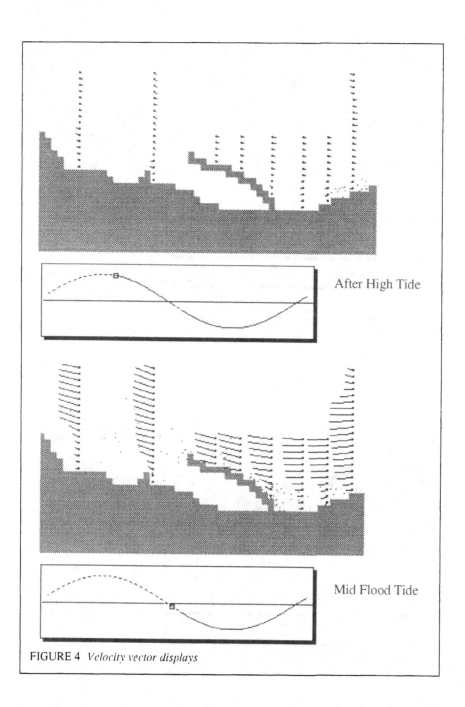

After High Tide

Mid Flood Tide

FIGURE 4 *Velocity vector displays*

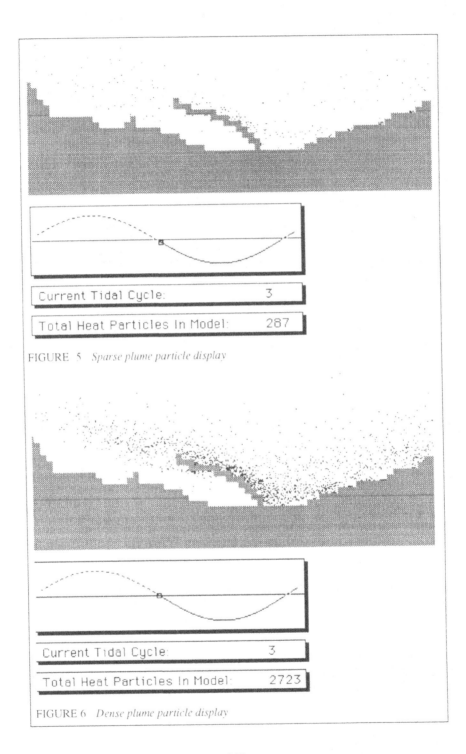

Current Tidal Cycle: 3

Total Heat Particles In Model: 287

FIGURE 5 *Sparse plume particle display*

Current Tidal Cycle: 3

Total Heat Particles In Model: 2723

FIGURE 6 *Dense plume particle display*

42 Ecosystem modelling utilising desktop parallel computer technology: The contributions of freshwater and sediment to habitat succession in the wetlands of Lousiana

M. L. White, J. W. Day Jr, T. Maxwell, R. Costanza and F. Sklar

Abstract

We have developed a spatial modeling workstation, consisting of hardware and software tools to allow development, implementation and testing of spatial ecosystem models in a desktop environment. In this paper we report on the modeling workstation development and subsequent application to the wetlands of southern Louisiana.

The modeling workstation utilizes output from a geographic information system (GIS) for managing spatial data and a general dynamic simulation language for developing unit models to produce ecological simulations framed in a spatial format. The model computer code utilizes parallel processors resident in a desktop microcomputer (Macintosh™). Time series maps of the state variables and the predicted wetland habitat that are generated from model simulations are readable by the GIS system for further display, analysis, and animation.

Using this workstation we have simulated the habitat changes, in numbers of hectares, that are predicted to occur in a wetland during the next fifty years due to the introduction of fresh Mississippi River water and associated sediment. By variously adjusting the volumes and timing of fresh water in the simulation, an optimum management strategy to enhance fisheries and preserve subsiding wetlands can be devised.

Introduction

Low elevation coastal wetlands, particularly in southern Louisiana, U.S.A.., are literally under siege from both human induced activities and natural disturbances. The ability to predict the indirect impacts of these activities and disturbances is important because the suggested solutions have far-reaching economic and social implications. Modeling is often

501

used as a tool to understand such a system, (Hall & Day, 1990), (Morris, Houghton, & Botkin, 1984) and in particular have been used in research to investigate processes in southern Louisiana (Penland, Suter, & Boyd, 1985), (Bahr, Day, Gayle, Hopkinson, Smith, et al., 1977). It can be useful in the planning and management of natural resources as well. However it has had limited use in management in the wetlands of southern Louisiana due to many factors. Information transfer from the scientist to the resource manager is an ongoing problem with whole conferences devoted to methods of information exchange (Janes & Hotchkiss, 1990). Recently advances in desktop technology have been explored as a possible solution to this problem (Rayes, Day, & White, 1992 submitted). However the greatest hindrance to the utilization of modeling in resource management is the overwhelming complexity of the systems to be modeled. The spatial variability of the landscape and the interactions of the physical and biological processes have placed limits on what can be predicted with confidence. Recently some inroads have been made on this problem using dynamic spatial modeling applied to the Terrebonne wetlands of southern Louisiana (Costanza, Sklar, & White, 1990).

We have found that landscape models offer the comprehensive ability to summarize spatial and temporal events into a single framework thus, they can be used to investigate the effects of general land and site-specific impacts on an ecosystem's development and change. The broad utility of such models for both research, management and policy purposes illustrates their effectiveness and value for ecological study and resource management. A model of the wetland landscape in western Terrebonne parish, Louisiana has simulated existing wetland conditions with up to 90% accuracy and has projected indirect impacts up to 50 years into the future (Costanza, et al., 1990). Development of models such as this has been limited in the past by the large amount of input data required and the difficulty of even large mainframe serial computers in dealing with large spatial arrays. These two limitations have begun to erode with the increasing availability of remotely sensed data and the Geographical Information Systems (GIS) to manipulate them, and the computer technology has advanced, particularly with the development of parallel computer systems which allow computation of large, complex, spatial arrays.

Background

In this paper we focus on the development of process-based landscape models to predict ecological succession in coastal wetlands. This type of model simulates spatial structure by first compartmentalizing the landscape into a geometric design and then describing the flows within compartments and spatial processes between compartments according to location-specific algorithms. Examples of process-based, spatially articulate landscape models include wetland models (Sklar, et al., 1985; Costanza, et al., 1986; Kadlec; 1988 Boumans and Sklar, 1991; Costanza, et al., 1990), oceanic plankton models (Show, 1979), coral reef growth models (Maguire and Porter, 1977) and fire ecosystem models (Kessell, 1977).

The Coastal Ecological Landscape Spatial Simulation (CELSS) model is a process-based spatial simulation model consisting of 2,479 interconnected cells, each representing $1km^2$ which was constructed for the Atchafalaya/Terrebonne marsh/estuarine complex in south Louisiana (Sklar, et al., 1985; Costanza, et al., 1990). Each cell in the model contains a dynamic, nonlinear ecosystem simulation model with six state variables. State variables which are tracked by mass balance are water volume, salinity, nitrogen, primary productivity, detrital material, and suspended sediment. The model is generic in structure and can represent one of six habitat types (i.e. fresh marsh, swamp, saline marsh) by assigning unique coefficients. Each cell is connected to adjacent cells by the exchange of water and materials such as salt, sediment, and/or nutrients. Habitat succession occurs in a cell when it's environmental variables (i.e. salinity, elevation, water level, productivity, etc.) fall outside the ranges for its designated habitat type. Succession means that the cell habitat type and all the associated coefficients are switched to a set representative of the changed conditions.

The CELSS model was written in standard FORTRAN (3021 lines of code) and was run with a time step of one week on a variety of computers including a VAX 11/780, IBM 3034, IBM 3090 and CRAY X/MP. A typical 22 year simulation run takes about 24 hours of CPU time on the VAX, 2 hours on the IBM 3034 , 30 minutes on the IBM 3090 and 15 minutes on the CRAY. This model took four people about 4 years to fully develop and implement using a supercomputer. The model has proved to be very effective at helping us understand complex ecosystem behavior and at guiding policy and research. We are now concerned with reducing the time involved for both developing and running this type of model and moving the modeling approach to smaller, less expensive computers.

General Spatial Modelling Package

Computer systems are much more effective tools for research and education if they are easy to use. Our goal was to utilize new developments in parallel computer architecture to bring useful spatial ecosystem modeling to an easy to use microcomputer platform. Toward this end we developed a spatial ecosystem modeling system to run on Apple Macintosh II™ computers with the possibility of exporting code to run on a Connection Machine for very large problems. The system consists of three major software components shown diagrammatically in figure 1.

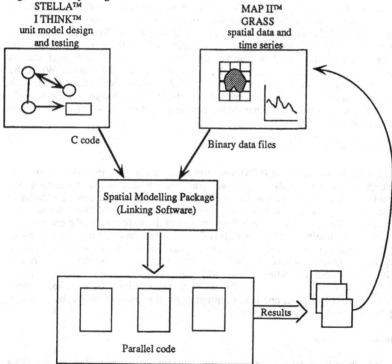

FIGURE 1 *Major software components of the spatial ecosystem modeling system. Components consist of (1) unit model development, (2) data assembly, and (3) linking 1 and 2 in time and space utilizing available parallel processors. Results are exported in GIS readable format.*

503

The major phases of developing a spatial model are: (1) unit model development and testing; (2) data assembly and manipulation; and (3) linking the data and unit model to run in space and time. In our system the unit model development and testing is done using STELLA ™ and ITHINK™, general commercially available dynamic simulation model development packages that are very easy to learn and use (Bogen, 1989). The data assembly and manipulation is done using one of two Geographic Information Systems: GRASS for more elaborate and complex operations, or MAP II™ for more moderate applications. These pieces are linked together with a Spatial Modeling Package (SMP). The SMP prepares code to be run on either the Connection Machine, Levco transputer parallel boards installed in the Mac II or the native Macintosh II series 680x0 processors. The system generates output files which can be read and manipulated using one of the GIS packages.

Unit model

The unit model is the local ecosystem site models which can be linked through horizontal flow of nutrients, sediments, etc. to form spatial ecosystem models. In our case the individual unit models are developed using STELLA, a software package for the Macintosh, designed to facilitate model construction (Costanza, 1987). It is not necessary to use STELLA for this purpose, since all that is required is to send to the next component of the system a set of difference equations, and there are many ways of arriving at these equations. But STELLA greatly facilitates this process and it is just this sort of computer facilitated improvement in speed and ease of use that we aim for in the overall system. STELLA uses symbols that are based on Jay Forrester's systems dynamics language (Forrester, 1961), which has become popular among modeling practitioners as a way to define and communicate a model's structure. Icons, representing stocks, flows and functional relationships are manipulated with the mouse to graphically build the model structure. The defining equations can then be typed in analytically, making use of numerous built-in functions, or entered graphically, using either a graph pad or a data table. When the definitions are complete, the model can be run. STELLA greatly increases the ease with which one can change the model and see the effects of those changes on the model's behavior. It allows the computer to handle the computational details and frees the user to concentrate on model development.

Spatial data assembly

Geographic Information Systems (GIS) are software systems used to collect, store and manipulate spatially referenced data. The development of spatial ecosystem models requires large quantities of spatial data, such as land use, habitat, and climate maps and the ability to manage these data. These maps will typically be collected from many different sources and come in widely differing formats. The GIS component of the SMP handles the crucial tasks of translation, manipulation and storage of this data. For the application presented in this paper we have used MAP II, an easy to use, raster-based GIS which runs on the Macintosh. It is adequate for most applications but it lacks (in the current version) the flexibility and extendibility of GRASS and other systems which are correspondingly more difficult to master. In general, however, we have tried to make the Spatial Modeling Workstation work with a number of GIS's so that one is not limited to those mentioned above, but can take advantage of new developments as they arise.

Spatial modeling package

The SMP consists of three components: (1) a translator for converting unit model equations into parallel C code (2) a grid manager for setting up and running the spatial array on the appropriate parallel computer and (3) a GIS interface for handling input and output from the

GIS systems. The functions of these components are described in detail elsewhere (Costanza and Maxwell, 1991) and for purposes of this paper will be summarized.

Translator The core of the SMP is a translator program which converts unit model equation output files to C code which can utilize either the Connection Machine parallel computer, Levco transputer boards installed in the Mac II, or the native Macintosh II series processor. This module is written in THINK C and runs as an application on the Macintosh family of computers. It converts scalar unit model variables into two dimensional arrays, provides the code for inter-grid cell communication and parallelizes the code for the type and number of parallel processors that will be utilized. In order to generalize the SMP, a naming convention is required for variables used in the unit model.

Grid manager This generic parallel program handles the programming details of simulating spatial dynamics using parallel processors. The task of decomposing the total gridded study area over the processors available and setting up the information exchange along the adjacent boundaries of sub-grids is handled in such a way that parallel modules can be added or subtracted and the grid manager will adjust accordingly.

GIS interface The interface between the GIS generated data and the SMP was designed to be simple and flexible, so that many different GIS systems could be accommodated. This module reads and writes a simple ASCII based raster array format for all spatial data since most existing GIS systems will also read and write data in some version of this format (usually with an appropriate header file). Thus, one would prepare the spatial data in the GIS system of choice and export it as ASCII raster files. The SMP reads these data files and produces results in the same format, which can be read back into the GIS system for display and further analysis.

Application to Barataria basin

Study Area

The area to be modeled includes the wetlands and estuaries of the Barataria Basin of southern Louisiana. This interdistributary estuarine-wetland system is bound on the north and east by the development levees which protect the urban areas along the Mississippi River. On the south and west the area is bound by the development levees which protect the urban areas along Bayou Lafourche and on the south by the Gulf of Mexico (Figure 2).

A brief but pertinent environmental description of the area is given in The Ecology of Barataria Basin, Conner and Day, 1987. "The basin is still an extremely dynamic system undergoing constant change because of geologic and human processes. The basin has been closed to river flow since the leveeing of the Mississippi River in the 1930-40's and the closing of the Bayou Lafourche-Mississippi River connection in 1902. A small amount of water enters the basin through the Intracoastal Canal via the locks in New Orleans. Precipitation provides the main source of freshwater for the basin. During periods of high water on the Mississippi River and given certain wind and sea conditions, freshwater from the river can exert some influence on the lower part of the basin." Additional water input to the basin consists of the storm water runoff that is pumped from urban areas. Tidal interchange takes place along the southern boundary of the study area and is characterized by the tidal record and salinity measurements recorded for Grand Isle, Louisiana.

Overview of Barataria model

The model consists of 5,355 interconnected cells, each representing 1 square kilometer. Each cell contains a dynamic simulation model and each cell is connected to its four nearest neighbors by the exchange of water and suspended materials (salts, nitrogen, phosphorus and suspended organic and inorganic sediments). The buildup of land or the development of open water in a cell depends on the balance between net inputs of sediments and local organic peat deposition on the one hand, and outputs due to erosion and subsidence on the other. The balance of sediment inputs and outputs is critical for predicting how marsh succession and productivity is affected by natural and human activities.

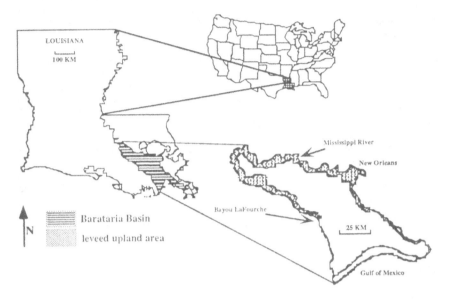

Figure 2. *Barataria basin of southern Louisiana, U.S.A. which is modeled by using the desktop ecosystem modeling package. The area is divided into 5355 one km2 cells.*

Forcing functions (inputs) are specified in the form of time series over the simulation period. Included are the weekly values of Gulf of Mexico salinity and sea level, sediments and nutrients input from the runoff at 22 pumping stations located around the basin, rainfall, temperature, and wind speed and direction. Water can exchange with adjacent cells via overland flow or it may be prevented from exchanging with adjacent cells by the presence of levees. A future version of the simulation will allow transport via canals and natural bayous and the model's canal and levee network will be updated each year during a simulation run.

The change in water level in each cell is determined in the model by water exchanges in and out of the cell across all four boundaries plus surplus rainfall (precipitation minus evapotranspiration). The hydrologic component models two-dimensional gravitational flow which is forced by wind. This approach does not simulate short term hydrodynamics since it neglects momentum transfers, vertical dynamics, and Coriolis forces, but it does approximate the major longer term effects fairly well in this very shallow system.

Water crossing from one cell to another carries both organic and inorganic sediments. This sediment is partitioned between being deposited, resuspended, lost due to subsidence, and carried to the next cell. The relative rates of each of these sediment exchanges in each location is a function of habitat type. Plants and nutrients within each cell also influence these exchanges and flows. Changes in other abiotic material concentrations (i.e., salts, nitrogen) are also a function of water flow between cells and concentration of materials in the cells, along with internal deposition and resuspension.

The primary production in a cell is related to flooding regimes, turbidity, temperature, and elevation. In addition, the relationship between weekly productivity and these variables is different for each of the habitat types. Maximum production occurs at "optimal" levels of inputs and production is reduced with increased deviation from the optimum. The response of primary production to different nitrogen concentrations was simulated with a Michaelis-Menten type rate equation, in which production continues to increase with increasing nitrogen, but at a decreasing rate.

Habitat succession occurs in the model when the physical conditions in a cell become more "appropriate" to a different habitat type. The state variables in each cell are monitored and compared to he physical environment (i.e., salinity, elevation, water level, etc.). If the values of the state variables change to the extent that the environment in the cell is outside the range for its currently designated habitat type then the cell's habitat type and all the associated parameter settings are switched to a new, better adapted set. For example, if salinity in a cell that is fresh marsh rises beyond a certain threshold value and remains at this high level long enough then the model converts the cell to brackish marsh habitat and changes all the associated parameters.

The starting conditions for the landscape of the Barataria basin model are the habitat data shown for 1955 from the U.S.. Fish and Wildlife Service maps. (Figure 4a). By 1978 and 1988 conditions had changed sufficiently to cause the habitats shown in Figures 4b and 4c. Timing comparisons for simulations on the various machines that the CELSS type model has been run are shown in Table 1. The execution timeper iteration of the code is only one order of magnitude larger on the Macintosh than it is on the CRAY while the estimated cost is over two orders of magnitude smaller.

Table 1. *Comparison of time per iteration of the CELSS model using various computers*

computer utilized	MacIIx	VAX 11/780	IBM 3033	IBM 3090 600	CRAY X/MP
# cells in simulation	5355	2749	2749	2749	2749
# iterations (22 years)	6126120	3144856	3144856	3144856	3144856
time /simulation (hours)	1.50	24.00	2.00	0.14	0.08
time/iteration (normalized)	9.24	288.	24.	1.66	1.
approx. $/computer	$20,000♣	**	$5,000,000***	$8,000,000Δ	$10,000,000

♣enhanced with 12 parallel processors   ** N/A   *** in 1978   Δ1988 at special university rate

Summary

Parallel computer hardware and software are now well developed enough to allow their use in ecological systems modeling. Parallel systems are particularly well-suited to spatial modeling, allowing relatively complex unit models to be executed over a relatively high resolution spatial array at reasonable cost and speed. When combined on a user friendly microcomputer , with dynamic model development software, one has a powerful yet easy to use spatial modeling workstation. We have demonstrated these techniques by applying them to a spatial modeling application involving simulation of the Barataria basin in southern Louisiana. These simulation have demonstrated that a desktop computer enhanced with relatively inexpensive parallel processing hardware can rival the CRAY supercomputer in

507

speed of computation. In terms of ease of use the desktop environment is far superior. The next generation of the workstation presented in this paper is currently being developed and details of its enhancements are available in DeBellevue and Costanza, 1990. This workstation as well as its successor should allow the application of advanced computer modeling techniques to a much broader range of problems, especially those involving detailed, spatially articulate ecosystem and landscape modeling.

Acknowledgments

The research was supported by the U. S. Fish and Wildlife Service at Louisiana State University, Baton Rouge, Louisiana, U.S.A. under Louisiana State University Cooperative agreement No. 14-16-0009-87-925 task 9 and at Chesapeake Biological Laboratory, University of Maryland, Solomons, Maryland, U.S.A. under purchase order #84110-89-00682 Titled: "Parallel processor computer modeling system development", R. Costanza, PI.

References

Bahr, L. M. J., Day, J. W. J., Gayle, T., J. Gosselink, G. J., Hopkinson, C. S., Smith, D., & Stellar, D. (1977). A Conceptual Model of the Chenier Plain Coastal Ecosystem of Texas and Louisiana Energy Resources Corporation, Cambridge, Mass. 115 pp.

Bogen, Daniel K.,1989, Simulation Software for the Macintosh, Science, Vol. 246, October 6, 1989 pp138-142

Boumans R. M. J., and F. H. Sklar, 1991, A polygon-based spatial model for simulating landscape change, Landscape Ecology, in press:

Conner W. H. and J. W. Day Jr., 1987, The Ecology of Barataria Basin, Louisiana: An Estuarine Profile, National Wetlands Research Center U. S. Fish and Wildlife Service; U. S. Department of the Interior Biological Report 85(7.13)

Costanza R., F. H. Sklar and J. W. Day Jr., D. A. Wolfe, 1986, Modeling spatial and temporal succession in the Atchafalaya/Terrebonne marsh/estuarine complex in south Louisiana, Estuarine Variability, Academic Press, New York: 387-404.

Costanza R., 1987, Simulation modeling on the Macintosh Using STELLA, BioScience, 37: 129-132.

Costanza R., F. H. Sklar and M. L. White, 1990, Modeling Coastal Landscape Dynamics, BioScience, 40: February 1990, 91-107.

Costanza R., T. Maxwell, 1991, Spatial ecosystem modelling using parallel processors, Ecological Modeling, in press:

DeBellevue, E. B., and R. Costanza, 1990, Unified Generic Ecosystem and Land Use Model: A modeling approach for simulation and evaluation of landscapes with an application to the Patuxent River watershed; Proceedings of Chesapeake Research Conference New Perspectives in the Chesapeake System; Chesapeake Biological Laboratory, Solomons, MD. 20688

Forrester J. W., 1961, Industrial Dynamics, MIT Press, Cambridge, MA.: 464pp.

Hall, C. A. S., & Day, J. W.,1990,. Ecosystem Modeling in Theory and Practice: an introduction with case histories Niwot, Colorado: The University Press of Colorado.

Janes, E. B., & Hotchkiss, W. R.,1990, Transferring Models to Users. In E. B. Janes & W. R. Hotchkiss (Ed.), (pp. 404). Denver, Colorado: American Water Resources Association.

508

Kadlec R. H., D. E. Hammer , 1988, Modeling nutrient behavior in wetlands, Ecological Modeling, 40: 37-66.

Kessell S. R., C. A. S. Hall and J. W. Day Jr., 1977, Gradient modeling: a new approach to fire modeling and resource management, Ecosystem modeling in theory and practice, Wiley-Interscience, New York: 576-605.

Maguire L.. A., J. W. Porter, 1977, A spatial model of growth and competition strategies in coral communities, Ecological Modeling, 3: 249-271.

Morris, J. T., Houghton, R. A., & Botkin, D. B.,1984,. Theoretical Limits of Belowground Production by *Spartina alterniflora*: An analysis Through Modelling. Ecological Modelling, 26, 155-175.

Penland, S., Suter, J. R., & Boyd, R.,1985,. Barrier Island Arcs along Abandoned Mississippi River Deltas. Marine Geology, 63, 197-233.

Rayes, E., Day, J. W., jr., & White, M. L.,1992 submitted, Ecological and Resource Management Information Transfer for Laguna de Terminos, Mexico: a computerized interface Coastal Management.

Show I. T. Jr., 1979, Plankton community and physical environment simulation for the Gulf of Mexico region, Summer Computer Simulation Conference, Society for Computer Simulation

Sklar F. H., R. Costanza and J. W. Day Jr., 1985, Dynamic spatial simulation modeling of coastal wetland habitat succession, Ecological Modeling, 29: 261-281.

43 The uses of intelligent systems and integrated design environments in water quality and hydraulic modelling

T. M. Slow and R. G. S. Matthew

ABSTRACT

An increasing important idea in engineering computing is that of applied Artificial Intelligence (AI) - Expert and Decision Support Systems (ES/DSS). A pioneering yet clearly synergistic concept is the fusing of the two concepts, the ES and the traditional engineering model into an advanced form of an 'integrated design environment'.

One AI application area that both improves upon the concept of the stand alone ES/DSS and aids integration is that of the 'intelligent system' (IS)- a combination of hypermedia, object orientation, expert systems and database technologies.

Such IS's are now beginning to be applied as the intelligent core to Executive Information Systems (EIS), and Management Support System (MSS) but they can offer improvements in the software engineering approaches to integrated design environments. Independently, an IS can improve efficiency and provide 'intelligent assistance', and a level of integration not previously available to engineers.

The construction of such a core IS, ESWHAM (Expert System for Water quality and Hydraulic Analysis Modelling) is described, with proposals of how this can be expanded into a full integrated and intelligent design environment for water quality and hydraulic modelling (using existing commercial or in-house models.)

The prototype IS operates in MS-Windows (using conventional DOS based models) on a 386 PC.

1. Introduction

The application of artificial intelligence (AI) to engineering, and water engineering in particular, is currently receiving a great deal of interest from researchers and consultants alike. This emanates from the success of the Alvey commission's Water Industry Expert Systems Club (WIESC) [6], and advances in the design of practical applied AI - Knowledge Based Systems (KBS), Decision Support Systems (DDS) and Expert Systems (ES). The majority of the engineering examples of such systems are extremely archaic when compared to their computing-research counterparts. Nevertheless, they represent simple practical examples of ground work needed before much more far reaching concepts can be achieved i.e. the more integrated use of (applied) artificial intelligence (water) engineering.

Abbott [1] describes this process, fifth generation modelling, as :-

'The fifth generation takes further the problem of conducting a discourse of average intelligibility by communicating in wide subsets of natural languages, alongside more conventional number descriptions, and it inseparably applies elements of intelligent behaviour during this operation. Thus fifth generation modelling in this field can be characterised as a fusion of Computational Hydraulics (CH) and work in the area commonly referred to nowadays as Artificial Intelligence (AI)'

$$\text{Fifth-generation modelling} = (CH) \cup (AI) \qquad (1)$$

At present, the areas of artificial intelligence and computational hydraulics have been separated into different disciplines. The computer researchers now consider that a 'data centred' or object oriented approach is more productive for the majority of conventional computing activities - from simple drawing packages to complex mathematical modelling. In computer-based hydraulic modelling, it is clear that any application of AI will in some way or another be based around one such information system - the mathematical model.

Existing applications of AI in water engineering have been stand-alone expert systems. These expert systems normally fall into one of two categories. Firstly, the application is based around a pure example of anarchical stand-alone expert system. This system is an expert system designed an 'intelligent' front end or 'input-data generator' based around one specific mathematical model. There are no live data communication links to the model, let alone any two way interactive links. The user interfaces with the expert system and then is expected to interface with the model separately. Secondly, the expert system is used to provide intelligent assistance to the user. This is perhaps best viewed as an intelligent help system or intelligent adviser. Once more, the ultimate goal of interfacing with the model, is ignored or poorly implemented, and therefore any level of integration is not achieved.

Clearly, this integration is the goal of any successful implementation of artificial intelligence in hydraulics, to confirm to the concept of fifth generation modelling [1] or any integrated design environment. The linking of the existing information sources - models, databases, Executive Information Systems (EIS), Management Support Systems

512

(MSS), Geographical Information Systems (GIS) and Computer Aided Drawing/Design (CAD) - with practical AI and other advanced computer technology, provides a system infinitely superior to the current constituents - the stand alone model and expert system. This approach can be achieved and enhanced through the use of an 'intelligent system'.

2.   Intelligent Systems

Most artificial intelligence applications have, at one time or another, been referred to as 'intelligent systems'. The traditional areas of AI : vision- and speech- recognition systems, neural networks and expert systems can all in their way be called intelligent systems - as they possess a varying degree of what is termed 'intelligence'.

However, a new term - the 'intelligent system' - is now being used to depict a system based around the integration of the existing technologies of expert systems, hypermedia, object oriented (or 'data-based approach'), and database technologies. Therefore using the same notation [1] as Equation 1 :

$$\text{Intelligent System} = (ES) \cup (HM) \cup (OOPS) \cup (IS) \tag{2}$$

where ES = Expert System, HM = Hypermedia, OOPS = Object Oriented Programming Techniques, IS = Information System/Database

The fusion of these technologies is synergistic and systems can be developed that not only complement each other, but can solve tasks that the constituent technologies could not independently.

For instance, the stand-alone expert system is geared to come to conclusions - reach goals - by manipulating facts and knowledge and by inferring new rules or relationships along the way in order to reach its original goal. The amount of knowledge based information (stored as rules or objects in advanced ES) that needs to be stored even for the smallest of ES is often extremely large. It is the size and complexity of such systems that has been a resulting factor in their independence. Clearly a way of manipulating this vast knowledge base is needed if any level of integration is to be achieved.

Schoen and Sykes [4] state :

'in many cases, the problem is not making the decision but rather in obtaining sufficient information on which to make a reasonable decision'

The use of hypermedia- a way of moving through data in a non-linear fashion - allows the expert system to use a smaller number of rules or concentrate on a pertinent specific area of rules - this allows the expert system to produce more tailored results more quickly. The converse can also apply, if the user moves through an information system using hypermedia until the number of possible options is reduced sufficiently then even the simplest of expert systems can be used to reach the final goal, which might be the

513

retrieval of a very specified piece of information - a kinetic constant, which is dependant on a vast number of factors, for instance.

There are many design strategies for these integrated 'intelligent systems', ([3],[5]), but a simplistic symbolic representation is put forward by Bielawski and Leward [2]. They propose three basic designs for the intelligent system. The first two are a simple representation of the co-operative effect of hypermedia and expert systems, as described briefly above. The third, however, is a far superior concept and incorporates our goal of tighter integration. This model illustrates the fusion of hypermedia and expert systems technology, with information systems. (The whole system is more flexible and powerful if object oriented techniques are used both in its construction and implementation/software engineering.)

Model 1 : Expert system with hypermedia support The first model approach, as shown schematically in Figure 1, for an intelligent system is the combination of an hypermedia component to supplement a main expert system module. The hypermedia component acts to aid the navigation of the expert system to reach its goals. The hypermedia system is simply a navigation tool for the expert system.

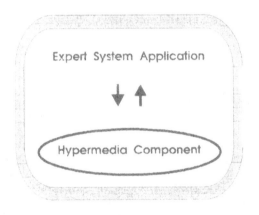

FIGURE 1 *Model 1 - Expert system-based intelligent system with hypermedia support [2]*

Model 2 : Hypermedia with expert system support The second model scenario, as shown in Figure 2, is the hypermedia system with expert system support. The hypermedia system navigates it way around the data in a non-linear format, when the information is narrowed down sufficiently, then the expert system can then narrow the search further using its rule based inference techniques.

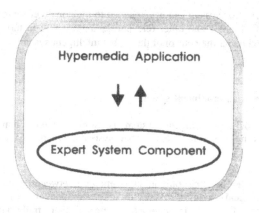

FIGURE 2   *Model 2 - Hypermedia-based intelligent system with expert system component [2]*

Model 3 : integrated intelligent system based around an information system The third model, as shown in Figure 3, is based on a different, almost opposite approach. At the heart of this system lies the database/information system, (i.e not the hypermedia/expert system modules). The two modules of hypermedia and expert system work together and then interact with the information source. This approach is 'data centred' as it is the existing information system that is the key to the system and not the expert system or hypermedia engine.

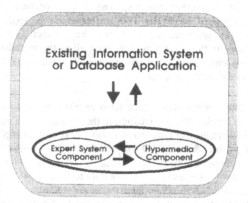

FIGURE 3   *Model 3-Integrating intelligent systems with existing information and data-base applications [2]*

This approach integrates the two information technology modules around the existing information source. The information system is the crux of the system and a synergy is formed between the three integrated interacting modules. Clearly, in the specialised case, of mathematical modelling , the existing information system is the mathematical model.

515

However, the information system can be expanded to include other sources of information (Note 1) such as GIS, CAD and other relevant databases, with in turn can act with the mathematical model under the control of the global intelligent system.

3. Construction of an intelligent system

The construction of an intelligent system is more complex than the traditional development of the individual modules - expert systems, hypermedia systems, databases, and models.

Interaction The system must not only allow interaction between the constituent modules but communicate and monitor each of the components over 'live' internal communication/data links. This interaction means that traditional programming approaches are restrictive. The conventional programming approach is to move through the programme step-by-step in a pre-ordained fashion determined by the programmer and software engineer. However, the continually changing modules of the intelligent system preordain a new conceptual approach - this approach can take many forms, but by far the most efficient is that of object orientation.

Object orientation The principle of object orientation can be pictured as a set of individual objects in a global sea. Each object has an event object oriented driven nature i.e. it receives a message from the global sea performs a task and then returns a message to the sea. Therefore the process is started by placing a message(s) in the sea, only the relevant object(s) respond to this message(s) and place another message(s) in the sea which are acted upon by another object(s). The process continues in a way that is inevitably different each time. Moreover, it is possible to use the application in ways that the software engineer/programmer may not even have exactly originally envisaged. Messages can be placed in the sea, by the user (as well as the modules of the intelligent system) by interacting with screen event objects through windows, text entry boxes, dials and all manner of other interface devices.

Intelligent Filters Intelligent systems must have interfacing units, or objects, to other information sources other than those contained in the system. These can be thought of as 'intelligent' versions of traditional data input/output filters (for example, those that exist for the exporting of graphics from a CAD package to a word processor). The 'intelligence' is the ability to communicate with the intelligent system modules. For example, a intelligent interface unit for survey data must be able to understand the format of the data input file, but also know when to act on the data, or more accurately where to go (Expert System component) to determine information relevant to the data (i.e. if some is to be ignored or modified or sent to another object). In short, the interfacing units must have a limited degree of 'dynamic' intelligence to allow them to interface with the intelligent system and their real world outside agency. The role of these 'intelligent filters' in an integrated intelligent system is shown schematically in Figure 4.

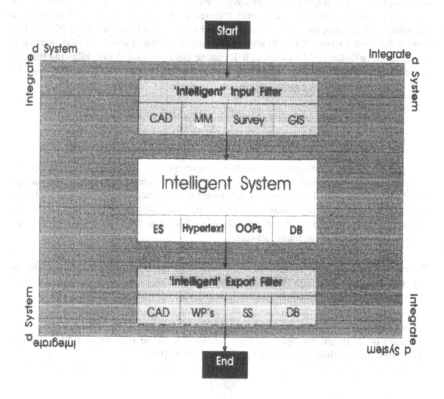

FIGURE 4 *Schematic View of Integrated Intelligent System*

PC Development The developer must also take into consideration the hardware platform on which such a system will be ultimately developed. Nearly all expert systems in hydraulics have been developed to run on workstations (i.e. powerful, and relatively expensive computers). The reasoning behind this is due to the fact that storage and need for speed in knowledge processing was not available to an acceptable level using PCs. Any systems developed were therefore to simple and limited to allow such systems to be practical. In fact, one of the final conclusions of the Alvey Commission [6] was to recommend development on such powerful hardware platforms. Nowadays, however, in many cases the mathematical models have been ported to the PC, but the separate expert systems remain on the more powerful computers.

Current technological advances in the power of PCs, now means that such systems that include aspects of AI can be developed practically on PCs. Furthermore, the object oriented development environments now exist on the PC to enable such systems to be constructed.

The evolution of the PC has for many years been retarded by the operating system. The current interest in and commercial success of Microsoft's Windows environment is proof

517

that things needed to improve. The MS Windows environment is an event driven but not fully object oriented system. If an object oriented development environment is added to the Windows base then the criteria for intelligent system development on the PC are met, with some restrictions. The Windows environment also allows for the two way conversation or linking of data - through 'Dynamic Data Exchange', thus allowing the task of intelligent links to be simplified. Also the traditional DOS models can be 'multi-tasked' with any intelligent system running inside the Windows environment.

4. Development of ESWHAM, a practical IS

The system under development has been developed to run on a PC using the MS-Windows environment. It is both an intelligent system and integrated design environment prototype. The roots of it developed are illustrated in the name, ESWHAM, an Expert System for Water quality and Hydraulic Analysis Modelling. The system uses the multi-tasking facilities of Windows and the Intel 3/486 to run DOS-based concurrently with the components of the main software.

The role of ESWHAM is to illustrate the practical possibilities of the intelligent system and provide a core on which the main foundations of a more complex intelligent system and integrated design environment call be constructed.

The format of ESWHAM is shown schematically in Figure 5. The intelligent system is composed of three internal applications : Knowledge Based Interactive Editor (KBIE), Knowledge Based Editor (KBE) and File Format Editor (FFE).

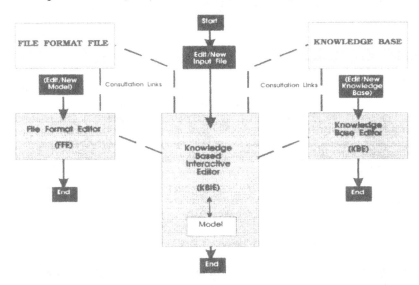

FIGURE 5 *Schematic View of ESWHAM*

The system around the revolves the Knowledge Based Interactive Editor (KBIE). This is the main facet of the system. It controls the expert system and hypermedia engines of the system. The user will work solely in the KBIE, under normal conditions. The KBIE consists of a knowledge based text editor (interchangeable with a graphical representation). This editor is configured or tailored for the specific model concerned by an internal 'consultation link' with the File Format File (*.FF). Furthermore, a model specific knowledge base is 'consulted' (Knowledge Base (*.KB)), this is used by the expert system and hypermedia components of the system to analyse the file, and suggest not only the correct format of the model input data (normally automatic), but make judgements on the specific modelling-linked nature of the data (with possible consultation with the user). This allows subtle errors to be noted and corrections suggested from the knowledge contained in the knowledge base. Once the model file(s) are deemed to be in a suitable state, by the system and the user, the model is run 'inside' ESWHAM, and the run-time errors and results processed.

5. Integrated design environments (IDE)

The design considerations of the integrated design environment are for the most part concerned with the nature of integration. The combination of different software applications, even across different hardware platforms is complex simply in the conversion sense. However, for the independent modules to be assembled into anything which can fairly be described as integrated, the links will need to process some level of dynamic intelligence.

This is where the real strength of the core intelligent system is evident. It is through the use of intelligent systems that integration can be achieved and combined with a level of knowledge based reasoning, hypermedia navigation, and object oriented techniques. The system can communicate internally with reason and infer assumptions when a number of unique scenarios arise. The hypermedia components allow data to be accessed in non-linear fashions which are more a kin to human thought processes. The object oriented data structured approach allows the concept of the 'reusable code object'. This allows the software to be modified and fine tuned without the traditional back to basics approach. Most important of all, the intelligent system as part of the integrated design environment, hold unconventional ground in that it places primary importance on the information source and its information. It is important to note that the majority of software holds performs the converse of considering the system to be constructed around independent modules, all possessing astonishing independent qualities but little or no harmony between the facets.

6. The extension of ESWHAM into a full IDE

ESWHAM has been tailored in its prototype stage to accept the most common file formats, used by the US EPA software suites. The use of the File Format File, and its respective editor, the File Format Editor (FFE), allows the number of model formats to be increased by creating new file formats, or in the more complex cases, the code level manipulation of the file format 'objects'. The respective knowledge base pertinent to each model can be created by the knowledge engineer or in exceptional cases by the end user, via the Knowledge Base Editor (KBE)

A number of 'intelligent' filters have been written to illustrate the effectiveness of this techniques and show the concepts of integration. The primary example of this is the filter that links with Microsoft's spreadsheet, Excel, this allows data to be exchanged between ESWHAM and Excel, and uses another two way link using Dynamic Data Exchange (DDE) to chart data from ESWHAM model input/output files in Excel. Further development of these intelligent links will allow the extension of ESWHAM into a fully integrated design environment.

7. Conclusion

Clearly intelligent design environments are an improvement over the traditional collections of independent software packages. The synergy created when the independent facets are combined is self evident and the relative merits of knowledge based reasoning, hypermedia, and object oriented techniques mixed with the original information system by way of the intelligent system can be seen in practical terms through the system, ESWHAM.

Intelligent systems and integrated design environments are already viable, and the flexibility of such systems allows them to remain both practical and productive.

8. Glossary

Artificial Intelligence (AI) The branch of computer science that investigates computer systems that exhibit human-like intelligence - simply a way of making computers think intelligently.

Decision Support Systems (DSS) Computer-based information systems that combine models and data in an attempt to solve non-structured problems with extensive user involvement

Executive Information Systems (EIS) Computerized systems that are specifically designed to support executive work.

Expert System (ES) The application of AI - an interactive program that incorporates judgement and rules of thumb, intuition and other expertise to provide knowledgeable advice about a variety of tasks in a specific area.

Hypermedia A relational data-base structure that links and accesses different types of media, such as text (hypertext), graphics (hypergraphics), sound and film, in a non-linear way.

Intelligent System (IS)   The combination of expert systems, hypermedia and database technologies, using utilizing aspect of object orientation.

Object Oriented (OOPs)   Term for programming practices or compilers that collect individual programming elements into heirarchies of classes, allowing program objects to share access to data and procedures without redefinition.

Management Support Systems (MSS)   A business information system designed to provide past, present, and future information appropriate for planning, organizing, and controlling the operations of the organisation.

9.   Notes

1. Information is defined for our purposes as data, knowledge, graphical or vector based spatial data. Data is defined here as numerical or ASCII data as one might find in a traditional database.

10.   References

(1)   Abbott, M.B., (1991), *Hydroinformatics*, Avebury Technical.

(2)   Bielawski, L., & Leward, R., (1991), *Intelligent Systems Design : integrating Expert Systems, Hypermedia, and Database Technologies*, Wiley.

(3)   Gillies, A.C., (1991), *The Integration of Expert systems into Mainstream Software*, Chapman & Hall Computing.

(4)   Schoen, S., & Sykes, S., (1987) ,*Putting Artificial Intelligence to Work* , Wiley.

(5)   Shafer, D., (1989), *Designing Intelligent Front Ends For Business Software*, Wiley.

(6)   Walker, R.S. (editor), (1988),*Water Industry Expert Systems Club - Final Report*, Wrc Swindon.